Brian H. Kaye

Golf Balls, Boomerangs and Asteroids

The Kaye Collection from VCH

Science and the Detective
Selected Reading in Forencic Science
Hardcover, ISBN 3-527-29251-9
Softcover, ISBN 3-527-29252-7

A Random Walk Through Fractal Dimensions
Second edition
Softcover, ISBN 3-527-29078-8

Chaos & Complexity
Discovering the Surprising Patterns of Science and Technology
Hardcover, ISBN 3-527-29039-7
Softcover, ISBN 3-527-29007-9

© VCH Verlagsgesellschaft mbH. D-69451 Weinheim (Bundesrepublik Deutschland), 1996

Distribution:
VCH, P. O. Box 10 11 61, D-69451 Weinheim (Federal Republic of Germany)
Switzerland: VCH, P. O. Box, CH-4020 Basel (Switzerland)
United Kingdom and Ireland: VCH (UK) Ltd., 8 Wellington Court, Cambridge CB1 1HZ (England)
USA and Canada: VCH Publishers, Inc., 337 7th Avenue, New York, NY 10001 (USA)
Japan: VCH, Eikow Building, 10-9 Hongo 1-chome, Bunkyo-ku, Tokyo 113 (Japan)

ISBN 3-527-29322-1 (hardcover)
ISBN 3-527-29323-X (softcover)

Brian H. Kaye

Golf Balls, Boomerangs and Asteroids

The Impact of Missiles on Society

VCH

Weinheim · New York · Basel · Cambridge · Tokyo

Professor Brian H. Kaye
Laurentian University
Ramsey Lake Road
Sudbury, Ontario P3E 2C6

> This book was carefully produced. Nevertheless, author and publishers do not warrant the information contained therein to be free of errors. Readers are advised to keep in mind that statements, data, illustrations, procedural details or other items may inadvertently be inaccurate.

Published jointly by
VCH Verlagsgesellschaft mbH, Weinheim (Federal Republic of Germany)
VCH Publishers, Inc., New York, NY (USA)

Editorial Director: Dr. Peter Gregory, Dr. Jörn Ritterbusch
Production Manager: Dipl.-Ing. (FH) Hans Jörg Maier

Every effort has been made to trace the owner of copyrighted material; however, in some cases this has proved impossible. We take this opportunity to offer our apologies to any copyright holders whose right we may have unwittingly infringed.

Cover Illustration by Phil Harms, London, England.

Library of Congress Card No. applied for.

A catalogue record for this book is available from the British Library.

Deutsche Bibliothek Cataloguing-in-Publication Data:

Kaye, Brian H.:
Golfballs, boomerangs and asteroids : the impact of missiles on society / Brian H. Kaye. –
Weinheim ; New York ; Basel ; Cambridge ; Tokyo : VCH, 1996
  ISBN 3-527-29323-X kart.
  ISBN 3-527-29322-1 Gb.

© VCH Verlagsgesellschaft mbH, D-69451 Weinheim (Federal Republic of Germany), 1996

Printed on acid-free and chlorine-free paper.

All rights reserved (including those of translation into other languages). No part of this book may be reproduced in any form – by photoprinting, microfilm, or any other means – nor transmitted or translated into a machine language without written permission from the publishers. Registered names, trademarks, etc. used in this book, even when not specifically marked as such, are not to be considered unprotected by law.

Composition: Filmsatz Unger & Sommer GmbH, D-69469 Weinheim. Printing: betz-druck gmbh, D-64291 Darmstadt. Bookbinding: J. Schäffer GmbH & Co. KG, D-67269 Grünstadt

Printed in the Federal Republic of Germany

Biography

Dr. Brian Kaye was born in Hull, Yorkshire, England, in 1932. He obtained his B.Sc., M.Sc., and Ph.D. degrees from London University after studying at the University College of Hull, where he was a George Fredrick Grant Memorial Scholar. After working as a scientific officer at the British Atomic Weapons Research Establishment (Aldermaston) he taught physics at Nottingham Technical College from 1959 to 1963. He then moved to Chicago, where he was a Senior Physicist in the Chemistry Division of the IIT Research Institute (the Research Institute of the Illinois Institute of Technology). There he studied problems as different as why dirt sticks to the fibers of carpet to the design of better propellants for space rockets.

Since 1968 he has been Professor of Physics at Laurentian University in Sudbury, Ontario. He specializes in powder technology, which deals with the manufacture and properties of cosmetics, explosives, powdered metal pigments, drug powders, food powders, and abrasives. He has written a standard text on characterizing powders and authored over 100 scientific papers.

In 1977 his interest in the complex structure of soot involved him in the new subject of fractal geometry, an interest that led to the books "A Random Walk Through Fractal Dimensions", which is now in its second edition, and "Chaos & Complexity. Discovering the Surprising Patterns of Science and Technology", both published by VCH. He is also interested in forensic science which resulted in the book "Science and the Detective", also published by VCH. The philosophical side of science has always interested him and has been complemented by his activities as a methodist local preacher in the Sudbury region of Ontario, Canada. He is just as likely to be found holding a service in a protestant church as he is to be lecturing on fractal geometry and chaos theory at the University.

Preface

This book grew out of a need for a general science course at the first year level for students of the liberal science program and as an elective for arts students who had to take one science credit in their studies. As I attempted to develop this general science course, I found myself having to research the practical science problems that had not been covered in my own degree studies. For example when a student asked me how they knew how many dimples to put on a golf ball I had to confess that I did not know the answer. Other questions that sent me scurrying to the textbooks were; "why does a boomerang boomerang"? and "what were the prospects of asteroids hitting the Earth in the next century"?

Basically the book evolved into a study of missile physics ranging from bows and arrows, through the story of David with his sling shot, to how genetic engineers were firing miniscule gold bullets into the heart of a cell to change its genetic structure. Extensive teaching experience established the fact that many students did not have problems with science but with the vocabulary of science. Students will not stand up in class and ask what is meant by torqueshear and work in the scientific sense. The stress on building up vocabulary in this book has grown out of experience in the classroom and not just the fact that my hobby is lexicography — (the writing of dictionaries).

The need for the development of vocabulary was particularly important since many of the students using this book were pursuing the course by correspondence and they often were located in small communities in sparsely populated areas of Northern Ontario with very little access to reference libraries.

This book is the second in a series aimed at completing a triad of courses for liberal science majors at Laurentian University. The second year text, Science & The Detective, has already been published. I hope to complete a third book entitled Soot, Dust and Mist (which looks at the way that dust affects out lives all the way from crop spraying to variations in the climate due to dust in the spiral arms of the galaxies) in the next year or so. Many people contributed to the ideals which are finally embodied in the text und I hope that other readers will enjoy this exploration of the science of missiles both in our everyday life and in outerspace that surrounds us. My students tell me that the study of Golf Balls and other missiles has helped them enjoy the science around them in everyday life. I hope this will be the experience of the readers as they study the impact of missiles around them.

B. H. Kaye
Professor of Physics
Laurentian University

Acknowledgements

I wish to acknowledge the many students who debated topics in class. I would also like to thank Garry Clark who worked extensively on the diagrams; my two daughters Sharon and Alison who typed the majority of the script along with Julie Gratton-Liimatainen who also worked on the text from time to time. Sharon Kaye undertook the difficult task of clearing the copyright for all of the information used in the book. My editors in Germany, Dr. Ute Anton and Dr. Peter Gregory have always been supportive of this work and I wish to acknowledge their help in arriving at a final script.

Dedication of the Book

To the memory of Auntie Alice who was a second Mother to me.

Table of Contents

Word Finder XIII

Chapter 1	**Why do Golf Balls have Dimples?**	1
1.1	Warning: Golf Balls are Potentially Lethal!	3
1.2	Featheries, Gutties, and Composite Balls	4
1.3	Bouncing Balls and Leaping Athletes	8
1.4	Golf Balls with Perfect Centres of Gravity	13
1.5	What turns an Innocent Dimpled White Ball into a Lethal Missile	16
1.6	Golf Ball Dynamics: Impulsive Deformation and a Lift from Dr. Magnus	19
1.7	The Scientific Study of Dimples	23
1.8	Floating Golf Balls and Aggressive Crocodiles	27
1.9	Golf Balls in the 21st Century	27
	References	28

Chapter 2	**The Science of Bows and Arrows**	31
2.1	Using Potential Energy to Launch a Missile	33
2.2	Working at Storing Potential Energy in a Bow	33
2.3	Graphical Illustration of the Energy Stored in a Shielded Bow .	41
2.4	Vane Strategies	43
2.5	The Composite Bow	45
2.6	The Crossbow and William Tell	48
2.7	How Big should an Arrow Be?	49
2.8	The Bow and Arrow in Military History	50
2.9	Pulleys for Bows	54
2.10	Metaphorical Missiles and Metamorphosed Bows	58
	References	62

Chapter 3	**Racketeering Missiles**	63
3.1	Love and Tennis	65
3.2	Tennis Balls: Flannel Wrap and Gas!	67
3.3	Vibrations, Sweet Spots, and Space Age Rackets	69
3.4	Serving the Ball	77
3.5	How Do They Measure the Velocity of those Speeding Balls? .	78
3.6	Slow Down, You Move Too Fast	80

3.7	Carbon Feathers and Plastic Skirts for Battered Birds	81
3.8	Are You Being Servied? (The Robot is Here)	82
	References	83

Chapter 4	**Bolas, Boomerangs and Bouncing Bombs**	85
4.1	Gauchos, Bolas and Spinning Tops	87
4.2	Boxcar Integrators and Lasers	101
4.3	Some Circumspect Vocabulary	104
4.4	Killing Giants and Catching Fish	108
4.5	Dancing Aborigines and Skipping Stones	118
4.6	Deadly Missiles of Cricket and Baseball	122
4.7	Keep Your Eye *Away from* the Ball?	130
	References	134

Chapter 5	**Darts, Stone Disks and Boomerangs**	137
5.1	Javelins and Snow Snakes	139
5.2	Bernoulli's Principle, Venturi Throats and Pitot Tubes	143
5.3	Stone Disks and Flying Dish Pans?	152
5.4	Killing Sticks and Boomerangs	156
5.5	Flying Toys of Tomorrow	159
	References	160

Chapter 6	**Pea Shooters, Rockets and Rifles**	163
6.1	Peashooters and Blowpipes	165
6.2	From Muskets to Machine Guns	181
6.3	Shrapnel, Dumdums and Devastators	192
6.4	Laser Rifles and Swords of Light	199
6.5	Moon Shots	202
6.6	Tit for Tat in Missile Development	205
6.7	Manufacturing with Missiles	214
6.8	Fatal Fiesta Frolicking?	215
	References	216

Chapter 7	**Rockets: From Fireworks to Trans-Galactic Missiles**	219
7.1	Rockets and Newton's Third Law of Motion	221
7.2	Getting Rockets off the Ground	231
7.3	What Goes Up, Must Come Down (Most of the Time!)	238
7.4	Heat Transfer Mechanisms	245
7.5	Designing Heat Protection Shields for Returning Space Missiles and Capsules	262

7.6	Did Astronauts See Shooting Starts in Their Eyes?	263
7.7	Circulating Missile Messangers	269
7.8	Interplanetary and Trans-Galactic Missiles with a Message	276
7.9	The Future of Fireworks Displays	282
	References	283

Chapter 8	**Cosmic Collisions**	**285**
8.1	Target Earth	287
8.2	The Dynamics of Asteroid Collisions on the Surface of the Earth	292
8.3	The Tunguska Attack: Cosmic Missile or Alien Spaceship?	299
8.4	Hairy Stars and Telltale Tails	304
8.5	Apprehending Cosmic Drifters	312
8.6	Cosmic Missiles and the Disappearing Dinosaurs	315
8.7	Moon Struck	321
8.8	Space Junk	326
8.9	What are the Chances of a Person being hit by a Cosmic Missile	327
	References	328

Chapter 9	**Some Down to Earth Missiles**	**331**
9.1	Ice Bullets and Hailstones	333
9.2	Latent Heat and Nucleated Ice	334
9.3	Rainmaking and Hailstone Abatement	339
9.4	In Search of Nothing?	344
	References	350

Chapter 10	**Humans as Missiles and Targets**	**351**
10.1	Belly Flops and Bungee Jumping	353
10.2	Sheep Skin Jackets and Metal Helmets	355
10.3	Crash Helmets and Safety Visors	361
10.4	Dementia Pugilistica	366
10.5	Automobile: Convenience or Deadly Missile?	371
10.6	Sports Aerodynamics	379
10.7	Roller Coasters as Missile Systems	382
	References	383

Chapter 11	**Micro and Miscellaneous Missiles**	**385**
11.1	Volcanic Ash and Birds	387
11.2	Deer Flies and Bumblebees	389
11.3	The Nuts and Bolts of Abrasive Cleaning	394

11.4	Measuring the Size of Dust	395
11.5	Hammers without Handles	401
11.6	Fat Bodies and Magic Bullets	406
	References	409
Index		411

Word Finder

A
ablate 237
ablation 234
absolute temperature scale 242
accelerometer 373
acetylene 171
acrobat 102
acronym 102
Acropolis 102
active satellite 273
activity 224
actor 224
Aerobie ring 155
aerodynamic diameter 396
aerogel 248
agenda 224
agent 224
agile 224
agitate 224
airfoil 151
aluminum powder 231
ambush 195
Ammon, Egyptian god 179
ammonia 179
ammonium perchlorate 231
amorphous 59
Anabasis 115
anachronism 271
Andromeda 279
aneurysm 150
angle 104
angle of attack 152
Ångstrom 176
angular velocity 89
anneal 357
antenna 273
antinodes of oscillation 74
anvil of a hail cloud 340
apogee 273
Apollo 59
apostle 273
apostrophe 273
apparatus 102
archaeologist 116
Archemedes 182
Archemedes principle 211
Armortek (ceramic armor) 207
aromatic compound 171
arrows of Apollo and Artemis 59

arrows of desire 57
Artemis 59
artificial gravity 92
asteroids 292
asthenosphere 260
astronomical unit (distance) 292
astronomy 17
astrophysics 296
atmosphere 143
atmosphere of the Earth 261
atmospheric pressure 144
atom 177
attitude angle 152
autogenous pulverization 405
automobile airbag 373
available redundancy 368
aversion 239
Avoirdupois 123
axis of rotation 89
azide 375
azote 375

B
backspin on a ball, effect 21
bacteria 399
badminton 81
ball-in-socket joint 96
ballast 211
ballistic armor 207
ballistic missile 230
ballistic pendulum 188
balsam 173
Barnard's Star, possible planets 280
Barringer crater 287
basal melt pool 296
basalt 296
baseball 122
battering ram (ballistic pendulum) 190
bazooka (anti-tank rocket) 228
Bermuda grass 12
Bernoulli's principle 143
Bernoulli, Daniel 143
biodegradable bullet 196
black hole (astronomical) 258
black powder (gun powder) 166

black-body radiator 250
blasting gelatin 175
blowpipes 165
bola/bolas 87
bolide 299
bolt for the crossbow 48
bomb 169
bombardier 170
bombardier beetle, spray defence 392
boomerang 156
bow machine 54
bowled out 125
bowler 125
boxcar 101
boxcar integrator 101
boyers 45
Brahe, Tyco (astronomer) 306
breach (break) 183
breccia lens 296
brecciated rock 290
breech 183
breeches (clothing) 183
Bren gun 191
britches (clothing) 183
Brownlee, David (cosmic dust) 312
bucket brigade detector 101
bull's eye 61
bullet 183
bullet pass (football) 100
bungee jumping 355
buoy 212
buoyant weight 211
bursera tree 392

C
caddies 5
calibrate 183
calibre of a bullet 183
caloric 39
calorific content of food 40
calorimeter 39
camouflage 185
candidate 238
candle 238
cannon (gun) 183
canon (church official) 183
carbon char 263
carbon dioxide 171

cartridge 183
caseless ammunition 191
CAT scanning 368
catapult (see also slingshot) 111
cataract 237
cattie 111
caught out 125
caustic soda 102
cellulose 178
Celsius temperature scale 39
Celsius, Andros 39
Cenozoic 316
Centaurs 50
center of gravity 13
Centigrade temperature scale 39
centrifugal force 89
centripetal force 88
chaos 267
charcoal 166
Charles law, for gases 244
Charles, Jacques (properties of gases) 242
chondrite 314
chondrules 314
chromosome 408
ciliary muscles 267
circuit 106
circumference 105
clap 169
climate 239
climax 239
clinic 239
coefficient of restitution 8
coherent light 199
cold front 339
Colt revolver (autoloading gun) 191
comet 273
composite bow 45
compound bow 54
Computerized Axial Tomography (CAT scan) 368
computerized tomography 368
concrete 349
confectioners sugar 177
Congreve, Sir William (rocket weapons) 227
conservation of angular momentum 89
constrict 175
convection current 248
convection 248
convective growth 339
convert 239
cordite 179

cork 123
corpuscles of light 264
cosine wave 106
cosmic dust 309
cosmic rays 266
cosmos 267
cotton 177
counter-intuitive 3
court tennis (indoor) 67
crack 169
crater 290
Creighton Fault 291
Cretaceous 317
cricket 122
cross-linking 28
cryoblation 237
cryogenic fuels (liquid propellant) 236
cryogenic grinding (cryo-grinding) 237
cryogenic pulverization 237
cryogenics 236
cryonics 237
crystal 236
crystal ball 236
crystal chandelier 236
crystal spheres 307
crystalline 236
crystallographer 236
cul-de-sac 223
curare 165
cybernetic organism 212
cybernetics 207
cyborg 212
cycles per second (cps) 106

D
Dazer 201
dead comet 301
dead soft 357
Death Valley Days (television duster) 358
decimal 104
decline 239
decoy 195
deduction 247
degree of angle 104
degree, temperature 39
demented 366
dementia 366
dementia pugilistica (punch drunk) 366
density of a substance 27
deoxyribonucleic acid (DNA) 408

detonate 171
deuterium (isotope of hydrogen) 348
deuterium oxide 348
devastator bullet 196
Dewar flask 258
Dewar, Sir James 258
diameter 105
diamond pipes 171
diamonds 171
Diana 59
diatom 177
diatomaceous earth (kieselguhr) 174
disappearing filament pyrometer 256
discus 152
distaff 225
divert 239
dolomite 349
dolphin 96
Doppler effect 78
dozen 105
Dragon Lady (U2) 313
dry ice 338
duck 65
ductile metal 356
dynamic pressure 148
dynamics 18
dynamite 174
dyne 36

E
echoic 169
education 247
elastic 10
electrical circuit 106
electro-optics 199
electron 345
element of a fluid 23
empirical knowledge 6
energy 18
Enfield rifles 185
entomologist 388
eponym 104
erg 38
ergonomics 17
erotic emotions 57
ether 264
ethereal 264
ethnology 156
etymology 65
Euclid 403
Euclidean geometry 403
experience 7

experiment 6
expert 7
explicit 254
extrapolation 242

F
Fahrenheit temperature scale 39
Fahrenheit, Gabriel Daniel 39
fast bowler 125
Faucault, Jean 94
Fawkes, Guy 225
feed-forward control 209
feedback control 209
Fernsprecher (phone German) 103
field performance indicator 12
fire damp (methane) 171
Fizeau, Armand (speed of light) 264
flak 358
flywheel 91
focal point 169
footprint of a cosmic missile 310
footprint of a satellite 271
force 49
forensic science 188
fossil 315
fractal geometry 402
Frisbee 153
fugitive 89
fungi 392
fuse 169
fuselage 227
fusileer 169

G
Gaia 103
galaxies 279
Gatling gun (autoloading gun) 191
Gay-Lussac's, law for gases 244
Gemini 96
Geographos (asteroid) 320
geography 103
geologist 315
geostationary orbit 271
geosynchronous orbit 271
geyser 337
gimbal 95
Giotto de Bondone (artist) 308
glucose 177
Goddard, Robert (rocketry) 228
golf 5
golf links 4

good humor 268
govern 208
government 208
governor (political) 208
graphite 173
gravitational attraction 35
green paper 252
greenhouse effect (carbon dioxide) 256
greenhouse 256
grenada 170
gross (one dozen dozen) 105
gun 181
guncotton 173
guttapercha 5
guttie 5
gyroscope 95
gyroscopic compass 96

H
Halley, Edmond 307
hard vacuum 146
hard-court tennis 67
heavy water (D_2O) 345
Hess, Victor Francis (cosmic rays) 266
heuristic programming 209
high compression balls 8
high explosive, example 171
hollow-point bullet 197
holocaust 102
Holocene 316
holographic manuscript 101
holography 101
home plate 127
horizon 239
horizontal 239
humorous 268
Hussein missile 193
hydrazine rocket fuel 236
hydrodynamic focusing 395
hydrofoil 151
hydrogen oxide 348
hydroponic culture of plants 291
hydrosphere 261
hypersonic wind tunnel 240

I
icing sugar 177
ideogram 167
igneous (volcanic) rock 296
ignite 296
impasse 223
implicit 254

implode 145
impulsive force 20
incandescent 238
incendiary bullet 193
inclination 239
incoherent light 199
Indian Mutiny 184
induction 247
inelastic 10
infrared radiation (heat) 248
insulate 245
insulin 247
intangible 88
intercontinental rocket 230
interpolation 244
intuition 3
invert 239
ionosphere 261
ions 261
iris 267
Islands of Langerhans (insulin) 247
isotopes 192

J
javelin 139
Joule, James Prescott 38
joule 38
Joule-Kelvin effect 234
Jules Verne Launcher (JVL) 203

K
Kamikaze 212
Kelvin temperature scale 242
Kentucky Bluegrass 12
Kevlar 358
Khaki (camouflage) 185
killing stick 156
kilojoule 40
kiloPascal 146
kinematics expert 18
kinetic energy 18
Kylies 156

L
laminar flow, defined 23
Langmuir, Irving (deerfly investigation) 390
laser range finder 202
laser rifle 200
laser, acronym 102
latent fingerprint 334
latent heat 334
latent image 334

lawn tennis 65
"lead" pencil 173
lens 168
lenticular 157
lentil 168
Leslie's cube (heat transfer demonstration) 250
Ley, Willy (rocketry) 229
LIDAR (light radar) 193
lipid 407
liposome, medication 407
liquid oxygen 231
liquid propellant 231
liquified natural gas 237
lithography 260
lithos 260
lithosphere 143
logarithmic scale 241
logarithms 241
Lord Kelvin 244
love 65
low Latin 169
Lucretius 103
luge, aerodynamics 382
luminiferous ether (medium for light) 264

M

Mach number 188
Mach, Ernst 188
machine 17
macroherbivores 392
magazine 183
magic bullet 59
Magnus effect 21
maiden over 125
man-machine interface 18
mandrel 231
manometer 145
Maralinga, Australia 156
massage 296
Maxim gun (autoloading gun) 191
mercury 39
mesosphere 261
Mesozoic 316
metamorphosis 59
metaphor 57
meteor 253
meteorite 253
meteorologist 253
methane 171
method 59
methyl group 173
metrologist 253

Michelson, Albert 266
Michelson-Morley experiment 266
Microabrasion Foil Experiment (MFE) 309
microcapsules 407
microherbivores 392
micrometer 176
microphone 71
microscope 72
microtome 360
Milikan, Robert Andrew (cosmic rays) 267
milk-stool rocket 104
milliwatt 200
molecular drizzle 314
Molotov cocktail 198
momentum 221
monochromatic light 199
monolith 260
monomer 178
Morpheus 59
morphine 59
morphology 59
music of the spheres 307
musket 183
musketeers 183
muzzle 182
mythonym 104

N

nanometer 176
Napier, John (logarithms) 241
neutral buoyant state 211
neutrino 344
neutron 346
Newton, Sir Isaac 35
Newton's First Law of Motion 35
Newton's Second Law of Motion 36
Newton's Third Law 223
newton 38
niobium laser 201
nitrocellulose 177
nitroglycerin 173
Nobel prizes 175
nodes of oscillation 74
norgestone 197
nuclear physicist 345
nuclear reactor 224
nuclear winter 319
nucleating technology 337
nucleus (atomic) 345

O

old dynamite 174
omen 306
one radian 106
onomatopeic word 169
Oort cloud 308
Oort, Jan (astronomer) 308
orbit 269
orbital trajectory 269
organ 16
ornithologist 388
oscilloscope 72
over 125
oxidizer (rocket fuel) 231
ozone sickness 262
ozone 261

P

π (pi) 106
Paleozoic, Earth history 316
Parkinson's disease 366
Parthian shot 51
Pascal 146
passenger cage 373
passive satellite 273
Patriot missile 193
peashooter 165
percussion cap 171
perfectly black body 251
perfectly elastic collision 10
perigee 273
perihelion 304
perimeter 105
period of rotation 106
periodic table of the elements 346
periphery 105
periscope 72
pervert 239
PETN 180
phalanx 114
philistine 108
phonetics 103
photocell 71
photography 102
photomultiplier 349
photosynthesis 177
phytoplankton 175
pictograph 167
pill 124
Pioneer 10 (spacecraft) 278
pitch 125
pitcher 125
Pitot tube 150

Planck, Max (quantum theory) 254
planet 176
plastic bullet 197
plastic explosive 180
plasticizer 180
Pleistocene 316
Pluto 296
plutonic rocks 296
polar orbit 272
Polara golf ball 25
politicians 102
polymer 178
polysaccharide 178
polyurethane 231
pomegranate 170
potential energy 18
precession of a top 93
pressure 49
Principia, Newton 269
propellant 165
proton 345
proximity fuse 212
pseudo 237
pseudonym 237
pub 140
pugilist 366
pyrophoric 233

Q

quantum theory 250
Questra, soccer ball 12
quoit 139

R

racket 65
racketeer 66
racquet 65
"radar gun" 79
radian measure 106
radiation 106
radiometer 387
radius 105
railgun 205
railway wagon 101
range 79
Rayleigh, Lord 254
RDX 180
reaction 223
reactor 223
reagent 223
recline 239
recoil 223
reduction 247

relativity, special theory of 266
resolving applied forces 37
resonate 61
restitution 8
revert 239
revolutions per minute (rpm) 106
riblet tape (fluid drag) 97
Richter Scale 301
Richter, Charles F. 301
rocket plume 194
round of ammunition 183
royal tennis 67
rubber bullet 197
Rutherford, Ernest 143

S

saccharine 178
Sadarm 206
Sagger missile 206
Sagittarius 35
sal ammonia 179
saltpeter 166
satellite 269
scent spray bottle 148
Schonbein 177
scientific method 59
Scud missile 193
seismogram 301
seismograph 301
semantic trap 395
Semtex 180
serendipity 174
set-up 102
shark 96
shatter cone 288
shock metamorphic features 296
shooting star (meteor) 253
Shrapnel, Henry 195
shuttle cock construction 81
silica gel 248
sine wave 106
sinuses 106
smart mine 206
smelling salts 179
snow snake 140
Snowdrop 258
Soddy, Frederick (isotopes) 348
soft vacuum 146
solar wind 304
solid carbon dioxide 338
solid propellant 231
specific heat 263
spectacles (glasses) 168

spectator 250
spectrum 249
spent uranium 192
spin bowler 126
spindle 225
spindly 227
spinster 227
splash 169
spores 392
sports ergonomics 18
Springfield repeating rifle 191
star 252
static pressure, Bernoulli's principle 148
Stefan's Law 252
Stefan, Josef 252
stick sling 114
sticky wicket 126
stimpmeter 11
Stingray 201
Stonehenge 104
strategy 195
stratosphere 144
streamline 23
striated surface 288
strike zone 127
stroboscope 19
stumped 125
Sudbury Neutrino Observatory (SNO) 345
sulfurcrete (SNO) 349
Super High Altitude Research Project (SHARP) 203
supercooled liquid water 336
supersonic wind tunnel 240
sweet spot 71
synchronized 271

T

tactic 213
tangent to a circle 88
tangible 88
target 61
targeted delivery 406
targeted drug 61
Teflon 100
telescope 72
tendon 45
tendonitis 70
tennis 65
tennis elbow 70
thermally efficient flowers (Snowdrop) 258
thermoforesis 248
thunder clap 186

Titan rockets 103
toluene 173
Tomahawk, guided missile 193
tome 360
topspin on a ball, effect 21
torch 89
torpedo 96
torpid 96
torque 89
Torr, unit of pressure 146
Torricelli, Evangelista 145
Torricellian vacuum 145
tortional pendulum 90
toxicology 51
toxophilite 51
tracer bullet 193
transducer, defined 71
translational velocity 87
transparency 256
transponder 273
triangle 107
trigonometry 107
trinitrotoluene (TNT) 173
tritium 348
troposphere 143
tuning fork 72
tunnel vision, Goliath 114
turbulent flow 23

U
U-tube manometer 145
ultraviolet catastrophe 254
ultraviolet radiation 248
umbilical cord 104
umpire 65
universal joint 96
universe 239
university 239
uranium 103
Uranus 103

V
vacuum 144
Vacuum Monster 313
Venturi throat 143
vertical 239
vibration 71
vitreous humor 267
von Braun, Werner (rocketry) 229
vortex 239
vulgar Latin 169

W
water tunnel 377
Watt, James 199
Watt 199
wavelength 73
weightlessness 91
Wein's Law 254
Wein, Wilhelm 254
Whipple Meteor Bumper 309
Whipple, Fred (nature of comets) 308
white hot 238
Whitehead, Robert (torpedo) 96
whitewash 251
Wide Area Mine (WAM) 206
William Thompson 244
wind angle 157
work 36
work hardening of a metal 356

X
xylophone 74

Y
yo-yo 98
Yucatan 319

Z
Zeppelin 140
Zeroth Law of thermodynamics 255
zoom lens 169

Chapter 1

Why do Golf Balls have Dimples?

Chapter 1

Why do Golf Balls have Dimples?

Section 1.1 Warning: Golf Balls are Potentially Lethal!

Walking one day by a golf course in Wisconsin I was startled to hear a sharp crack as a golf ball narrowly missed my head and hit a tree. My companion cheerfully remarked

> That could have killed you, you know. A man was killed last week at a golf course in Illinois when he was hit by a stray golf ball traveling at high velocity.

As I picked up the innocent looking little white ball that had rolled away from its encounter with the tree I noticed its structure with a new respect. As I held up this potentially lethal missile, my companion quickly moved on from his morbid commentary to pose an interesting question

> How come a golf ball has dimples? Surely common sense would say that a ball should have a smooth surface if you want it to travel as far as possible.

My companion's statement reminded me of a comment once made by Einstein, probably the greatest physicist of the 20th century. He pointed out that common sense is actually nothing more than a collection of prejudices laid down in the mind prior to the age of 18 [1]. Often when discussing things like the shape of a golf ball one makes the statement that "intuitively one feels it should be smooth." The dictionary definition of *intuition* is that it is

> the power of the mind by which it immediately perceives the truth of things without reasoning or analysis.

As we will discover many times in this exploration of the science of missiles the truths encountered in science are often *counter-intuitive*; that is the real truth is the opposite of what our intuition would suggest to our mind as the truth. It is this fact that often makes it difficult to grasp and understand scientific truths. I must admit that when my companion said that "surely a smooth ball would go further than a rough one", my instinct was to agree with the statement. I had to admit to my colleague that in my three years of study for a science degree I had never acquired an understanding of why a golf ball had dimples. I promised that as soon as I returned to my study I would explore the reasons for the dimples on the golf ball.

As I searched for information on the dimpled appearance of the golf ball, I started to find some fascinating information. I found that a golf ball driven by a tournament professional leaves the tee at about 70 meters per second or 250 kilometers per hour (156 miles per hour. Note: we will give metric measures of speeds etc. but since "miles" still dominate in the USA, the equivalent classical quantities will be quoted immediately after the metric units.) I found that a driven golf ball accelerates 10 000 times faster than the fastest acceleration available in the most powerful of sports cars. My physics textbook told me that the dimpling of the golf ball decreased the zone of turbulence behind the ball thereby decreasing the resistance experienced by the ball at high velocities; we will discover what that means later in this chapter [2, 3]. I also discovered that most golf balls have 336 dimples spread evenly over their surface, but that in some more recently suggested designs for better golf balls, the number, the pattern, and depth of the dimples are varied [4–6]. As I contemplated the dimples on the near lethal missile from Wisconsin which lay quietly on my desk in front of me, I resolved to gather together all the science I could find about this interesting missile and all the other missiles used in ball games and related games involving the throwing of objects such as the hammer, the boomerang, and the Frisbee. I also began to study more dangerous missiles used in warfare, starting with bows and arrows, and working my way up to guided missiles and antitank shells. Later, as my explorations expanded, I found myself looking into the physics of how biologists introduce small foreign bodies into living cells to develop techniques used in the field of genetic engineering, and at the potential damage inflicted by fast, miniature meteorites on spacecraft moving through cosmic dust. I also looked at the potential effect of the earth being hit by asteroids. Hopefully, the resulting essays on the science of a myriad of missiles gathered together in this text will prove as interesting to the reader as the exploration of the facts proved to be for me.

Section 1.2 Featheries, Gutties, and Composite Balls

To discover how the golf ball gained its dimples, we need to begin with a brief history of the game of golf [2–4]. There is some controversy over who invented golf, but the reference books seem to come down on the side of the Scots [5]. It appears that they started to play this game along the sand dunes of the coast where the hazards of the game included losing the small ball in rough patches of grass on a sand dune, on a sand dune itself, or in pools of water on the shore. To this day the modern builders of golf courses imitate the problems met along the seashore by building sand traps and water hazards into golf courses. The word *links*, as in *golf links*, appears to come from an Old English word for shoreline. The aim of the game of golf is to hit the ball around the golf links, or golf course as it is often called, lodging the ball in a series of small cups on each of several greens with the minimum number of strokes of the club. The word *golf* itself seems to come from an old Scots word for club. Golf was so popular amongst the Scottish people that in 1457 the King of Scotland, concerned that golf was taking up too much time and distracting people from the practice of archery which he deemed essential for the defense of Scotland, published a royal decree in which he demanded that "golfe be utterly cryed downe." However, Scots continued to play their favorite game. It was introduced into France by Mary Queen of Scots. The people who were learning to play the game in France were called "cadets." This word is pronounced, in French, "cad-day." For this reason the young boys and girls who carry the clubs around after the golfers on modern golf links are called *caddies*. The first golf balls were made out of untanned cowhide that was stitched to the desired shape leaving a small hole in the side [6]. The ball was then turned inside out and stuffed with as many boiled feathers as could be pushed through the hole. After the hole had been closed, the feathers were left to dry. Finally, the craftsman pounded the ball round and painted it white.

The first major change in the design of golf balls occurred in the middle of the 19th century. It is said that an ardent golf-playing professor at St. Andrew's University in Scotland received a statue of the Hindu god Vishnu that had been shipped by sea from India. This fragile statue had been packed in a cradle of a rubbery substance called *guttapercha*. This substance is the solidified juice of various Malaysian trees. The name comes from the Malaysian words "getah" meaning gum and "percha", a tree. We are told that the gum, as it oozes from the tree, hardens to form a gummy-type substance not unlike a hard rubber. The professor must have noticed that when this material fell on the floor it bounced because we are told that he quickly fashioned a golf ball out of the guttapercha. The original guttapercha balls were made of sheets of the dried gum wound into a ball which was boiled and pressed into shape. The popular name for this type of ball was *guttie*. The gutties greatly reduced the cost of golf balls but they had a problem: they became brittle in cold weather and were prone to split on impact.

We are told the Scottish professor noticed that as he played with his new ball that the more nicks and depressions his golf clubs cut into the surface of the ball, the further the ball traveled. The professor then experimented with the deliberate dimpling of the ball to improve its flight characteristics. Because of the careful observations of the golf-loving professor, today's dimpled golf ball evolved from its damaged, gummy ancestors. Different companies making golf balls experimented with different patterns of dimples and finally settled on today's pattern of 336 symmetrically placed holes [6].

The modern golf ball was born in 1902 when Coburn Haskell of Akron, Ohio made a ball by winding elastic thread on a spherical, hard core, adding a final casing of guttapercha. In more recent times, a solid ball made out of a suitable plastic has become available. A section through a modern traditional golf ball is shown in Figure 1.1.

Some readers may be surprised to find that the dimpled golf ball does not owe its origin to the wonders of modern science but evolved from astute observations on the aerodynamics of damaged balls by people who knew virtually nothing about physics.

a)

b)

Figure 1.1 The modern golf ball is descended from a feather-filled hand-sewn leather ball, and may have various internal constructions. a) A modern golf ball usually has 336 dimples. b) The golf ball can be made from a rubber pellet surrounded by elastic, then covered with the outside case, or may have a solid plastic core.

In science, when we discover something that works by *experiment*, even when we do not understand the theory behind our innovation, the knowledge is described as *empirical knowledge*. This word comes from a Latin word "empiricus" which means some

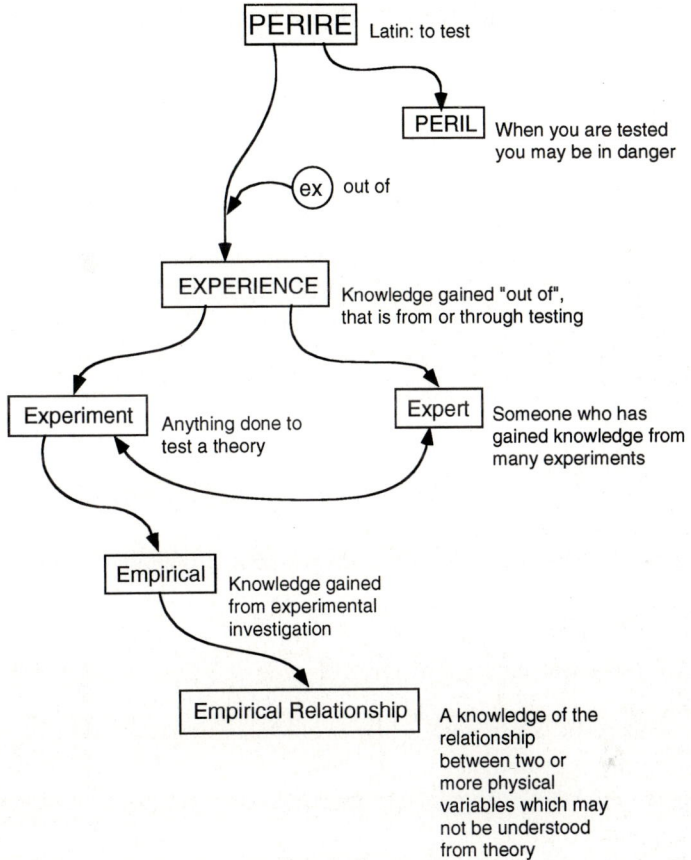

Figure 1.2 Empirical knowledge is gained by experience.

thing gained by trial and error. This word empirical is closely related to the words *experience*, *experiment*, and *expert*, as illustrated by the wordweb of Figure 1.2. As modern scientists start to apply their knowledge of aerodynamics to the design of golf balls and clubs, they are finding that the old players using trial and error have managed to evolve a very efficient system of ball and club for obtaining near optimum dynamics. However, there are still innovations to be made in design, but professional golfers' associations are resisting some of the new efficiently designed golf balls because they feel that it may become too easy for the relatively unskilled player to play good golf [6]. Thus, as we shall discover, some of the more modern designs of golf balls cannot be used in professional golf tournaments because they have a self-correcting action which counteracts the effect of a miss hit with the club (see the discussion of the Polara golf ball later in this chapter).

Section 1.3 Bouncing Balls and Leaping Athletes

An important property of the balls used in different types of ball games is the amount of energy it they retain after bouncing off a surface such as the golf club or the tennis racket. In everyday speech people talk about the "bounce of the ball". A professional golf player can sometimes be observed to check the bounce of his ball by bouncing it on a smooth surface before starting a golf game. Scientists measure the "bounceability" of a ball by means of a quantity defined as the *coefficient of restitution* [7].

The word *restitution* is used in everyday speech to describe the type of payment made to replace a damaged object or to make good loss of income caused by the act of the person making the payment of restitution. This meaning of the word does not help us much to understand what is meant by restitution in a game of golf! The term restitution comes from two Latin root words: "re" meaning again and "stature", to make, to stand (hence the word statue). To make restitution then literally means to restore an object or a situation to its original condition. We all know from experience how the return height of a bouncing ball decays with repeated bounces as shown in Figure 1.3(a). The fact that the ball fails to reach its original release height is a failure of the system to achieve restitution of the original height. The coefficient of restitution is a quantitative measure of the loss in height at each bounce. When released from the same height, a ball made of material with a low coefficient of restitution would not bounce back to the height achieved by a ball with a higher coefficient of restitution.

The term coefficient came into use in science when scientists were measuring many properties of material and listing these measured properties before theories were known. The word was coined by joining "co" which means together, and "efficient" which originally meant capable of doing something. Efficient came from Latin root words "ex" meaning out and "facere" meaning to make. Thus, if we knew the coefficient of restitution, we would know how things worked together to make the ball bounce. Scientists exploring the properties of different materials would make balls of all sorts of different materials and measure the bounce of each ball. They would then plot the coefficient of restitution against the substance in the ball. For example, when the early manufacturers of golf balls experimented with how tightly they wound the elastic thread around the central core, they found that the tighter they wound the thread, the higher the coefficient of restitution of the bouncing ball. These better bouncing balls are called *high compression balls*.

We will meet the coefficient of restitution in any game where the ball is hit with a bat or a racket or when the ball bounces off a wall or other hard surfaces. It can be shown that the scientific measure of the coefficient of restitution is given by the formula shown in Figure 1.3(a). When we bounce an ordinary rubber ball off a surface, we learn through experience that if we wish to catch the ball at the same height at which we release the ball from our own hand, we must give the ball a little bit of extra energy as we throw it down on the surface. This means we release the ball from our hand such

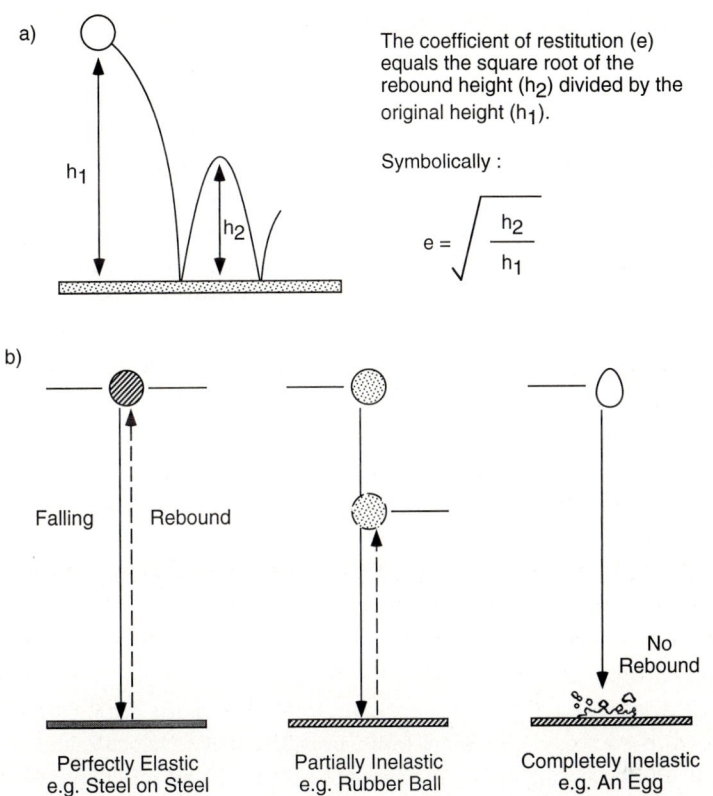

Figure 1.3 The coefficient of restitution is used to quantitatively describe how high something will bounce after hitting a hard surface. a) A method of determining the coefficeint of restitution. b) A perfectly elastic collision results in the object returning to its original height. A perfectly inelastic collision will result in no bounce.

that it is already traveling at the speed it would have gathered if it had been released from a greater height. Then the first bounce height will be back up to the position of our hand. A few years ago scientists working with rubber discovered that they could make a particular type of rubber that had a very high coefficient of restitution and that balls made of this material would bounce higher than ordinary rubber balls. When played with, these "super balls" appeared to bounce in a very lively manner. Because people threw them down with the same energy they would give to ordinary rubber balls, the super balls bounced back higher than anticipated causing surprise to most observers.

It can be seen that the higher the coefficient of restitution describing the interaction of a ball with a given surface, the higher the ball will bounce. If the ball were to lose no energy at all, bouncing back to exactly the height from which it was dropped, the coefficient of restitution would be equal to 1. In scientific terms, the collision of the ball with a surface from which it rebounded without loss of energy is described as a *perfectly elastic collision*. I can remember, when I was taking my physics courses at high school, that I was greatly confused by the terms *elastic* and *inelastic* as they were used by my physics teacher. I remember being very confused as he demonstrated that a steel ball falling on a steel plate was almost a perfectly elastic collision whereas a soft rubber ball falling onto a surface had a relatively low coefficient of restitution, and a collision between two rubber balls was a relatively *inelastic collision*. I now realize that in science the word elastic has a second meaning that seems to have nothing to do with its usual meaning in everyday speech To an ordinary person the word elastic means stretchable, easily deformed with rapid recovery of shape. After all, one puts an elastic band around the wad of papers, exploiting what one describes as the elasticity of the rubber. The word elastic comes from a Greek word "elastikos" meaning to drive. The idea was that an elastic body could be made into a spring which could be used to drive machines or throw rocks. Today we make simple toys for children driven by elastic bands that are twisted before release of the toy. Confusion has arisen because in everyday speech we have focused on the springiness of material as meaning elasticity whereas the scientist uses the word elasticity also to describe the conservation of energy during a collision. It just so happens that rubber which is very stretchy and hence "elastic" in everyday speech absorbs energy when stretched and so it will not undergo an elastic collision. [See Figure 1.3(b)].

Some of the words having very specific meanings in science which are not the same as those used in everyday speech are force, pressure, speed, velocity, and work. As we develop our understanding of the behavior of golf balls and other interesting missiles, we will have to define carefully each of these words to avoid confusion caused by ideas we have acquired in everyday communication.

In the USA regulations state that a golf ball must not attain more than a certain velocity when fired from an official machine designed to simulate the blow received from a ball in the full drive from a golf tee. Indirectly this restricts the permissible value of the coefficient of restitution. The typical commercially available golf ball in the 1980s had a coefficient of restitution of about 0.7 [3]. Some of the new solid golf balls have higher coefficients of restitution than the classical golf ball. For example, in an advertisement for a golf ball known as the Guidestar, it is claimed that this ball has 21% more rebound power than other commercially available balls. As already mentioned, the Professional Golfers' Association is concerned that new developments in the technology of making balls may make it too easy for a mediocre player to achieve good golf scores. Daish [3] suggests that any move to improve the coefficient of restitution beyond the value of 0.7 would be banned by the golfing associations.

When the golf ball is hit by a club the coefficient of restitution is essentially that exhibited by the golf ball when it is bounced off a hard concrete surface. However, once the ball arrives on the green the physical dynamics of the ball are determined by the surface of the green. (Assuming the unfortunate golfer does not hit a sand trap or water hazard!) Professional golfers are very concerned with the physical properties of the surface of the green and the people preparing a surface of the golf links take great care with the grass growing on the green. However the physical properties of the grass surface can vary considerably from course to course [8, 9].

Apparently there are two major types of golf course surfaces favored by different groups of golfers. The original golf course, which is claimed to be the father of all the golf courses in St. Andrew's, Scotland, favors one type of surface whereas in North America golfers prefer what is known as the *"Augusta dream"* after a famous American golf course in Georgia. In the words of Pearce [9]

> *The Augusta green courses have wide fertilized and gorged fairways, narrow close-cut roughs and consistently irrigated pesticide treated greens. To recreate the "Augusta dream" American Bermuda grass with its rapacious demands for water and chemicals has become the essential ingredient for top flight courses.*

Pearce goes on to tell us that

> *The traditional British links is now widely seen (by professional golfers) as a quaint anachronism,*

and that

> *Foreign golfers such as the American Jack Nicklaus hate it when the British open comes to Royal St. George's in St. Andrew's.*

When one starts to look at the literature on the type of grass favored for a golf course one soon finds oneself dealing with names such as "agrostis tenuis" and a grass known as "red-fecucha rubia." The properties of turf used in different sports is so important that there is actually a "Sports Turf Research Institute" in Bingley, Yorkshire, England. An employee of that institute, Mike Canaway, tells us [9] that

> *Many factors influence the quality of greens, the use of fertilizers and irrigation, wear, weather, animal and fungal pests, soil and damage, as well as the actual grass.*

The *speed of a green*, how fast a ball tends to roll across it during putting, is particularly important to players. Researchers estimate this value by sending balls across the green from a standardized metal ramp called a *stimpmeter* to measure how far they roll. Other instruments are used to measure the hardness of the green and the coefficient of restitution when the ball is dropped from a standard height. Another machine is used to fire golf balls at greens at realistic speeds, angles, and spin. Photographs are taken as they hit the turf.

Moisture on the greens surface also slows the ball. In Britain some of the greens, which are dominated by what is known as annual meadow grass, can become so spongy that they are known as sponge puddings.

The way in which balls bounce on the turf has also become a major problem in World Cup soccer matches. In preparation for the World Cup series, played in the spring of 1994, two specialists in the discipline of turf science studied the bouncing of footballs on each of the pitches where the various matches were played [10]. With the information that they collected they tried to make sure that the grass surface of every stadium, used in the competition around the United States, was as similar as possible. The experts found that the type of playing surface varied over the nine different locations stretching from Massachusetts to Florida and California where the games were played. In the warmer climates a type of grass known as *Bermuda grass* turf is used. This gives a surface not unlike the putting green of a golf course. In contrast more northerly pitches use *Kentucky Blue Grass* which is softer. It cannot be cut so short as the Bermuda grass and this offers more resistance to the ball as it rolls and bounces. In the World Cup games fields with artificial turf, had to be overlaid with sod because the World Cup rules required that all games be played on real grass. Just before the games started the experts used a device known as a *field performance indicator* to make sure the grass was according to standards. In the operation of the field performance indicator, a football is rolled down a ramp inclined at an angle of just less than 45 degrees. When the ball hit the turf, a gauge measured how high the ball bounced and the distance rolled by the ball was measured. The Rosebowl turf in Passadena was taken to be the standard for the games. To make other turfs match the Rosebowl turf the experts adjusted the cutting height for the mower or used selective watering and rolling procedures before the soccer match [10].

Commentators watching the World Cup matches in 1994 reported that some players consistently kicked passes too hard as if they were underestimating how far and fast the ball would travel [11]. According to reports, Italian, Belgian, and Norwegian players all complained about the ball used in the World Cup. For this particular series of soccer matches a special ball had been developed which when kicked received more energy than traditional soccer balls. This ball had been especially developed by Addidas and was known as the *Questra*. It had been specially designed in an attempt to improve the liveliness of the soccer game for the benefit of the spectators. In the Football Associations bounce test a ball is dropped from two meters onto a hard surface and the time it takes to reach its maximum height is measured. The Questra rebound was 5% quicker than 5 balls produced by rival manufacturers. The Questra ball had a special polyurethane coating which reduced friction. On the inside of this layer Addidas added a rubbery layer that allowed players to get more "bang from their boots". A spokesman said if you strike the ball hard it flies very rapidly off the foot because all of the energy goes into kinetic energy and none is wasted on the ball, i.e. in scientific terms the coefficient of restitution was higher. If however the ball is kicked softly the material has time to deform and so it is easy to control. The new ball was made available to the competitors in the World Cup series three months prior to the tournament but it appears that some players were unable to modify their playing skills to cope with the new ball [11].

The energy interaction between bouncing balls and the playing surface is not the only area of athletic competition where the coefficient of restitution is important. Thus, purists argue that Carl Lewis was able to record the fastest ever hundred meters in the World

Athletics Championships in Tokyo, 1991, because the track was unusually hard. In the words of Coghlan [12]:

> Tests on samples of the Tokyo track showed that the track was unusually hard, a property that suits sprinters and jumpers who want as little as possible of the power unleashed by their leg muscles to be dissipated by the track's surface – the harder the surface the greater the reaction forces imposed on the track.

Coghlan goes on to point out that, although such hard surfaces favored long-jump athletes and sprinters, running for long times on such surfaces can cause problems for long distance runners. Thus in the Tokyo games Liz McColgan, the ten thousand meter champion, complained that the track in Tokyo had given her sore limbs because of the prolonged jolting they received on the hard track. In these days of modern technology the authorities have started to specify the frictional and rebound characteristics of a surface. In the Barcelona Olympics a synthetic rubber known as polychlorophene, which came in the form of prefabricated sheets, were laid down to create uniform running surfaces. The reader interested in the development of standard surfaces and the measurement of the properties of these surfaces can find details in the articles by Coghlan and Brody [12, 13].

Studying the physical behavior of bouncing objects has some surprising applications. Thus in Holland, Joost Vander Burg is studying the bouncing behavior of pea seeds. In this research, he drops the peas from a height of a few millimeters onto the sensor which measures the forces which act on the surface during the impact. He has shown that the best seeds have a high coefficient of restitution whereas damaged peas, which would rot in the ground when planted, give a different type of interaction with the surface. Not only could his device be useful for selecting prime seeds but in the article describing his work it is said that he could also select the best peas for making pea soup [14].

Section 1.4   Golf Balls with Perfect Centers of Gravity

In the advertisement for the Guidestar patented golf ball it is said that the Guidestar has a center of gravity which is 97.5% perfect compared to 58% for manufacturer A, 28% for manufacturer B, and worse for other manufacturers. To understand what is meant by the center of gravity, let us consider what happens when we hang an irregularly shaped piece of wood from a nail. Consider the piece of wood shown in Figure 1.4 in which three holes at points A, B, and C have been drilled. If we hang the piece of wood from a nail at each of these holes, we would find that a plumb line drawn from the point of suspension would cross the profile as illustrated in Figure 1.4(a). The point at which the 3 lines met would be called the *center of gravity* of the board. It is that

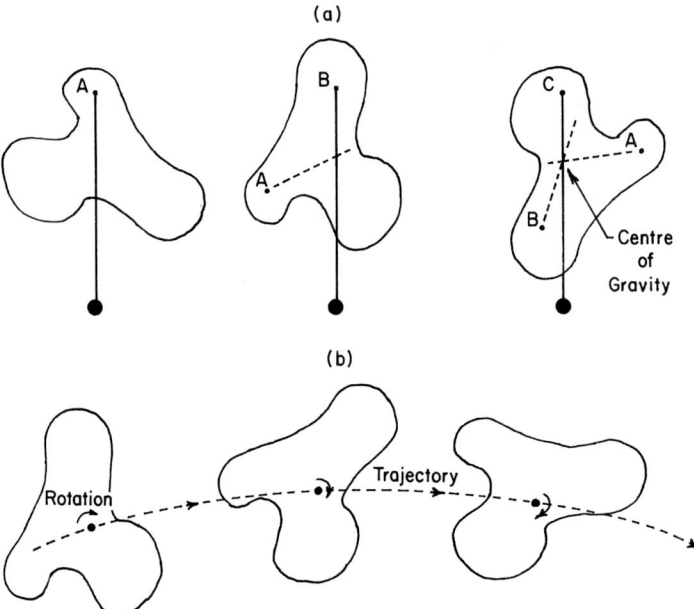

Figure 1.4 The center of gravity of an object is the point at which all of its mass can be considered to be concentrated. a) The center of gravity of a thin piece of wood can be found using a plumb line as shown. b) When an irregular shaped object is thrown, its trajectory follows the path of the center of gravity.

point at which the weight of the object can be considered to act as if all the mass of the object were concentrated at that point [7, 15]. Thus if we cut out a shape such as that of Figure 1.4(a) we would find that it would balance on a needle placed at that point. If we then took the piece of wood and threw it through the air, we would find that the trajectory of the piece of wood would be the same as if all of the mass of the piece of wood were concentrated in a tiny ball located at the center of gravity of the piece of wood (for the meaning of the word trajectory see Figure 1.5). If we had thrown the piece of wood with a spinning action, we would also have found that it rotated about the center of gravity as it flew through the air as shown in Figure 1.4 (b).

Let us consider what would happen if we studied a disc with a uniform distribution of material in its construction. If we checked on its center of gravity using the plumb line, as for the odd shape in Figure 1.4(a), we would find that the three points met at the center of the circle as shown in Figure 1.6(a). If we then modeled the construction of a golf ball in two dimensions with the central disc having a hard core which was slightly off center, then the system would be as shown in Figure 1.6(b). For this system, attempts to measure the center of gravity would result in the three lines meeting at a point which is off center. If the disc of Figure 1.6(a) is thrown through the air with a spinning action, it would spin about its center of gravity and, in the words of a golfer, it would fly true. If, however, the disc of Figure 1.6(b) were to be thrown in a similar

1.4 Golf Balls with Perfect Centers of Gravity

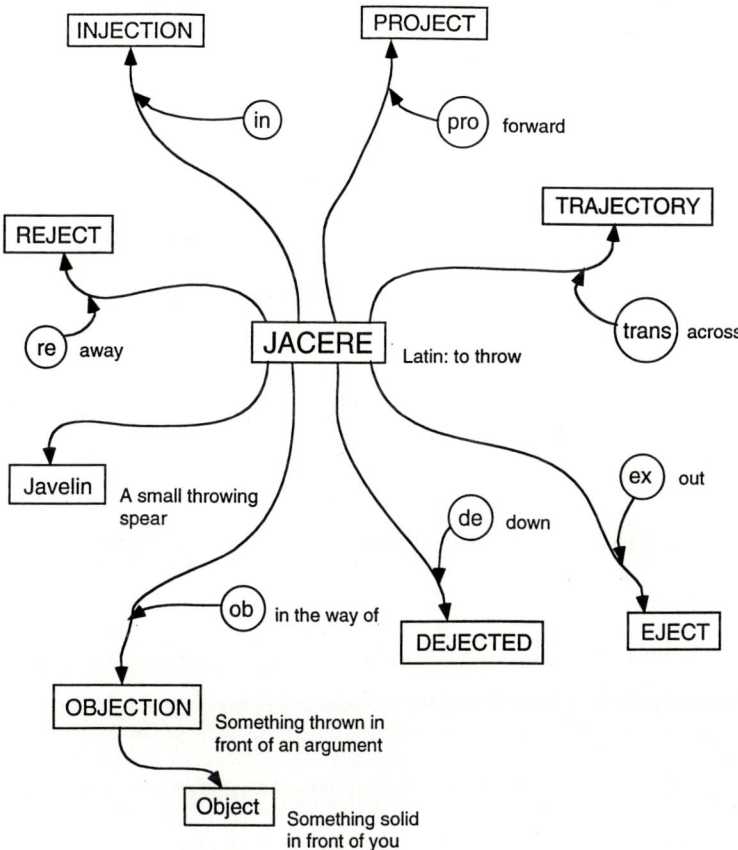

Figure 1.5 The technical term for the path traced out by a thrown object is "trajectory."

way, it would rotate about its actual center of gravity as it flew and would appear to wobble in space as shown by the sequence of Figure 1.6(c). Thus, a golf ball whose center of gravity does not coincide with the geometric center appears to wobble in flight and does not fly "true". We can be sure that manufacturers try to minimize the deviations of the center of gravity from the center of their golf balls However, as the advertisement for the Guidestar ball indicated, traditional balls are not always perfect with respect to the location of the center of gravity. After a ball has been hit a few times it may become sufficiently deformed for its center of gravity to be displaced from the geometric center of the golf ball causing the ball trajectory to deviate from what the golfer calls true flight. This is why a professional golfer will change balls after a few hits during a game.

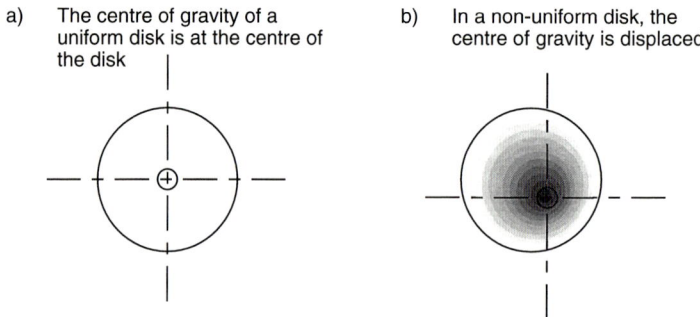

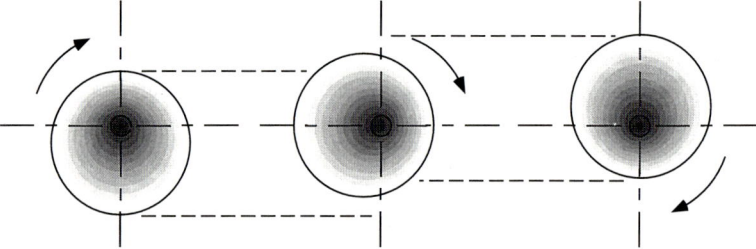

Figure 1.6 In a composite golf ball a non-centered "center of gavity" (also known as the center of mass) causes the ball to wobble in flight as it spins. a) The center of gravity of a uniform disc is the center of the disc. b) In a non-uniform disk, the center of gravity may be displaced. c) If the center of gravity is offset from the physical center of the disc, the disc would wobble as it moved through the air spinning about its center of gravity.

Section 1.5 What turns an Innocent Dimpled White Ball into a Lethal Missile?

The answer to the question posed in the above title is energy. When the club hits the ball, energy is transferred to the ball. The amount of energy stored in the traveling ball determines how dangerous it is on impact. From a scientific point of view, the energy of an object is its capacity to do work. The term energy comes from the Greek word "ergon" meaning work. This word entered everyday English in a slightly altered form to give us the word *organ*, for a machine capable of work. Thus an organ of the body is organized tissue that can work to achieve tasks such as pumping blood around the body. We normally think of a machine as being something complicated with rotating meshing gears and knobs. A machine, however, can be something as simple as a lever or a wedge (see discussion of pulley in Chapter 2). The word *machine* is defined as "any

1.5 What turns an Innocent Dimpled White Ball into a Lethal Missile?

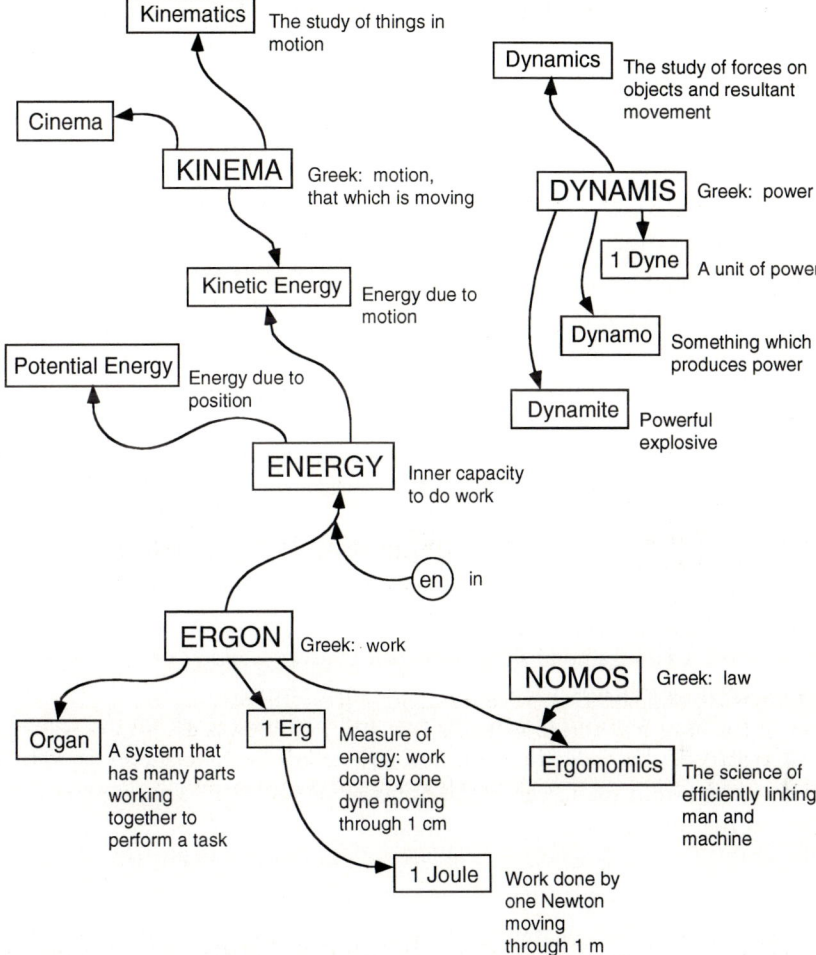

Figure 1.7 The energy invested in making a golf ball move can transform it from a dimpled mass into a lethal missile.

instrument for the conversion of motion." Ergon has been used recently to create a new word to describe a new science: *ergonomics*. As can be seen from the wordweb of Figure 1.7, this word was created by linking ergon to the root word "nomos" meaning law or set of rules. This root word is found in many scientific terms such as *astronomy*; the laws governing the behavior of the stars coined from the Greek root word "astron", a star. The term economics, originally the rules governing the running of a large house, comes from "oikos", Greek for a house. Ergonomics is the science of linking man and machine efficiently so that each performs at optimum levels. A branch of ergonomics is *sports ergonomics* concerned with achieving optimum sports performance by efficient design of the link between the operator, the machine, and their environment. A golf club is a lever, a simple machine in the strict sense of the word. A sports ergonomics specialist is concerned with designing the club so that, with the greatest comfort and ease, the golfer can transfer the maximum energy to the golf ball with the minimum of effort. The linkage between the golfer and club is defined by the ergonomist as the *man–machine interface*.

The sports ergonomics specialist must consider not only the interaction between the golf club and the ball but also how the golfer is positioned in his environment. Thus two mechanical engineering students studied the playing of 14 different golfers as part of a project carried out at the Massachusetts Institute of Technology (MIT). They concluded that the stiffness characteristic and spike patterns of the left and right shoes of the golfer could be designed differently to make the player more stable and improve the efficiency with which the ball is hit [16]. Thus the way in which a golfer's feet grip the ground affects the way in which he is able to swing the club and hit the ball. In these days of ever-improving sports technology, the sports ergonomist is becoming an important specialist [17].

The word energy was made by adding "en", meaning "in", to "ergon". Thus *energy* is the ability to do work inherent or hidden in a system. Thus we say that the moving golf ball has energy because in essence we could use the impact of the ball with an object to achieve work. In the game of skittles the ball's energy is used to knock over the skittles. In the game of billiards, the energy of one ball hitting another ball is used to put the second ball into motion; that is, the first ball performs work on the second. The energy available in a body due to its motion is described as its *kinetic energy*. This technical term comes from the Greek word "kinema" for motion. This same Greek term has given us the word cinema, a place where we watch moving pictures. The specialist who describes the motion of various objects is a *kinematics expert*. If he is concerned with the forces putting something into motion and subsequent changes in that motion, then the subject he is studying is described as *dynamics* from the Greek word "dynamis" for power (a word which has been used to coin the word dynamite for a particular powerful explosive).

If we consider an object at rest on the top of a cliff, we say that it has *potential energy* because of its position. If it is then pushed over the edge, this potential energy progressively becomes kinetic energy as the object is attracted by the gravitational forces of the earth. In some sports we store energy in a system which will then fire the missile by a process that converts potential energy into kinetic energy. Thus, when the archer draws the bow, energy is stored in the stretched bow. This potential energy is then converted into the kinetic energy of the arrow as it leaps from the bow. It can be shown that the kinetic energy of an object moving with a velocity v is equal to half the mass of the body multiplied by its velocity squared. The term "squared", meaning "multiplied by itself", is expressed mathematically by adding a small raised "2" after the symbol representing the squared quantity. Thus we can also write

$$KE = \tfrac{1}{2} m v^2$$

where KE = kinetic energy, m = mass of an object, and v = velocity of the object.

Although mathematical formulas, such as this equation for kinetic energy, might initially paralyze some readers with fear, one can see, by comparing the equation with the long winded sentence, that there is an advantage in using symbolic summaries of physical relationships. Remember the "2" written just above and to the right of the v indicates that the operation "velocity times velocity" must be carried out to calculate the kinetic energy of the moving ball.

Very often in the discussion of the physics of sports we use the words mass and weight interchangeably. Strictly speaking weight is the force exerted when a mass is acted on by the gravity that the body experiences in its environment. Since ordinarily we are only talking about playing golf upon the surface of the Earth, we use the words weight and mass interchangeably. Note however that one of the Apollo astronauts played a little golf on the surface of the Moon for the sake of demonstrating the effect of the weaker gravity on the Moon's surface. The mass of the golf ball was the same on the surface of the Moon as on the surface of the Earth, but the weight of the golf ball created by the gravitational attraction of the Moon was much less than the weight of the golf ball on the Earth because the mass, and thus gravitational attraction, of the Earth is much greater than the Moon's. A full discussion of the difference between mass and weight is beyond the scope of this essay but interested readers should consult standard physics textbooks [7, 15, 18].

Section 1.6 Golf Ball Dynamics; Impulsive Deformation and a Lift from Dr. Magnus

As the golf ball moves through the air, it must use some of its energy to overcome the opposition of air friction. At the same time, it is dragged downwards by the gravitational pull of the earth. The flight pattern of the golf ball is determined by the relative effects of these forces acting on the ball.

Many physics textbooks contain a picture of the behavior of a golf ball during the short time that it is in contact with the club. Many of these pictures come from the laboratory of Dr. Edgerton at MIT, who specialized in the high speed photography of moving objects [19, 20]. He developed an instrument for studying the dynamics of fast moving objects called a *stroboscope*. This word was coined by Dr. Edgerton from the two Greek root words "strobos" meaning twisting or turning, and "scopeein" to view or see. Originally a stroboscope was a light designed to flash at the same speed or some multiple of the speed of rotation of something like a phonographic disc. When illuminated by a regularly flashing light the rotating object appears stationary. The stroboscope is now almost universally called simply a "strobe."

When discussing high speed photos of a club hitting a ball attention is usually drawn to the fact that the golf ball, which to the ordinary touch seems to be a hard, not easily deformed object, is flattened by the club during the moment of impact. Such a picture illustrates vividly that forces acting during high speed impact are much higher than the forces that can be exerted on a ball in a static situation. It is often shown in a worked example in physics textbooks that to deform a ball by the amount seen in a high speed photograph, a static load equivalent to a ton weight would have to be applied to the ball [15]. The transient forces that occur during high speed impact are very much higher than most people imagine and are described as *impulsive forces*. The origin of the word impulsive is illustrated in Figure 1.8. When a car hits a tree, during the short period in which the car is brought to rest, the forces acting on the car are many times higher than the weight of the car.

The energy used in deforming the golf ball results in a rise in temperature of the ball in the same way that a nail gets hot when it is repeatedly hit to drive it into a wall. It can be shown that the temperature of a golf ball rises on average by $1°$ Fahrenheit when it is driven off the tee. In games such as squash or tennis, where the ball is hit repeatedly, the ball gets noticeably hotter during the game. This results in altered dynamics of the ball during play.

The solid golf ball moulded in one piece has a tendency to split as a result of the high impulsive forces acting on it as it accelerates off the tee. During a good drive, the club head is traveling at 175 kilometers per hour (110 miles per hour) when it hits the ball. High speed photographs have shown that irrespective of the style of the golfer and his swing technique, the time of contact between club and ball is very short and always about the same magnitude: that is 0.5 millisecond. This is one half of a thousandth of a second! A careful analysis of a golf shot shown by a series of high speed photographs of the moving ball and club shows that the ball and club are only in contact for the first part of the sequence of pictures and that in fact for the last club position, the ball has disappeared. It has been shown that it takes 0.6 millisecond for the impact produced in the club by the collision with the ball to travel as a mechanical disturbance up the shaft of the club. Therefore, the ball has already left the club before the hand feels that contact has been made. There would be a further two-thirds of a second delay before the brain could signal the hand to adjust the the club if the player sensed that the blow was less than a perfect golf stroke. This time is required for the nerve signal to travel up to the brain to be processed before a subsequent signal is sent back to the hand for action. There is therefore no way that a golfer can really affect his stroke by corrective action if a bad swing has been made [3].

Another empirical piece of information that golfers worked out for themselves by trial and error is that, in order to make the ball travel as far as possible, the golfer should hit the ball with *backspin*. This type of spin creates lift in the ball by an effect known as the *Magnus effect*. Spin in the opposite direction is known as *topspin*. Topspin is virtually unknown in golf although widely used in such sports as tennis and table tennis. Topspin, because of the Magnus effect, causes the ball to dive. The physical cause of the

1.6 Golf Ball Dynamics; Impulsive Deformation and a Lift from Dr. Magnus

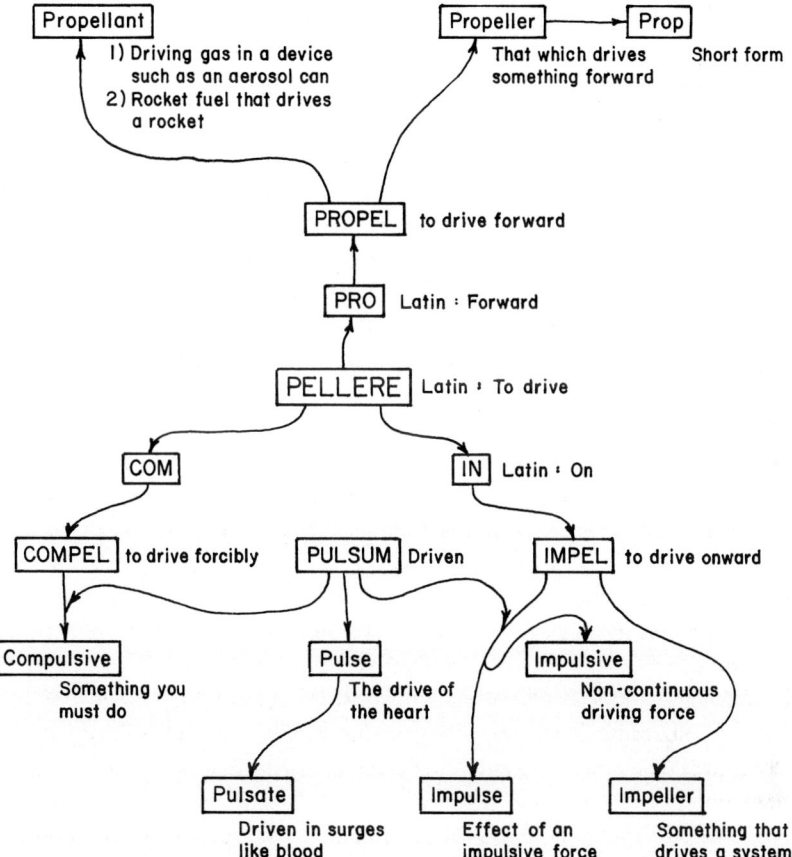

Figure 1.8 Impulsive forces can drive an object forward.

Magnus effect can be visualized by considering the sketch given in Figure 1.9(a). Scientists imagine that smooth lines can be drawn in non-turbulent flowing fluid to indicate the movement of elements (small pieces) of the fluid. (Note that scientists use the word "fluid" to include both liquids and gases — i.e., things that flow.) Dynamically this shows that the flow of fluid past a stationary spinning ball is exactly the same physically as the movement of the spinning ball through static fluid. As the ball moves through the static fluid, the flow lines in a fluid, such as air, are imagined as moving around the ball as shown in Figure 1.9(a). The flow lines followed by a small piece of the fluid can be made visible by trails of smoke in a device known as a wind tunnel. The sketches of Figure 1.9 are based upon flow patterns observed in a wind tunnel. The spinning of the ball causes the flow lines to build up on one side of the ball as illustrated in Figure 1.9(a). The physical effect of the pileup of the air at the top side of the ball is to create an upwards force as indicated in the diagram. The Magnus effect was demonstrated physically long before it was understood. It is related to the Bernoulli effect discussed later

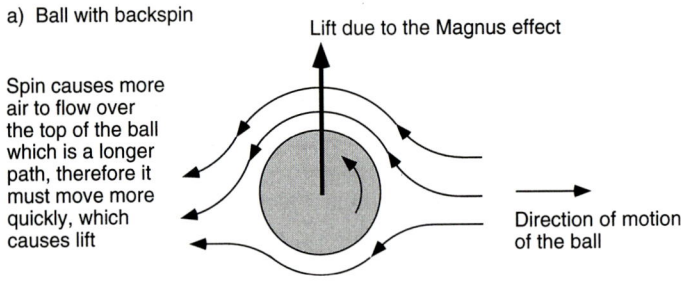

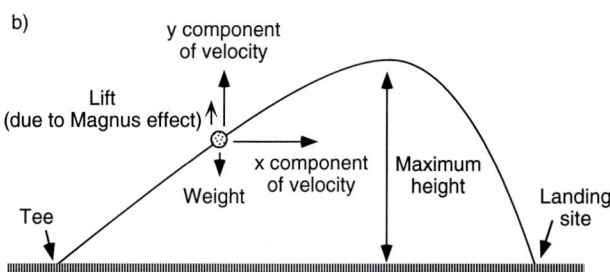

Figure 1.9 A golf ball with a backspin travels much farther than one hit without spin. This is a result of the Magnus effect. a) A spinning ball experiences a lift at right angles to the direction of travel due to the airflow created around the ball by the spinning action. b) Due to the Magnus effect, a golf ball with backspin will follow an extended path until it loses enough spin to follow a normal falling trajectory.

in Chapter 5. The Magnus effect is another counter-intuitive idea in that most people might expect that the pileup of air molecules on top of the ball would cause the opposite effect, that is, downward pressure on the ball. One way to think about the Magnus effect is that air molecules are so busy rushing past the top of the ball and each other that they do not have time to press down on the ball with the same force as exerted by air molecules on the under side of the ball. Thus because the molecules are busy hitting each other there is a net force upwards from the slower molecules below the ball.

The spin rates given to a golf ball are very high. The golfer uses different golf clubs with heads placed at different angles to achieve different spin rates in the ball. Thus with a club known as a driver, a spin of more than 50 revolutions per second is imparted so that the ball leaves the tee rotating at about the same speed as the motor in a typical vacuum cleaner. With a club known as a number 5 iron, the spin can be about 100 revolutions per second whereas with the number 7 iron, the rate can be 130 revolutions per second (approximately 8000 revolutions per minute). Experimental measurements have shown that up to 80% of the spinning is still present in the ball when it hits the ground. Again, calculations and photographs of real balls in flight have shown that for the first part of a lofted hit the ball appears to be almost weightless because the Magnus effect is so large that it almost counteracts the weight of the ball. Thus, for the first part of the flight from the tee to the green the path of the ball is almost a straight line. However, as the ball slows down because of air friction, the net flow rate over the ball drops sufficiently so that the weight of the ball becomes the dominant force and the ball drops steeply as indicated in Figure 1.9(b).

Section 1.7 The Scientific Study of Dimples

To understand the physical way in which dimples affect the flight of a golf ball we must learn something about the types of flow which can occur in a fluid. Scientists distinguish between two main types of flow in a fluid: laminar and turbulent. The origins and meaning of these two words are illustrated in Figure 1.10. In a moving fluid undergoing *laminar flow* one can imagine sheets of fluid which glide over each other in the same way that a pack of cards will slip sideways by card-over-card sliding when the pack is pushed on the top. When the velocity of a moving fluid is low, the flow conditions are laminar. Laminar flow of fluid around a smooth ball is shown in Figure 1.11(a). When discussing this type of fluid motion the scientist describes the movement of the fluid by focusing on the behavior of a small piece of the fluid which he calls an *element of the fluid*. He then describes the motion of the element of the fluid in laminar flow as one of quiet progression along a *streamline*. These streamlines can be made visible in moving liquids by means of aluminum pigments in water or smoke in air (see discussion of wind tunnels and liquid flow tunnels in Chapter 10). In laminar flow, an element of the fluid glides out of the way of the ball and back into a position similar to that occupied before the ball passed by. In *turbulent flow* the elements of the fluid are in chaotic, apparently random, motion. When the fluid encountered by a ball moves past it at high speed, a chaotic turbulent zone develops behind the ball as shown in Figure 1.11(b). Energy is dissipated in the turbulence and this energy has to come from the ball as it moves through the fluid. Therefore, the energy of the ball is lost more quickly when turbulence develops behind the moving ball. It can be shown that at the speed with which a driven golf ball moves through the air under normal conditions, a wide chaotic type zone as shown in Figure 1.11(b) will develop if the ball is smooth. By carrying out the appropriate experiments, scientists have discovered that the rough texture of the dimpled golf ball delays the development of the chaotic conditions behind the ball so that a much narrower turbulent zone exists [2–4,6,21]. This means that less energy is absorbed by the turbulent zone as the ball moves through the air. It seems that the rough surface of the ball makes the air cling longer so that it does not shed from the dimpled ball into turbulent conditions as quickly as it does with a smooth ball.

Once scientists started to use wind tunnels and liquid flow test chambers, they began to experiment to see if they could improve on the aerodynamics of the ball by altering

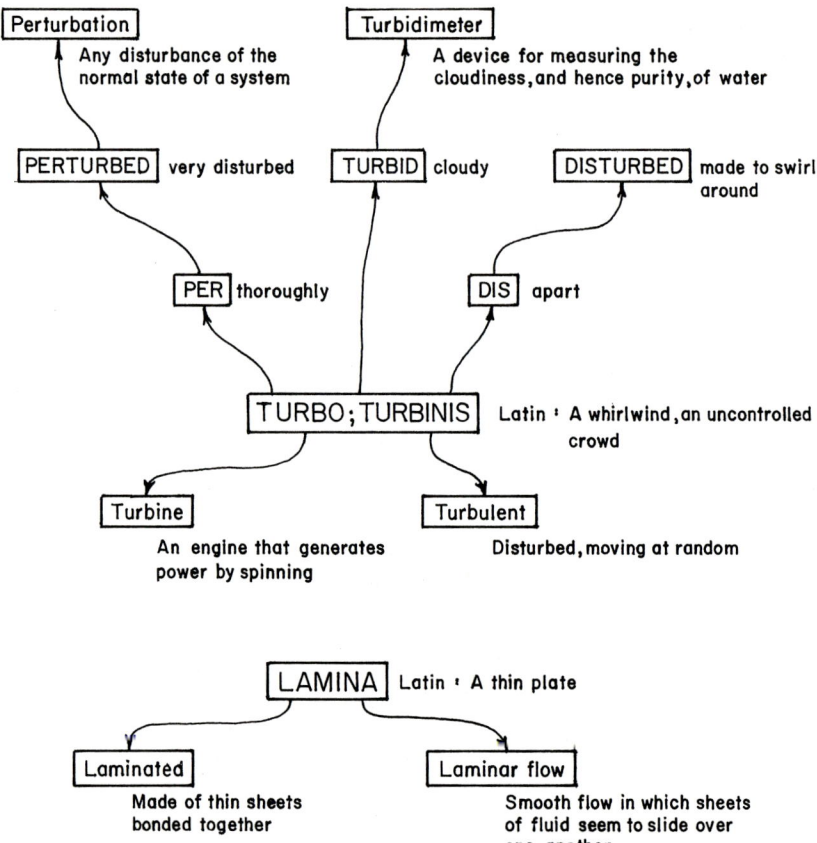

Figure 1.10 Scientists distinguish between two very different types of fluid motion: laminar and turbulent.

the pattern and depth of the dimples. A ball with a dimpled pattern filling a hexagonal region on the surface of the ball was introduced by Dunlop. It is claimed that this ball follows a truer path when moving through the air [6]. Another ball was developed in a series of experiments carried out by Fred Holmstrom, a physicist at San Jose State University and Daniel Nepela, a chemist consulting with IBM in 1974. They modified the ball by slowly filling in some of the dimples and driving the ball on real courses to see how the changes affected the flight characteristics. In their new ball, which is called a *Polara*, only 50% of the ball's surface is covered with conventional dimples. These are located on a band around the middle of the ball. The other dimples are much shallower. The ball is placed on the tee with the deeper dimple band in a vertical plane [6]. Apparently, when the ball is hit from this position, the flow of air around the dimples interacts with the air flow on the smoother sides to produce a gyroscopic effect that causes the ball to correct any miss hit which would send it to one side or the other of

1.7 The Scientific Study of Dimples

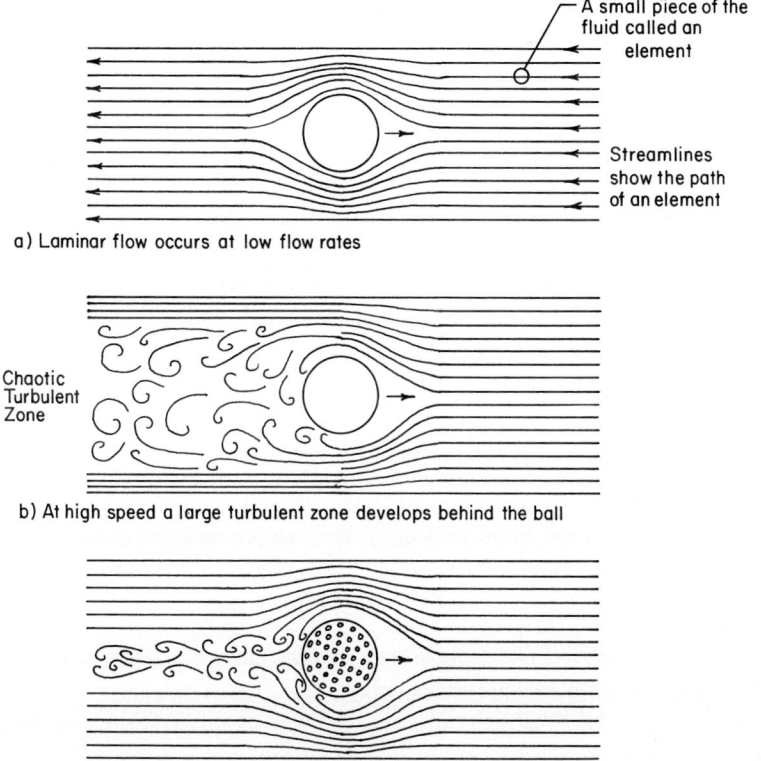

Figure 1.11 Dimples on the surface of the golf ball reduce the size of the turbulent zone created behind the ball at high speeds.

the desired trajectory (see discussion of Gyroscopes in Chapter 4). After players started to use the Polara in tournaments, the United States Golf Association adopted an amendment to its rules stating that a golf ball must be spherical in shape and be designed to have equal aerodynamic properties and equal moments of inertia (the meaning of this term will be discussed in Chapter 4) about all axes through its center. The new restriction rules out the Polara from tournament play. This ruling also stops manufacturers of golf balls from achieving the same kind of stabilization of flight by creating a golf ball in which the hard core is made into a disc that stores energy as in a flywheel when the golf ball is hit off the tee.

In 1994 an aeronautics instructor at the Massachusetts Institute of Technology, Jeffrey C. DiTullio invented a dimpled baseball bat. He tells us that

> *While I was sitting in traffic on the way home it occurred to me that if we could reduce the drag on a baseball bat then hitters could hit faster.*

With bats donated by the Boston Red Sox, DiTullio created three prototypes with dimples. He tested them in the wind tunnel at MIT and he found that the dimpled bats showed 60% less drag against 112 kilometers per hour (70 miles per hour) wind than did smooth bats. He worked out that this should allow a professional batter to swing a dimpled bat 5% faster adding 15 feet to a hit. Field trials with players from the Red Sox farm team showed a less dramatic improvement of only 3%. However, that would be sufficient to make a difference in a professional ball game. It will be interesting to see if the dimpled baseball bat becomes as common as the dimpled golf ball [22].

It is estimated that until 1989 golfers were driving about half a million golf balls a month off the decks of cruise liners. Many of these balls ended up in the stomachs of turtles, whales, and dolphins. In an attempt to solve this problem Patrick Kane and three associates patented a ball that looks and hits like a standard ball yet is completely biodegradable [23].

In describing his new ball Kane pointed out that

> *Normally golf balls have a plastic skin and a solid or rubber core. In our new ball, the outer skin is of paper, pulp, gelatine, or seaweed bound together by a water-soluble adhesive. The core is a mixture of sodium barcarbonate and sodium citrate — the ingredients that make Alka Seltzer fizz.*

The balls are intended to be used once on a driving range but Kane had his new ball, Aquaflyte, tested by a professional golfer who found that it can survive several drives. The drawback for the new ball is that it travels only about two-thirds of the distance of an ordinary ball. However for on-ship practice this may not be a disadvantage. Obviously if a dolphin swallows one of these balls it might even help him if he has an upset stomach!

Section 1.8   Floating Golf Balls and Aggressive Crocodiles

As already mentioned, the golf association controls the weight and minimum size of a golf ball at 45.927 grams with a minimum diameter of 4.26 centimeters. (These odd numbers represent the standardization of the typical sized ball at the time that the standard was set up.) A quantity which we will have occasion to meet in several chapters of this book on missiles is the average density of a missile or the density of a substance used in the making of a missile. The *density of a substance* is defined as the mass of the substance divided by its volume. In other words, it is the average mass per cubic centimeter of the material. If we work out the volume of the golf ball, it turns out to be 40.47 cubic centimeters. This is usually written 40.47 cm^3. If we take the mass of the golf ball and divide it by this volume, we get an average density of 1.13 g/cm^3. Water is 1.00 g/cm^3. The fact that the golf ball is more dense than water means that the ball sinks in water. Apparently so many golf balls are lost in the water hazards of Florida golf courses that there is a company that makes its living by sending scuba divers into the water hazards to recover golf balls and sell them back to the golfers. But, if we increase the radius of the golf ball by a small amount and keep the same legal weight limit, we can decrease the density of the golf ball so that it floats. For example, if we increase the diameter of the ball from 4.26 cm to 4.50 cm, which is an increase of about six percent, the density of the ball drops to 0.96 which means that it will float. Some companies are already offering floating golf balls. For the professional golfer this slightly larger ball would have different dynamics because the ball would experience more air resistance as it moved through the air. However, for the weekend golfer, the floating golf ball may bring a desirable economic change in his playing equipment. It was recently reported that one of the scuba divers attempting to recover golf balls from a water hazard in Florida was attacked by a crocodile. Scuba divers may also be in favor of floating golf balls!

Section 1.9   Golf Balls in the 21st Century?

Toward the end of the 1980s it was suggested that because of the increasing costs of maintaining a large golf course in a city area there was something to be said for introducing a lighter weight ball to be used on new shorter length golf courses. The lighter ball would mean that if the golfer exerted the same force with the club on the ball it would have less kinetic energy even though it left the tee with the same speed. Remember, since kinetic energy equals $1/2\ mv^2$, the smaller mass of such a golf ball would mean the kinetic energy would be lower which would mean the distance traveled would be shorter and, therefore, the golf course land required, cheaper.

With modern electronics, it is quite possible to implant a small transmitter inside the golf ball to transmit location signals when it is lost in the rough grass. We could probably even build robotic dogs to retrieve the beeping golf balls, provided that is they are not inside crocodiles!

A news story in 1992 indicated that the atomic age has brought about a new type of golf ball. In Manitoba there is a laboratory run by Atomic Energy of Canada. The scientists at the laboratory were curious to know what would happen if they treated a few golf balls with a high intensity beam of electrons. After the golf balls had been bombarded with the high energy electrons they traveled farther when hit; that is, their coefficient of restitution was increased. Starting in 1992 you could pay $37 (Canadian, including postage), to have 12 of your golf balls treated in this way. The scientist who carried out the experiments, Shewchukz, said "We make no claims that it will work." Apparently in the summer of 1992 the laboratories were receiving orders for 50 to 60 dozen balls a day and had a one-month backlog. One of the celebrities who hit the new golf balls was Peter Mansbridge, who was well known in Canada for being the anchor man on the Canadian Broadcasting Corporations "The National" and "Prime Time" news. Mansbridge claims that his "atomic ball" went 25 yards farther than a normal one. In fact, however, there is a relatively simple explanation for what was happening. In a plastic there are long molecular chains which are often side by side in the structure of an ordinary rubber or plastic used in making golf balls. When the electrons rip into the ball they disrupt the bonds of the molecular chains so that these actually join up with adjacent chains, a process known as *cross linking*. This cross linking of the rubber or plastic molecules can result in a more resilient form of the material which leads to a higher coefficient of restitution in the core material of the golf ball. The individuals paying to have the golf balls treated with high energy electron beams could have achieved a similar increase in coefficient of restitution by having golf balls made with the same substance that goes into super balls. Remember, however, that earlier in this chapter we quoted Daish as saying that "if we could increase the coefficient of restitution of a golf ball it would be probably be banned by the golfing authorities", so the atomic golf ball does not appear to have much of a future in competitive golf.

References

[1] Lincoln Barnett, *The Universe and Doctor Einstein*, Harper and Row. New York, 1948.
[2] Angelo Armenti, Jr. (Ed.), *The Physics of Sports*, The American Institute of Physics, P.O. Box 20, Williston, Vermont, 05495, 1992, p. 260.
[3] C.B. Daish, *Learn Science Through Ball Games*, American Edition, Sterling Publishing Company, New York, 1972.
[4] T.P. Jorgensen, *The Physics of Golf*, The American Institute of Physics, 1994. (See also the review by R.E. Fornes in *American Scientist*, March/April 1995, p. 7)
[5] A. Chase, "A Slice of Golf," *Science*, July/August.1981, pp. 90,91.
[6] L. Rubenstein, "Golf Balls, The Long Search for Perfection," *The Globe and Mail*, Toronto, Saturday, June 27,1981, p. S4.
[7] J.D. Cutnel and K.W. Johnson, *Physics*, 3rd Edition, John Wiley & Sons Inc., New York, 1995.
[8] See the article "Speedy Greens" in "Science and Sport", a *New Scientist* supplement to the issue of October 9, 1993, p. 16.
[9] The reader interested not only in the different types of grass surfaces used on golf courses but in the general ecological impact of the growing number of golf courses around the world, will find interesting information in the article by F. Pearce, "How Green is Your Golf?" *New Scientist*, September 25, 1993, pp. 30–35.
[10] B. Holmes, "World Cup Fields of Dream", *New Scientist*, May 14, 1994, p. 5.
[11] E. Coghalen, "World Cup Players Face a Whole New Ball Game," *New Scientist*, July 9, 1994, p. 4.
[12] A. Coghlan, "On the Right Track", *New Scientist*, August 1, 1992, pp. 34–45.
[13] H. Brody, "Measuring the Softness of an Athletic Surface", *The Physics Teacher*, Volume 30, January 1992, pp. 28–31.
[14] H.C. Cremers, "Drop Test Picks Out Best Peas for Soggy Soil", *New Scientist*, March 6, 1993, p. 20.

[15] J. Orear, *Physics*, MacMillian Company Inc., New York, 1979.
[16] See news item "Not Golfing Well? Blame Your Shoes," *Research & Development*, June 1993, p. 80.
[17] A. Turnbull, "Making All The Right Moves – Computer Models are Offering Top Athletes the Chance to Fine Tune Their Performances", *New Scientist*, July 25, 1992, pp. 23–27.
[18] D.H. Fender, *General Physics and Sound To Advanced and Scholarship Level,* "Cambridge University Press, 1957.
[19] D.R. Goodwin, "The Sorcerer of Strobe Alley", *Science*, June 1982, pp. 37–45.
[20] E. Zwingle, "Doc Edgerton, the Man Who Made Time Stand Still", *National Geographic*, October 1987, pp. 464–483.
[21] J.M. Davies, "The Aerodynamics of Golf Balls", *Journal of Applied Physics*, Volume 20, Number 9, September 1949, pp. 821–828.
[22] W. Gibbs, "Dimpled Baseball Bat", News item in *Scientific American*, July 1994, p. 98.
[23] J. Beard, "Dissolving Golf Balls Ends Dolphins' Distress", *New Scientist*, April 15, 1992.

Chapter 2

Robin Hood, William Tell, and Which Way did they go?

Chapter 2

Robin Hood, William Tell, and Which Way did they go?

Section 2.1 Using Potential Energy to Launch a Missile

In golf, energy is stored in the moving club enabling it to launch the golf ball as a missile by transforming the kinetic energy of the club into the kinetic energy of the ball. In archery the energy used to launch the missile is stored in the stretched bow as potential energy. When the bow is released its potential energy is used to create kinetic energy in the arrow. The design of the bow focuses on efficient transfer of the maximum amount of energy which can be stored in the stretched bow into kinetic energy of the arrow launched by the bow.

Section 2.2 Working at Storing Potential Energy in a Bow

McEwen [1] describes a bow as

> a two armed spring spanned and held under tension by a chord. Drawing the bow places the back, or outside curve, under tensile stress and the belly, or inside curve, under compressive forces. Any bow must adapt to these forces to avoid breaking and to propel the arrow successfully when the chord is released. When fully drawn the bow stores potential energy in its limbs. Releasing the bow string transfers this energy to the arrow throwing it into flight.

McEwan tells us that the various kinds of bows did not appear suddenly. He states [1]:

> Bow design seems to be a part of a gradual process of modification spanning many millennia and prehistoric cultures.

The various terms used in describing the physical structure of a bow are shown in Figure 2.1. The type of bow shown, known as a "longbow", is the type that was used by Robin Hood who dwelt in Nottingham forest during the reign of King Richard the

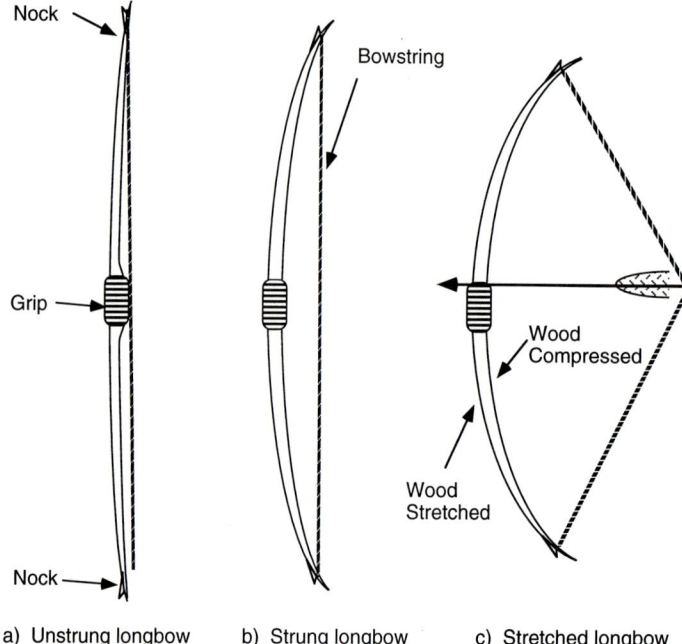

Figure 2.1 The longbow, as used by Robin Hood, is described technically as a "self-bow" from the fact that it is made from a single material.

First, also known as the Richard the Lionheart. In the North of England, close to my hometown, there is a small town called Robin Hood's Bay. Legend has it that Robin Hood came to this small sea port to help the local inhabitants fight off the attack of pirates. It is said that Robin Hood could fire an arrow across the bay. Anyone who has visited Robin Hood's Bay knows that firing such an arrow would be a super human feat.

When researching material for this chapter I was interested to find in a dictionary that the phrase "to draw the longbow" or "to longbow" meant to exaggerate and make extravagant statements [2]. Obviously archers used to tell the same stories about their arrows as fishermen do today about the fish that got away.

To understand why the flight of real arrows has physical limitations and to understand the design features of different types of bows we have to study the physics of work and of the storage of potential energy in strained systems. As in our study of the sophistication of the dimpled golf ball we will discover that, by trial and error, the bow makers of history have evolved a very sophisticated design which is hard to beat even with all our modern materials and computers. The sophistication of ancient bows was brought into dramatic focus in September 1991 when hikers in the Tyrolean Alps (on the borders between Italy and Austria) discovered, in the melting ice of a glacier, the mummified body of a man. This mummified body has come to be known as "The Iceman" [3–5]. Radiocarbon dating has established that the mummified body is 5100 years old. Alongside the body of the Iceman, trapped in the ice, were an unfinished bow and a quiver of arrows. In the words of Jaroff [4]:

> *The bow, which had not yet been notched for a bow string, was made of yew. Egg, an expert studying the remains tells us that yew is the best wood in central Europe for bow making and although yew is relatively rare in the alps the iceman had searched out the best material.*

To the archaeologist, as Jaroff states,

> *The iceman's quiver is an even rarer prize. It is the only quiver from the stoneage period found in the whole world.*

Egg [4] states:

> *The cargo of feathered arrows marks another first. Carved from viburnum and dogwood branches, a dozen of them were unfinished but two were primed for shooting with flint points and feathers. The feathers had been affixed with a resin like glue at an angle that would cause spin in flight and help maintain a true course.*

Another expert Notdurifter says "It is significant that ballistic principles were known and applied." The quiver also held an untreated sinew that could be made into a bow string and a ball of fibrous chord.

The antiquity of the archer as a military specialist is shown by the fact that Sagittarius is one of the figures of the zodiac. The term *Sagittarius* comes from the Greek word for arrow. To understand the dynamics of a bow and other systems for sending missiles on their way, we need to grasp the basic scientific concepts of work and energy. To start with we must study some very important laws of motion which were first set out by the British scientist *Sir Isaac Newton* who lived from 1642 to 1727. Legend has it that he gained his insight into the movement of the earth and planets and other celestial objects when he was seated under a tree on his mother's farm. It is reported that he noticed an apple falling from the tree and realized that, from a different perspective, it could be said that the earth was moving to the apple rather than the apple moving to the earth. Or even more fundamentally, that the two were moving toward each other by means of a mutually attractive force which became known as *gravitational attraction.*

Newton's first law of motion states that; a body in a state of rest or in uniform motion will continue in that state provided no forces act upon it. This seems to be very obvious today, but the Greek philosophers taught the opposite. They had worked it out that the natural state of all objects was no motion and that an object only moved when pushed. This, in a way, was a very reasonable observation since everything on the earth sooner or later comes to rest if nothing keeps pushing it. However, in an age of interplanetary space travel, we know that a rocket fired into outer space will keep on traveling forever if it experiences no other forces after the rocket fuel runs out. One of the nightmares of scientists working on space research is that one day an astronaut walking in space will have an accident in which a rocket thrust drives him away from the spacecraft snapping his tethering cord. If he has no other rockets to return him to the spacecraft he could travel into outer space forever at the same speed that he had when the thrust rocket driving him away from the spacecraft is exhausted.

Newton's *second law of motion* states that; a force acting on a load equals the mass times the observed acceleration of the body. This is written symbolically in the form

$$F = m \times a$$

where F = force, m = mass, and a = observed acceleration.

When an object falls toward the Earth we can measure that it experiences an acceleration of 981 centimeters per second per second (usually written as cm/s^2 or cm s^{-2}). This rate of acceleration is denoted by the symbol g. The observed acceleration is caused by the force due to gravity on the object. This force is identical to the weight of the body. From Newton's second law it follows that the weight, w, of the body equals the mass, m, times g. Written symbolically,

$$w = m \times g \text{ or simply } w = mg$$

A very important alternative form of this formula is

$$\text{mass} = \text{weight}/g$$

In the system of units we are using here, a system known as the centimeter-gram-second system (written cgs system), g is about 1000 cm s^{-2}. When I was a student working on problems in dynamics, I sometimes found when I looked at the answers at the back of the book that my answers were too great by a factor of a thousand. Fortunately, I realized that I must have forgotten to divide the weight by the acceleration due to gravity to get the mass before calculating the effect of a force on a body. As a professor teaching physics I have found this mistake — a failure to convert a weight to a mass — to be the most common mistake made by students studying mechanics and dynamics.

In the cgs system of units, the basic unit of force is the dyne. A *dyne* is the force that will accelerate a mass of 1 gram at a rate of 1 centimeter per second per second. Thus, if we apply a force of 1 dyne for 1 second, at the end of the second, the 1 gram body will be moving with a velocity of 1 centimeter per second.

Work is another one of those words which has a different meaning in science to that in everyday speech. To understand the difference between the scientific and everyday use of the term work, consider the systems sketched in Figure 2.2. The strict scientific definition of work carried out on a system is that the work done equals the product of the size of the force and the distance moved in the direction of the force. Thus, for the force exerted on the ball in Figure 2.2(a), the work done in accelerating the ball from zero to the velocity v over the distance L is given by the relation:

$$\text{work} = \text{force} \times \text{distance}$$

For this simple system, the work invested in accelerating the ball, assuming that the pushing of the ball occurs in free space where there are no frictional forces, equals the

2.2 Working at Storing Potential Energy in a Bow

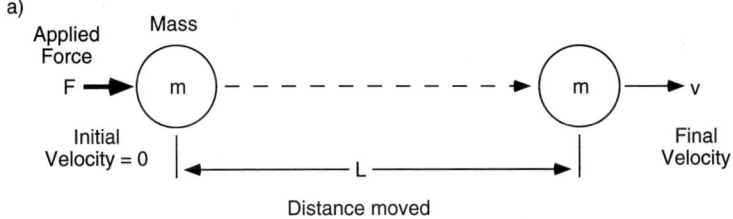

Work (W) equals the applied Force (F) times the Distance (L) moved in the direction of the force, symbolically:

$$W = FL = \tfrac{1}{2} m v^2$$

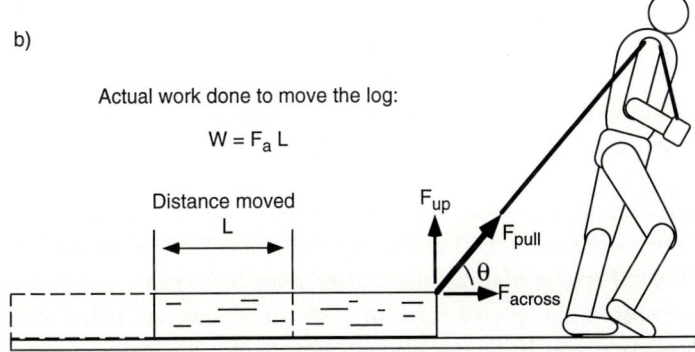

Figure 2.2 A scientist's calculation of the work performed by a laborer is often different from the laborer's own sensation of work. a) Scientific definition of work. b) A man pulling a log feels he is working harder than the amount of work he performs according to the scientific definition.

kinetic energy acquired by the ball. This fact is summarized symbolically in Figure 2.2(a). Consider now the case of a lumberjack pulling the log as sketched in Figure 2.2(b) over rough ground. The lumberjack pulls with force F_{pull} and moves the log a distance L. However, the scientific estimate of the work done on the log is not $F_{pull} \times L$ since the log is not moving in the direction of the force. In this situation, it can be shown that the effectiveness of the force applied to the rope by the lumberjack can be split into two forces acting at right angles to each other. These component forces are F_{across} along the direction in which the log is moved and F_{up}, an upwards force tending to lift the log. In scientific terms the splitting of the applied force into its components is termed *resolving the applied forces*. The term "resolve" and the related term "resolution" are again words which have different meanings in everyday speech and scientific terminology. It can be shown that the magnitude of the force along the direction of motion is given by the mathematical formula:

$$F_{across} = F_{pull} \cos \theta$$

where $\cos\theta$ is a mathematical function of the angle θ made by the rope with respect to the direction of the motion of the log. The value of $\cos\theta$ can be obtained by pressing the appropriate button built into many calculators or it can be looked up in books of mathematical tables. (An explanation of the origin and significance of mathematical functions is beyond the scope of this book.) The scientific estimate of work done on the log is

$$F_{\text{across}} \times L$$

The muscles of the lumberjack have actually done more work than this, since they are not 100% efficient, and the lumberjack will have used up more food calories than this calculation indicates; but from the scientific point of view, this is the only measure of work performed on the log! Furthermore, the log has not gained any kinetic energy after it has been moved. The work done, or the energy expended on the log, has been used up in overcoming the friction opposing the motion of the log.

In the cgs system of units, the basic unit of work is the *erg* which is defined as 1 dyne working over a distance of 1 centimeter. The erg and the dyne are very small quantities. In another system of units known as the SI system which we will use for most of this book, force is defined by studying what happens to one kilogram of mass. (Remember that a kilogram is 1000 grams.) Newton was honored by having this unit of force named after him. A force of 1 *newton* will accelerate 1 kilogram of mass at 1 meter per second per second. In the SI system of units, the unit of work is the *joule*. The joule represents the amount of work done when a force of 1 Newton moves an object over 1 meter. One joule of energy equals 10 million ergs, also written in the form 10^7 erg. Just as the notation 10^2 means multiply 10×10, so 10^7 means multiply 10 by itself 7 times to give the product of 10 million. The joule is named after a British scientist, James Prescott Joule who lived from 1818 to 1889. Joule pioneered the studies of the way in which work is transformed into heat [9].

Joule realized that water moving down a waterfall transforms the potential energy that it had by virtue of its position at the top of the falls into kinetic energy at the bottom of the waterfall. It is this kinetic energy gained by the falling water which is used to drive turbines in a hydro-electric power station. In an ordinary waterfall however, the kinetic energy that builds up in the falling water is lost in the churning turbulence of the disturbed water at the bottom of the waterfall and is transformed into heat energy which means that the water at the bottom of the waterfall warms up. At the time that Joule was carrying out his work, scientists had to make their own accurate thermometers and these instruments were rather fragile. It is reported that on his honeymoon in Switzerland, Joule spent most of his time measuring the temperature of the water at the top and the bottom of various waterfalls. It is said that he was so concerned about his thermometer that he used to walk behind the carriage in which his wife rode carrying his thermometer so that the bumpy road would not break it. Unfortunately, we have no record of Mrs. Joule's observations on Mr. Joule's scientific pastimes! However, we do know that Joule was successful in proving that the rise in the temperature of water at the bottom of a waterfall was exactly what he had expected from his calculations of the way in which kinetic energy is transformed into heat.

For a long time people believed that heat was a fluid which flowed into an object to make it hot. The Roman word for heat was "calor" and people called this imagined fluid *caloric*. An instrument for measuring heat exchanges in an experiment is still called a *calorimeter*. Originally the unit of heat energy used in scientific studies, the calorie, was defined as the amount of heat required to raise the temperature of 1 gram of water by 1 *degree Celsius*. As a result, we know that 4.18 joules of energy is equal to one calorie.

When exploring the physics of sports, it is often necessary to read articles from both England and the United States. Unfortunately, these two countries use different sets of thermometer scales to measure the temperatures involved. In the United States in popular articles on sports, the Fahrenheit scale of temperature is still widely used. The *Fahrenheit temperature scale* was invented by a German/Dutch physicist Gabriel Daniel Fahrenheit (1686–1736). Fahrenheit earned his living in Amsterdam by making instruments for sailors and other people interested in the weather. Up until that time the main fluid used in thermometers was alcohol or an alcohol/water mixture. *Mercury*, a liquid metal, which is now widely used in thermometers, had not been available in a sufficiently pure form for it to be used in thermometers. Fahrenheit developed a method for cleaning the mercury so that it did not stick to the walls of the tube in which it was made to rise and fall to follow changes in temperature. In setting up his temperature scale, Fahrenheit wished to avoid negative temperatures and so he used as a reference point a mixture of salt and water that froze at a temperature he called 0° on the Fahrenheit scale. Unfortunately, temperatures in North America drop much lower than this so Fahrenheit's primary aim of avoiding negative temperatures has not stood the test of time.

Newton was the first to suggest that the freezing point of water and the boiling point of water should be used as two easily reproducible reference temperatures when calibrating thermometers. On the Fahrenheit scale, the temperature of boiling water came out to be 212 degrees. (The word *degree* means "moving by one step" from the Latin word "gradus", a step. Thus, you earn degrees at various steps in your education toward full qualifications). The Swedish astronomer *Andros Celsius*, a professor at the University of Upsala in Sweden from 1730 until his death at the age of 43 in 1744, was the first to actually set up a reference temperature scale using the freezing point of water as zero and the boiling point at a hundred degrees. In Latin, the word "centum" means one hundred. Since the temperature scale set up by Celsius has one hundred steps from the freezing point of water to the boiling point of water, it became widely known as the *centigrade* (100 steps) *temperature scale*. In 1948, at an international meeting of scientists, it was recommended that the centigrade temperature scale become known as the *Celsius temperature scale*. In the late 1980s both centigrade and Celsius could be found in the scientific literature, although Celsius appears now to be becoming the favored terminology. To help the reader convert temperatures from Fahrenheit to Celsius scale temperatures, a chart is given in Figure 2.3.

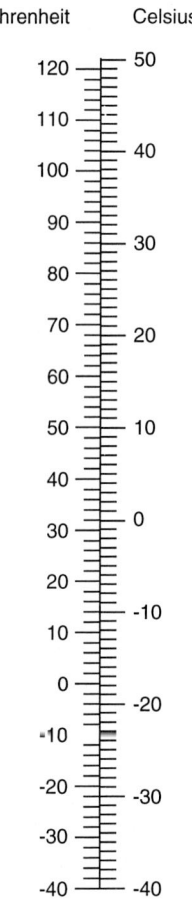

Figure 2.3 Comparison of the Fahrenheit and Celsius temperature scales.

It is not always realized that the Calorie mentioned in food product labelling is different from the scientific calorie. The Calories of food energy content are actually the same as kilocalories, which are 1000 times the basic *scientific calorific content of the food*. The Joule used on the food packet is again a *kilojoule* (1000 joules). Today the energy content of many food products is given on the label in both Calories and Joules.

Section 2.3   Graphical Illustration of the Energy Stored in a Stretched Bow

Now that we have developed the basic concepts of work and energy, we can describe in scientific terms how a bow is used to launch an arrow. The longbow is the simplest type of bow. Children often make this type of bow out of a length of springy wood which can be bent without breaking, with a string or cord stretched tight between its two ends. I can still remember the first bow made for me at age seven by a young man who carefully selected a long straight piece of ashwood from a hedge. He taught me how to string the bow by making a notch in each end of the stick. He then told me never to leave the bow strung overnight but always to unstring it when not in use so it would retain its springiness longer. For this reason all "self-bows" (bows made from a single piece of material, e.g., wood) are left unstrung when not in use. The English bows used by Robin Hood and his merry men were made out of the wood of the Yew tree. Gordon tells us that, contrary to popular belief, English longbows were not made of wood from English trees but from Spanish Yew [10]. The English government made it compulsory that merchants trading with Spain, importing Spanish wine, must always include in their return cargo some Yew tree staves suitable for making longbows. English archers using the yew bow rarely went south of the Alps or the Pyrenees (the mountains separating Spain from modern France). Henry Blyth, an associate of Dr. Gordon, has pointed out that the springiness of yew deteriorates rapidly with increasing temperature and that a yew bow cannot be used reliably above 35 degrees Celsius. Since the yew wood bow would probably snap if used in the Mediterranean regions, the longbow was not widely used by soldiers of the Mediterranean region, such as the Egyptians and the Greeks. They used the composite bow which will be described later.

Calculating the work carried out by an archer pulling on a simple bow is relatively easy: the arm exerts a force and the movement of the bow's string is along the line of action of the force. A tall, strong man can draw an arrow back about 0.6 meters (24 inches). Although the human reach from the nose to the hand is approximately 0.9 meters (3 feet), remember that the longbow must be bent to some extent to begin with to create initial tension in the bow and build up good tension by the time the bow is fully extended. The greatest force that a typical strong man could exert on a bow of this kind, when fully extended, is approximately 350 Newtons (about an 80 pounds pull). In some modern archery competitions the winner is the archer who can shoot the arrow the furthest. In these competitions, the archer uses a very powerful bow which he lays on the ground, placing his feet in it as he draws with his arms.

To understand how much energy is stored in a bow when we work at pulling the bow string it is useful to do a force distance type of graph of the type illustrated in Figure 2.4. This shows graphically the work done by the logger of Figure 2.2(b) as he drags the log with an effective force F_{across} over the distance L. It can be seen that the area

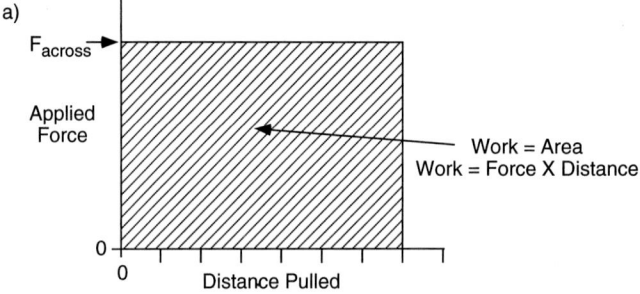

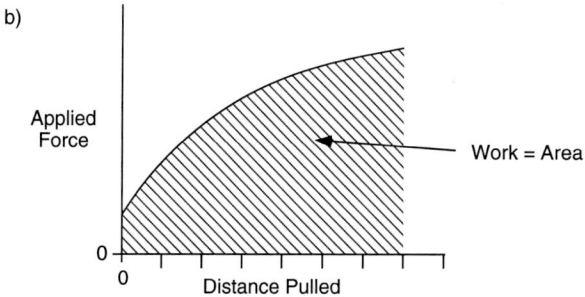

Figure 2.4 Force versus distance graphs can be used to determine the amount of work done in a given situation. a) When a person pulls a log, as in Figure 2.2(b), the area under the graph of force versus distance illustrates how much work is done. b) When pulling a bowstring, the force is continually increasing as the bow is stretched.

of the graph is force times distance. The area this represents is the work done in overcoming the friction between the log and the ground. The calculation of the work done using the bow string is more complicated because each time we move the bow string a short distance the force required to move it increases, as shown in Figure 2.4(b). However over the whole range of the movement of the bow string the force–distance graph still summarizes the work done in pulling the bow string and also represents the energy invested in the bow ready to launch the arrow [11].

Scientists at Reading University in Great Britain have studied the longbow using computers and they estimate that, in general, the energy stored in a longbow was just under one hundred joules. They have also found that both the longbow and the composite bow are surprisingly efficient in transferring the potential energy of the stretched bow into kinetic energy of the fired arrow. When the string of the bow is released by the archer, the arms of the bow start to leap forward and this tightens the string and pushes harder on the arrow. The process of transferring the energy from the bow to the arrow is a complex interaction of the moving arms, the bow, and the push of the string on the arrow itself [11]. You must never, never, never shoot a bow without a proper

Section 2.4   Vane Strategies

arrow in the bow. If this is attempted, there is no safe way of getting rid of the stored energy in the bow/stretched-string system and not only may the bow be broken by the wildly oscillating string, but the would-be archer runs a high risk of being injured by the uncontrolled bow string.

One aspect of the dynamics of an arrow leaving a bow which receives little attention in the physics literature is the role of the feather vanes at the end of the arrow. Apparently there is some discussion amongst archers as to whether one should use straight feather vanes or angled vanes. Some claim that the angling of the feathers, giving rotation to the arrow, stabilizes the flight of the arrow (the reasons why spinning stabilizes the flight of a missile will be discussed in detail in Chapter 4). Other archers claim that

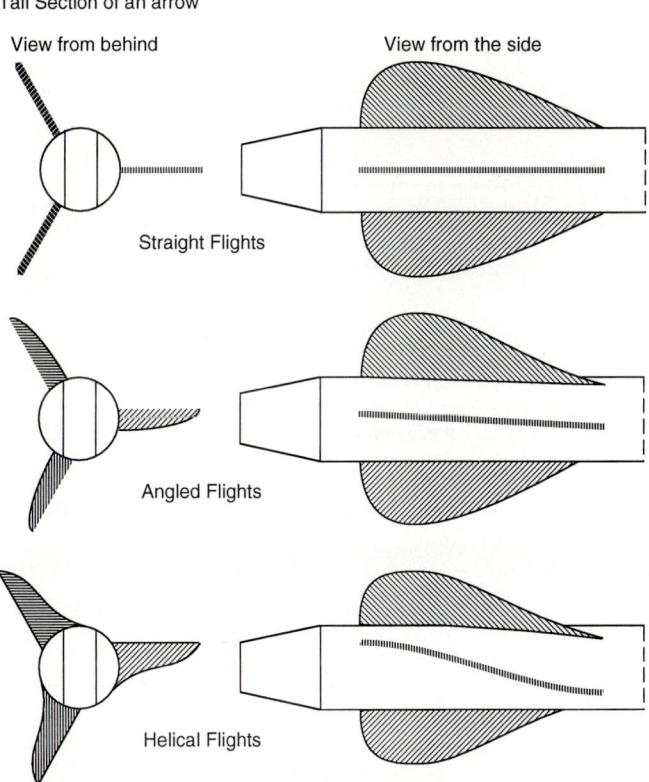

Figure 2.5 The feathers on an arrow can be arranged in various patterns. Some archers prefer straight feathers, while others like angled feathers which spin the arrow to stabilize its flight [6].

spinning the arrow wastes energy which detracts from the ability of the arrow to travel a long distance. In Figure 2.5 some of the configurations used in setting up the feathered vanes of an arrow are shown [12].

When researching the design of feathered arrows I discovered the explanation of a fact that had always mystified me. First of all I had always wondered why the feathers were not ripped off the arrow as the end of the arrow passed the center of the bow in its flight. The answer to this is partly that a real feather has some give so that any rubbing of the vane against the bow does not necessarily destroy the feather; however, the main effect is that the arrow, when it is fired from a bow, oscillates in the way illustrated in Figure 2.6 [12, 13]. In the words of Gareth Rees [13]

> *The arrow is not completely rigid, and the force exerted by the string does not pass through its center of mass. Therefore as soon as the archer releases the bow string the arrow begins to bend in flexural oscillations along its length. The frequencies of these*

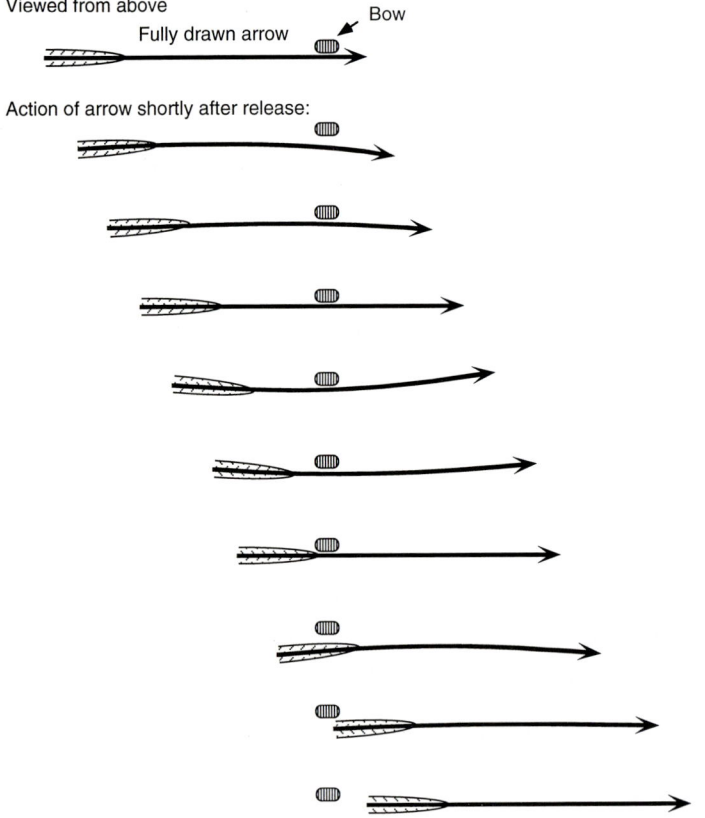

Figure 2.6 The stiffness of an arrow must be tuned to the bow so that it will oscillate in such a way as to minimize damage to the feathers [11, 12]

oscillations depend principally on the stiffness of the arrow. A correctly tuned arrow will bend away from the bow stave as it passes it. The oscillation period must be matched to the time taken for the arrow to be accelerated past the bow stave. Therefore there is a link between the draw weight of the bow and the properties of the arrow.

When discussing the effect of the oscillation of the arrow and the vanes at the end of the arrow the coaches manual for the Canadian Federation of Archers [12] makes the following statement

After the arrow has cleared the bow, and while the spine cycle (the technical term for the oscillation of the arrow) is still quite pronounced, the helical vanes begin turning the shaft making it necessary for the spine action to become circular. The drag factor is extremely high over this period and considerable loss of distance results. A less inclined vane angle is more effective as it takes over slowly, allowing most of the activated spine oscillations to be completed before rotation begins.

It is apparently left up to the individual archer to design his own compromise between the role of the vanes spinning the arrow and the flexing of the arrow to enable it to clear the bow without losing energy to the bow stave [12, 13].

Section 2.5   The Composite Bow

As already mentioned, the longbow made from yew wood would have its limitations in the dry climate of the Mediterranean area. The longbow also had several disadvantages in warfare in which the archer was on horseback or traveling in a chariot. For various reasons an alternate design of bow evolved, known as the *composite bow*. The composite bow demonstrates the surprisingly sophisticated technology that was developed by what we call primitive people. Consider for example a typical composite bow used by Asian archers who fought on horseback. As illustrated in Figure 2.1, when a bow is bent, the material in the front part of the bow is under tensile stress, that is, it is stretched. On the other hand, on the belly side of the bow the material is in a state of compressive stress. Some materials, such as stone and concrete, are strong under compressive stress whereas other materials are strong under tensile stress. This is why reinforced concrete has a wire mesh, which is strong in tension, to give overall strength to a reinforced concrete structure. Asian *bowyers* (the technical term for a maker of bows) knew that animal sinews had high tensile strength. (Sinews are the natural cords attaching muscles to bone in animals and man.) Another word for sinew is *tendon*. A dictionary definition of a tendon is

A tough whitish material consisting of numerous parallel bundles of collagen fibers. Tendons are stiff but flexible. They assist in concentrating the pull of the muscle on the small area of bone.

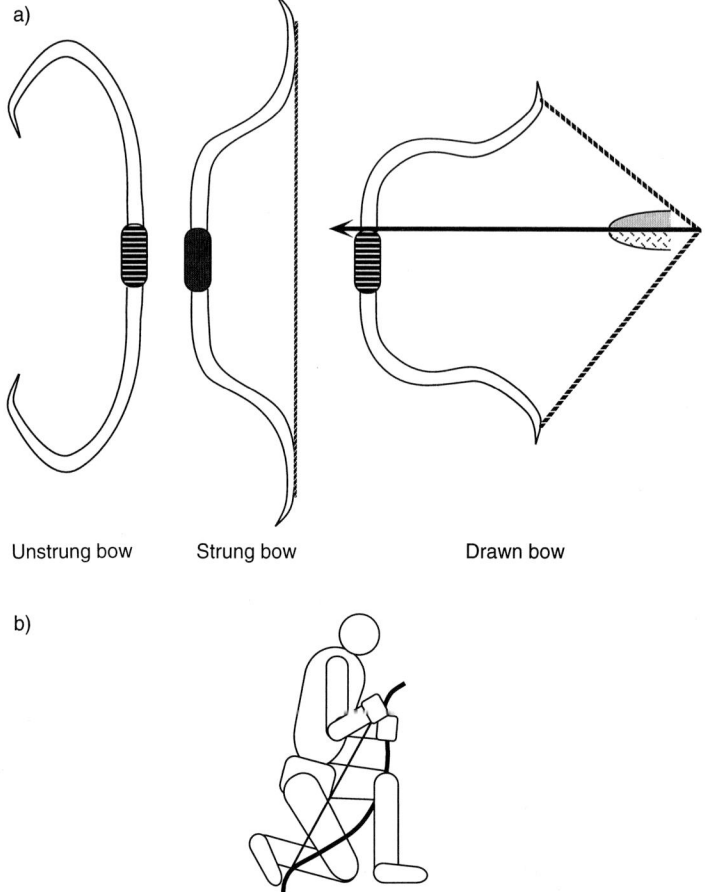

Figure 2.7 Composite bows exploit the natural properties of sinew and horn to enhance efficiency. a) Appearance of a composite bow. b) Technique used to string a composite bow.

Sinews have been widely used by primitive people as good natural chord for sewing animal skin and in making canoes. Asian bowyers used an adhesive derived from hide and fish swim bladder to glue animal sinew to the back of their bows. The type of composite bow shown in Figure 2.7 is referred to as a reflex bow, since, to string the bow, it has to be bent backwards. The bellies of these bows were fitted with a glued layer of animal horn. McEwan tells us that horn is twice as strong as hardwood under compression and that it has a high co-efficient of restitution so that it returns to its original shape giving back its stored energy to the arrow. McEwan also tells us us that another advantage of the composite bow is that it can be kept strung for prolonged periods without adverse effects. He reminds us that simple wooden self-bows and sinew-reinforced bows are usually kept unstrung between uses to avoid string deterioration and loss of power [1].

Experiments have shown that a tendon retains its useful mechanical properties up to approximately 55° C. However, since tendon is not good in damp conditions in the higher humidities of Northern Europe, this type of composite bow would lose its strength in climates such as those of England and Norway.

The technique for stringing a composite, reflex bow is illustrated in Figure 2.7(b). This picture shows how the leg is used to bend the bow as the string is placed in position. When first strung the tension in a composite bow is already much higher than in a longbow. Because of this initial tension in the string, the bowman has to pull the bow with a larger force to begin with. It has been estimated by scientists at Reading University that the composite bow can store approximately 170 joules of energy when stretched. Therefore, the arrow released from such a bow can have about 80 percent more kinetic energy than that fired from a typical longbow.

In Medieval times when knights fought battles man to man with heavy weapons, many stories were told about famous swords, swords so heavy that only the strongest could lift them and wield them in battle. But in the legends of ancient Greece, as told in the poetry of Homer, one hears not about famous swords but about special bows that only men of heroic strength could use. Thus, the great warrior Odysseus, who survived the Trojan war only to take twenty years to return to his island home in Ithaca, had a bow that required such a force to string it that only he was able to do it [14].

Because Odysseus took so long to return after the Trojan war, there were many who assumed he was dead and who would have liked to have married his presumed widow to inherit and control his fortune and his territory. These suitors descended on the house of Odysseus and besieged his wife Penelope with offers of marriage. Penelope, we are told, adopted two strategies to hold the suitors at bay as she waited for Odysseus to return. In one story we are told that she said that she could not marry until she finished the piece of cloth she was weaving. She wove all day and then at night secretly unwove the previous day's work so that the weaving took forever. In the other story Penelope tells the suitors [14],

> *I will bring you the great bow of the divine Odysseus and whosoever shall most easily string the bow with his hands and shoot through all the twelve axe rings with him shall I go and forsake this house of my marriage so beautiful and filled with fair things.*

By the time Penelope had settled on this final test, Odysseus had returned and was actually present in disguise among the suitors. Penelope had anticipated that none of the suitors could even string the bow. Then Odysseus revealed himself, strung the bow, and killed the suitors.

Section 2.6 The Crossbow and William Tell

Beginning in the 13th century AD, armorers began to build what came to be known as crossbows, which were bent with the aid of levers and gears. In this instrument, an example of which is shown in Figure 2.8, the potential energy is stored in a metal bow. The main purpose of the crossbow was originally to fire a metal arrow called a *bolt* with sufficient energy to pierce the heavy protective armor worn by the knights of the period. Gareth Rees points out that the crossbow, though much smaller than the longbow, is a more powerful weapon. It requires less skill and strength to operate than a longbow but the drawback is the time taken to draw and shoot (about a minute compared with as little as six seconds for a longbow). Rees tells us that by 1300 the crossbow had largely displaced the longbow on European battlefields despite being banned in 1139 by the Pope as "deathly and hateful to God and unfit to be used by Christians" [13].

William Tell is one of the legendary bowmen of history. According to tradition, he led the uprising of the local Swiss peasantry against the bailiff of the Austrian King. This bailiff, called Gessler, had put a hat on top of a pole and demanded that the local

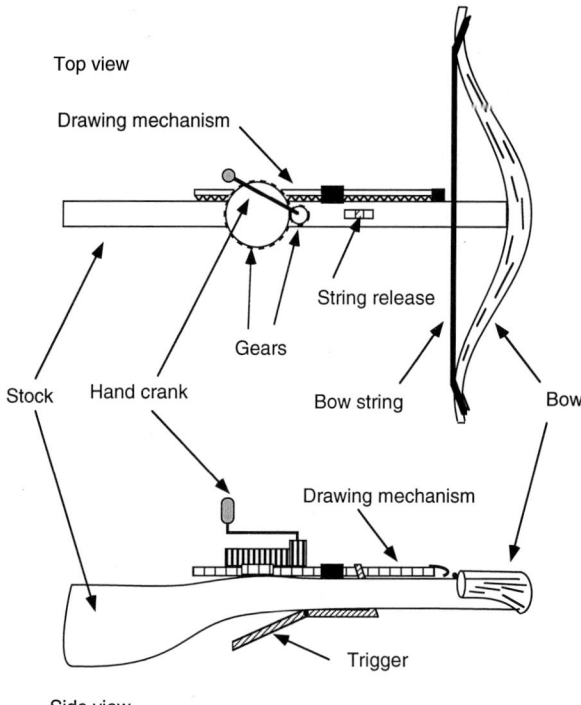

Figure 2.8 In a crossbow, potential energy is stored in a strong wood or metal bow which is usually cranked or levered into the drawn position. The metal "bolt", similar to a short arrow, can then be fired with enough energy to pierce armor [13].

people pay homage to the hat. When Tell refused to do this, he is reported to have been told by Gessler that the Austrians would listen to the Swiss demands if Tell would shoot an apple from the head of his young son using a bow and arrow. In illustrations of this story and in a movie made about it, William Tell uses a crossbow in this dramatic piece of archery.

Section 2.7 How Big Should an Arrow be?

The answer is: it depends on the purpose you had in mind when firing the arrow. Basically the transformation of the stored potential energy in a stretched bow into the kinetic energy of the flying arrow is a straightforward calculation since the kinetic energy of the arrow is simply

$$\tfrac{1}{2}mv^2$$

where m is the mass of the arrow.

Sometimes an arrow is fired in a trajectory such that it will go up into the air and then fall down. Blyth has pointed out that with respect to warfare there is no optimum design for an arrow. Sometimes, however, maximum energy for penetration is desired, in which case a relatively heavy arrow is fired at a large angle to the horizontal, so that it goes high in the air. It then builds up a good speed as it falls so that it hits the intended target with the greatest possible energy. It appears that this tactic was used by the Norman warriors of William the Conqueror when fighting the Saxon troops of King Harold of Great Britain in 1066. At the battle of Hastings, the Saxon King Harold was killed by an arrow that pierced his eye. Apparently the Saxons under Harold fought the Normans with the traditional massed infantry formation using weapons such as axes and broad swords for close-encounter battle. They probably did not know that arrows could inflict heavy damage on closely packed troops.

Penetration of an arrow through protective clothing depends not only on the kinetic energy of the arrow but also on the sharpness of the point. In everyday speech the terms pressure and force are often used interchangeably whereas in scientific language the two words have distinctly different meanings. In science, *force* is the magnitude of something which causes change. *Pressure* is the effectiveness of a force as it is applied over an area. We can explain the difference between force and pressure by considering what happens when we use a knife to cut through a piece of material. With our hand and the muscles of our arm, we can exert a given force on the knife. The effectiveness of that force is determined by the sharpness which is governed by the area of blade in contact with the material to be cut. The pressure is defined as the force divided by the area. The knife is said to cut into the material when the applied pressure is greater than the pressure which can be resisted by the material. A blunt knife that fails to cut into a material can be sharpened by reducing the area of the blade as illustrated in Figure 2.9. The increased

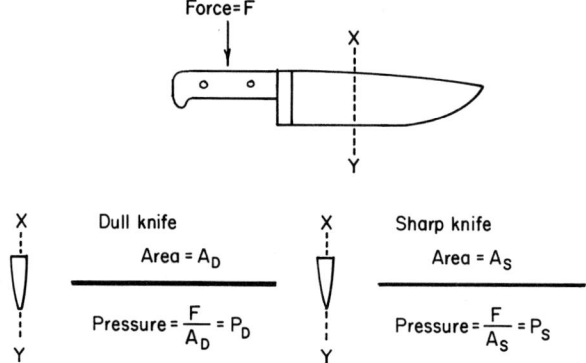

To sharpen a dull knife, rub away some of the edge to decrease its area.

Figure 2.9 A sharp knife is an efficient pressure amplifier.

pressure generated by the same force extended over the smaller blade area of the sharp knife will then enable the knife to cut, that is to break through the material. In the same way, whatever forces are exerted on an object by an arrow, the pressure available for penetration is increased if the area of the tip of the arrow is made smaller, or, in other words, if the tip is made sharper.

Section 2.8 The Bow and Arrow in Military History

The arrow has been used as an offensive missile in war for at least 45 centuries. Ancient texts which have come down to us from the Middle East show that as early as 2500 BC, the bow with arrows equipped with feathers was being used in warfare. The Assyrian Empire flourished from 745 AD until it fell to the Persians in the period 529–559 AD. Scholars attribute the superiority of the Persian army under Cyrus over the people of the Middle East to a battle technique adapted from the people of the plains of what is now Russia. These warriors of the plains could fire bows from horseback and move at great speeds. Apparently, up until that time, people of the Middle East had not mastered the technique of riding horses but used horse-drawn chariots in warfare. However, the Greeks must have been aware of these very able "horse-mounted" warriors, for in their mythology they talked about the *Centaurs*, pictured in these stories as half man and half horse. The idea of such creatures possibly arose when people without horses first met up with a raiding party of mounted men. If they had never seen men riding horses before, in their terror they could easily decide that the horse and man were a single creature. The fact that the Greeks associated these half-horse/half-man creatures with the barbarian north is emphasized by the fact that the Centaurs were usually depicted as wild, uncivilized creatures, always fighting with bows and arrows. In Greek mythology, however, one of them, Chiorn, is wise and gentle. It is said that he taught medicine to many of the Greek heroes. This story may have been generated by some historic events in which a person from the tribe of horsemen to the north, skilled in medicine may have migrated to live among the ancient Greeks. In mythology we are told that when Chiorn died, he was placed among the constellations of the stars and became the ninth sign of the zodiac. Modern astrological pictures often show the archer as an ordinary man pulling a bow; however older books on the star constellations show Sagittarius as a Centaur pulling his bow and aiming his arrow into other groups of stars.

It is interesting to note that modern English contains a figure of speech known as "*a Parthian shot.*" Brewer's *Dictionary of Phrase and Fable* [15] defines a Parthian shot as

> *a telling or wounding remark made on departure, giving an adversary no time to reply.*

This figure of speech is an allusion to the ancient practice of Parthian horsemen turning in flight to discharge arrows and other missiles at their pursuers [15].

It should be noticed that in Greek mythology there existed a race of female warriors living in Scythia, a region on the north side of the Black Sea, who did not allow men to remain in their tribe. The Greeks called this tribe "The Amazons." It is said that any sons born of unions with their neighbors were killed or sent to their fathers. It is also said that the girls had their right breasts burned off so that they might better draw the bow when indulging in warfare. The Greek word for "without a breast" is "amazon". The Amazon River in South America was named by the Spanish Explorer Franciscan de Orellana. In 1541, while sailing up the large South American river, he encountered a fierce tribe in which the women fought beside the men. For this reason he named the river the Amazon.

The Greek warriors did not always rely on the kinetic energy of their arrows being sufficient to cause the required damage to their enemies. As the history of the word toxicology shows, the Greeks and many other ancient people used to dip the tips of their arrows in poison to paralyze or kill their victims. This has led to the odd state of affairs in which the Greek word for bow has given us both the name *toxophilite,* for a lover of archery, and the word *toxicology,* for the study of poisons. The strange meanderings of the root word "toxon" are traced out in Figure 2.10.

Any student of military history soon becomes amazed at the stupidity of generals and the way in which they seem to fight a new war using the tactics and theories of the last war. For example, the generals of the Civil War in the United States managed to inflict enormous casualties on each other because they did not appreciate that the increased accuracy of new rifles made it stupid to march towards each other in massed ranks the way they had in earlier wars when guns were less effective. However, the generals of the American Civil War were no more stupid than the French generals of World

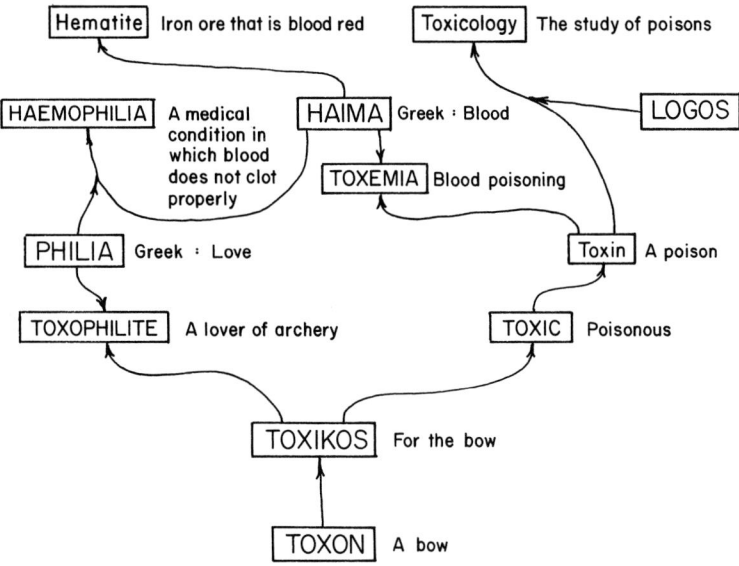

Figure 2.10 If kinetic energy did not make an arrow deadly enough, a quick dip in the poison pot usually did the trick!

War I who sent troops in red coats toward German machine guns. The French generals simply did not comprehend the devastation these new machine guns could inflict, especially upon highly visible soldiers. I have often pondered the amazing stupidity of military leaders. Upon consideration, the mind-set of an individual who makes a good soldier — one who obeys without question and with courage — is not best suited to creative innovation in the area of tactics. In all fairness, however, it should be pointed out that sometimes it is not easy to acquire the new tactics being used by an enemy. The Assyrians, when confronted with the mounted archers, may not even have been able to ride a horse let alone fire an arrow accurately when mounted.

In two famous battles between the English and the French, the English longbowmen were able to defeat and inflict heavy casualties upon superior French armies because of the use of wrong tactics by the French leaders. The first of these, the *battle of Crecy*, was fought in France on August the 26th, 1346, between Edward the Third of England and the French King, Phillip of Valois. The English force consisted of approximately 3900 knights, 11000 longbow archers, and 5000 light troops [13]. The French had at least 12000 knights, 6000 crossbowmen (mercenaries — paid soldiers from Genoa in Italy), and 20000 militia. There was also a Luxembourg division of cavalry under the King of Bohemia. Basically, the English won this battle because, although the crossbow could pierce armor, its range was less and its firing speed much slower than those of the longbow. The English King also used dismounted knights with foot soldiers to withstand the charge of mounted knights who came under fire from archers from each flank of the central core of dismounted knights. The English battle casualties were almost negligible whereas at least 1500 French knights were killed. This was a real example of how the crossbow was a good weapon used in the wrong circumstances. Long before the French could get within range of the knights they wanted to attack, they themselves were subjected to the longbow archers who could fire more arrows per minute from greater distances.

The second battle in which the fire power of the longbow played an important role was at the *battle of Agincourt* fought on October the 25th, 1415 between an English Army led by King Henry the Fifth, and a French Army led by Charles d'Albert Constable of France. At the time of the battle, Henry and his troops were actually trying to retreat to Calais to sail back to England. The English army consisted of 6000 lightly armed archers and they were intercepted by an army of 25 000 men consisting of armored cavalry and infantry. Before the fighting began, heavy rain turned the ground into a muddy field. In the words of one scholar,

> *The French generals' faulty tactics, notably the employment of massed formations against a mobile enemy, led to defeat.*

The French cavalry of knights in heavy armor became trapped in the muddy field where they became easy targets for the English archers. The English longbows could fire at a rate of 14 arrows a minute. It is calculated that about 6 000 000 arrows were shot at the battle of Agincourt. Up until this time there had been a steady increase in the amount of armor that the knights wore out of a need to protect themselves from crossbows and the heavy lances of charging armored knights. But too much protection became a liability. It is said that at Agincourt many of the knights died because they fell off their horses and were unable to stand up again let alone remount. Their assistants could not help because they were too busy fighting the English archers. Again, at Agincourt as at Crecy, the English casualties were very low while the French nobility suffered very heavy casualties [13].

The arrival of the gunpowder musket spelled the end of military archery. The success of the British infantry fire power against the Spanish Armada at the time of Queen Elizabeth was one of the first demonstrations of the superiority of the musket over the bow and arrow. We will explore the history of the development of the gun in Chapter 6.

The equipment of a modern archer competing in the Olympic Games includes a shield on the arm holding the bow and a special glove to protect the fingers. It is interesting to note that the woman wears a breast harness to avoid injury from the bow string. This is a somewhat easier solution to the problem than that adopted by the Amazons!

In present-day competitive sports such as the Olympics, modern technology is changing the equipment used by the athletes. Thus Bjerkile in an article on high technology used in Olympic sports [16] tells us that in 1993

> *Archery bows made from new materials are a far cry from the sleek wooden relicts that were still standard less than 30 years ago. The latest bows are made with a core of syn-*

thetic foam — a material composed of tiny glass beads embedded in a rigid foam matrix and wrapped in layers of carbon fibers and fiberglass. This was developed by Hoyt Archery. It is claimed that the Hoyt bow is lighter and more stable than wood core bows and is impervious to temperature changes.

An equipment expert Donald Rabska, quoted in [16], says

With wood you eventually get problems in the fibers and no matter what you put on it, wood absorbs moisture and is affected by temperature change This means that wood bows shoot fast in cold weather but get mushy when the temperature rises.

Rabska states that foam core bows return more of the released energy stored in the bow when the archer pulls back on the strings. Bjerklie tells us that bow strings and arrows have also been transformed.

Originally made of linen, strings are now made of a light-weight and low-stretch polyethylene that provides a velocity gain to arrows of several feet per second. The latest arrows are made from hollow aluminum with walls only .0006 inch thick. They are the lightest and therefore fastest yet. In fact they are about 20 feet per second faster than the aluminum arrows that were standard until the mid 1980s

In the article it is stated that such design changes have sent winner scores soaring. In a competition round, archers shoot a total of a 144 arrows aiming for a perfect score of 1440. Thirty years ago winning scores in major international tournament hovered around 1100. To win today an archer must shoot in the neighborhood of 1350 [16]. A recurve bow is defined as one in which the limbs, i.e., the part of the bow excluding the handle, are initially curved in the sense opposite to the curvature assumed by the limbs of a conventional flat bow with a bow string in place [17]. Note that the recreational bow of Figure 2.11 is fitted with stabilizers which minimize vibration when the bow is fired.

Section 2.9 Pullies for Bows

Figure 2.11 shows a new type of bow, used particularly by hunters. This type of bow is known as a *compound bow* or *bow machine*. It was invented by Wilber Allen of Billings, Missouri. As the story is told [18],

One day he was particularly frustrated when he tried to shoot a deer at a distance of 60 feet. The deer dodged every one of his well shot arrows and the hunter was convinced his arrows were just not flying fast enough.

As described in an article on the compound bow, the system that he evolved was an

odd looking combination of pulleys, wheels and criss-crossed cables.

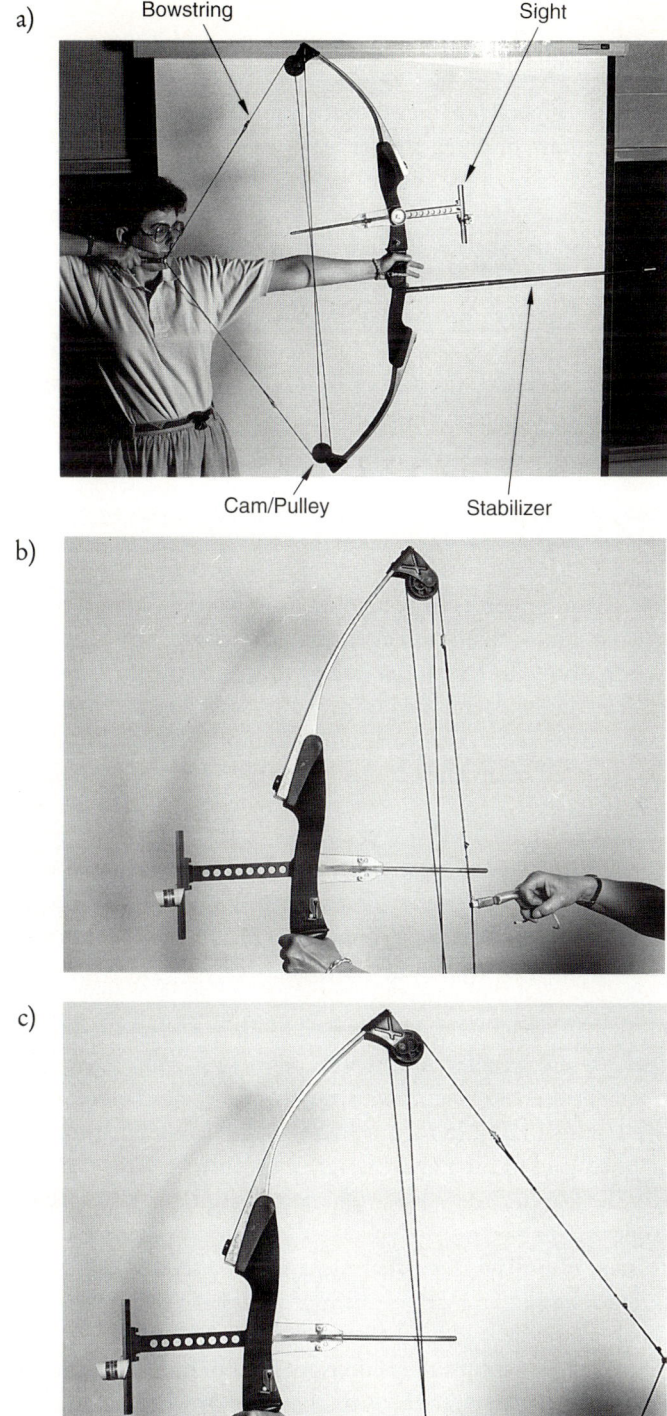

Figure 2.11 In modern competitive archery, several devices are used to improve the accuracy of the shot. a) An archer holding a drawn bow with pulley/cams, a stabilizer, and sight. b) Close-up view of a pulley/cams when the bow is undrawn. c) Close-up of a pulley/cam with the bow in the drawn position. Note the change in the orientation of the wheel with respect to b).

The bow machine design is an anathema to traditionalists who prefer the simple grace of standard bows; the bow machine shoots faster arrows. The International Archery Federation, official voice of olympic archery has refused to acknowledge the continuing evolution of the bow. They only permit recurve bows in their tournaments. As a consequence American compound shooters founded their own association, known as the International Field Archery Association to provide worldwide competition for the compound bow [18]. To understand the essential innovation introduced into bow design by Wilber Allen it is necessary to review the functioning of a device known as a pulley [19, 20]. Physics students learning the history of machines discover that ancient technology developed five basic machines. These are the lever, the wheel and axle, the pulley, the inclined plane and wedge, and the screw. The dictionary defines a pulley as

a wheel turning about an axis and having a groove on its rim in which runs a rope, chain or band — used for raising weights and changing the direction of pull.

Note that the word pulley comes from an old Greek word meaning a pivot or turning point. This somewhat limited definition does not really give us any hint about the powerful idea involved in using more than one pulley to reduce the force needed to achieve useful work. This idea can be appreciated from the systems sketched in Figure 2.12. Part (a) shows a simple pulley used to change the direction of an exerted force to lift an object. However, as recorded in an ancient manuscript,

If we want to move any weight whatever we tie a rope to this weight and pull the rope until we lift it For this is needing a power equal to the weight that we want to lift but if we untie the rope from the weight and tie one of its ends to a solid cross beam and pass its other end around a pulley fastened to the middle of the burden and draw on the rope our moving of that weight will be easier.

Note, however, that although the force required to lift the weight is half of the weight of the object when we use a pulley in this way, we now have to move the rope twice as far to lift the weight a given height; thus we still do the same amount of work to raise the body; see Figure 2.12(b). In Figure 2.12(c) we see a combination of pulleys that can reduce the applied load to lift a given weight even further. This type of pulley system was used by sailors to lift heavy loads on and off ships and also to lift the sails by pulling on the ropes.

By using the system of pulleys the archer using a bow machine can exert a moderate force which is in effect amplified by the system of pulleys and strings to enable archers to pull a bow which would be beyond their strength without the pulleys. Another feature of the bow machine is that by an ingenious method of mounting the pulley wheels off center and exploiting the way in which the bow bends when the strings are pulled the archer actually has to exert a lower force at the end of the pulling action as compared to the peak force that has to be applied during the pulling of the bow. The force–distance curve of a bow machine equipped with pulleys is shown in Figure 2.13. The bene-

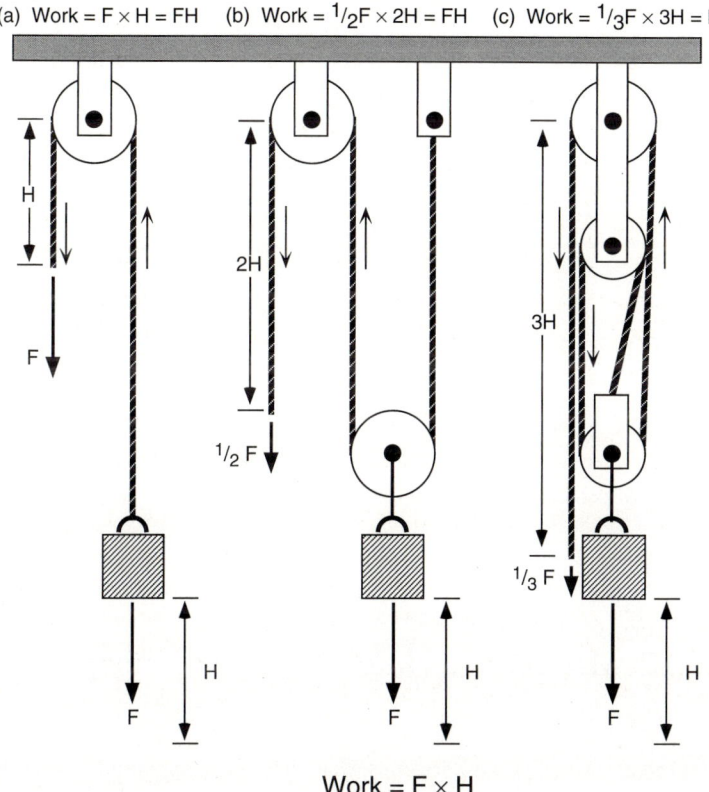

Figure 2.12 Pulleys can be used to increase the load that can be lifted by a given applied force. a) A simple pulley can be used to change the direction of a force. b) Two pullies can be used to halve the force needed to lift an object but the rope must be pulled twice as far. c) Three or more pullies can be used to lessen the applied force still further, but with correspondingly longer distances having to be pulled on the rope. The equipment shown here is known as a "block and tackle" and was widely used on sailing ships.

fit to the archer of this decrease in force towards to the end of the pull has been described by Cutnell and Johnson as follows:

> *One of the key features of the bow machine is that the force rises to a maximum as the strings are being drawn back and then falls to 60% of this maximum value when the string is fully drawn. This makes it much easier for the archer to hold the fully drawn bow while aiming the arrow, much easier than with an equivalent recurve bow.*

Cutnell and Johnson, have given a useful summary of the dynamics of different types of bows in their physics text [7].

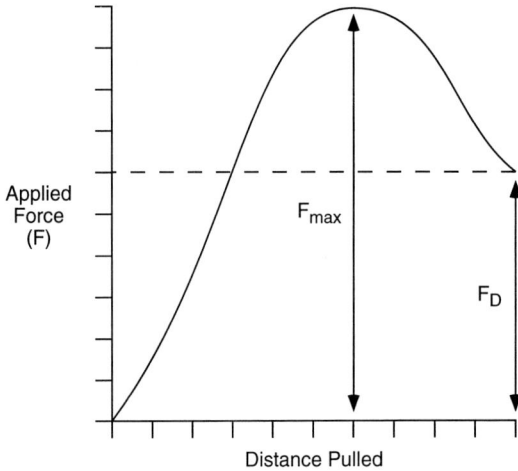

Figure 2.13 The force-versus-distance graph for a bow equipped with pullies is different from that of the self- or recurve bow.

Section 2.10 Metaphorical Missiles and Metamorphosed Bows

Students of the arts often protest strongly that they want nothing to do with technology; however, over the years, poets and artists have often embraced technology to create images with which to express their emotions and to create images of their theories and feelings. Thus St. Valentine's Day cards usually show Cupid the God of Love firing *arrows of desire* into the hearts of lovers. Here the arrow and bow have become metaphors for the act of falling in love or at least for the sensation of desire. The technical definition of a *metaphor* is that it is "a thing spoken of being that which it resembles." Thus, statements such as "she moved as swiftly as an arrow" or " she was as tense as a stretched bow" would be metaphorical. The origin of this word and related words is illustrated in Figure 2.14. The rootword in metaphor, from the Greek for to carry, is to be found in the child's name Christopher. This latter name comes from an old legend in which the person is said to carry the Christ child across the river and to received a new name for his endeavors [15].

Cupid is the Roman name for the God of Love. In Roman mythology he is said to be the son of Venus, the Goddess of Love. In Greek, Cupid was known as Eros and his arrows stimulated *erotic emotions* [15]. In most representations of Cupid or Eros, the figure is shown carrying a composite bow presumably because it would be too difficult to fly with a longbow! It is interesting to note that the target of the arrow of desire, the heart, is really inappropriate, chosen as a result of a basic mistake made by early Greeks in their study of the structure of the human body. The Greeks were well aware of the fact that the human brain had no pain nerve cells. They gained this knowledge when they treated deep head wounds. Since the skull is intended to be sufficient protection for the brain, there is no need for the brain to know by the activation of pain cells when it is in danger from an external force. Apparently the creator of the human body did not anticipate that human beings would develop weapons with which to damage the head sufficiently for the brain to be exposed. Because the brain had no feeling, the ancient Greeks assumed that it was basically a place where the blood was sent for cooling before returning to the rest of the body. They were also aware of the fact that the one organ of the body which they could not do without was the heart. In Greek thinking, the heart became the center of living and the location of emotions and thinking. Therefore, when Cupid wanted to stimulate desire with his arrows, he fired them at the heart of the unsuspecting human being. Much of Greek thinking was dominated by speculation rather than by experimentation. It was the development of the *scientific method* that cleared away such amazing misconceptions as the heart being the center of feeling and emotion.

As indicated in Figure 2.14, the word *method* incorporates the same rootword "meta" as metaphor, only now it is combined with the Greek word for road. A method was originally a set of instructions for reaching a new place from the present location. The *scientific method* is a way of approaching the world. In the scientific method, one first of all sets up a theory about how things work. You then decide what other things should also be true if your basic idea is correct. Finally, you test these predictions by experiment. If the ideas prove to be true, you say that your theory has been validated. If the results do not agree with predictions, you alter your theories about the world. In the scientific method, experiment is dominant and speculation is secondary. The scientific method of studying the body has shown that the heart is not a seat of emotion; it is a pump made out of meat.

In mythology books, the two gods Apollo and Artemis are always shown as youthful archers. In Greek mythology these two gods are said to be the twin children of Leto. *Apollo* is always shown as very handsome. He is the god of poetry and music. He is also said to be the god of the sun. When the Romans became the dominant force in the Mediterranean countries, they adopted many things from the Greek civilization and Apollo became one of their gods. In Greek mythology *Artemis* is known as the goddess of hunting; the Romans identified her with their own forest goddess *Diana* who was always out hunting. Artemis and Diana were also known as goddesses of the moon. In Greek mythology, the *arrows of Apollo and Artemis* were deadly missiles which spread disease. The Greeks used to explain epidemics of disease by supposing that these children of Leto were shooting at people at random. From this developed the idea that by praying to Apollo an epidemic might be halted. In this way Apollo became associated with the cure of disease. It is curious to note that in modern science, a new drug which is able to cure a disease is sometimes known as a *magic bullet*. Again, a mythical missile becomes an explanation of events in the health field.

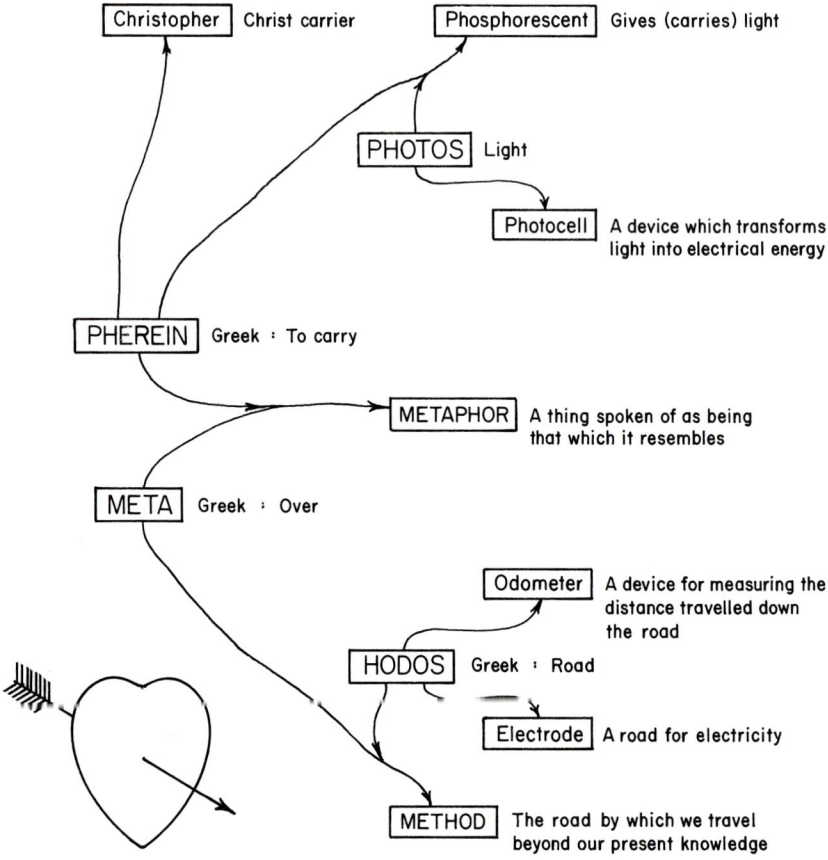

Figure 2.14 Arrows of desire are fired by Cupid in a misplaced metaphorical stimulation of erotic emotion.

Poets and religious leaders have always dreamed of an age in which there would be no warfare. When such an age dawns upon the earth, we are told in the words of the poet that

swords will be beaten into ploughshares and spears into pruning hooks.

In this phrase what is being said is that there will be a metamorphosis of weapons into things of everyday utility. One element of this word comes from the Greek word "morphe" meaning shape. *Morpheus* was the Roman god of dreams. The drug *morphine* is so called because it is said to give fantastic dreams before it destroys the person. *Morphology* is the scientific term for the study of shape and the word *amorphous* means "without shape." The root word "morphe" is combined with "meta", a word we met earlier in this section, to create the word *metamorphosis* which means a change of shape and sometimes of function in living creatures or other objects. Thus when a caterpillar becomes a butterfly, the change is described as a metamorphosis. The sword becoming a ploughshare is metamorphosed on the smith's anvil from a weapon of destruction to an implement of agriculture.

It is curious that poets in their imagery have never metamorphosed a bow into a harp. This type of metamorphosis appears to be the origin of all stringed instruments. Very early in the history of archery it was noticed that a plucked bow string made a pleasant musical note. Experiments would soon show that the note made by the string varied with the tension of the string and the material from which the string was made. An obvious innovation would be to put various strings side by side on the same bow to economize on the number of bows required to make music. Then one day it is probable that a music maker rested this multiple-stringed bow on a hollow box and noticed how the sound seemed to become louder. Again, experiment would show that the same effect was created whether the bow rested on the box with one end or with the center of the bow on the box. Through this series of observations, the craftsman of ancient times soon evolved an instrument known as a lyre with strings attached to a box stretched at various tensions to the arms of a bent bow. This was a metamorphosis from weapon launcher to music maker. We now know the reason the hollow box made the sound louder was that the vibration of the string was transmitted to the box in such a way that the surface of the box began to vibrate in sympathy with the string. In technical terms, the box is said to *resonate*, thereby becoming an amplifier for the vibrations of the string. All of the modern stringed instruments such as harps, cellos, and pianos are metamorphosed bows. A poet wanting to show the metamorphosis of a warlike society into a peaceful community could equally well use this imagery,

their bows will be made into harps as well as *their swords shall become ploughshares.*

In the word webs drawn to illustrate the origin and evolution of scientific words, we have used arrowheads on the lines to show directions. This abstract use of the arrow to show direction is so widespread that we hardly ever think about it. However, when scientists were starting to send out probes into outer space, they discussed what kind of plaque they could put on the spaceship to give a message to any alien intelligence that might capture the plaque in the outer realms of our space. Since the late 1980s, we have been so used to television shows in which aliens communicate effortlessly with each other that we do not always realize how difficult it would be to communicate with an alien intelligence. One of the suggested designs for the plaque used a stylized representation of the sun and its planets with an arrow pointing to the earth as being the origin of the spaceship. However, it was pointed out that a society that had never developed bows and arrows would not know how to interpret the meaning of an abstract arrow. Most certainly there was no guarantee an arrow would be interpreted as being an indicator of direction. It might instead be seen as a footprint of some alien bird on a distant planet and thereby be interpreted that these birds also inhabit the planet we call Earth.

English bowmen used to practice shooting at what we today call a *target*. The modern archery target has several rings whereas the early targets were probably a bale of hay with a single white circle drawn on it. This ring became a metaphorical *bull's eye*; a phrase still in use to describe targets in archery and darts. The word *targe* originally meant a shield. Interestingly enough, the bowman would not really be aiming at a shield, but rather, around the shield. The word has come into modern English through the French-speaking warriors of William the Conqueror. The French add "ette" to the end of the word to make something small. The original spelling of target was "targette." In modern North American English we use phrases such as "our expenditures are on target" meaning that they are moving towards the estimated figure arrived at when planning expenditure. In modern pharmaceutical technology a *targeted drug* is one that does not become activated until it has reached the part of the body which was intended by the doctor. Notice how in North American English, target has changed from a noun to a verb and now describes action rather than an object. Purist protectors of the English language would target such utilization for destruction, at least metaphorically!

References

[1] D. McEwan, R.L. Miller, C.A. Burgmen, "Early Bow Design and Construction," Scientific American, June 1991, pp. 76–82.
[2] Chambers Etymological English Dictionary, W. & R. Chambers Ltd., Edinburgh, Scotland, 1967 edition.
[3] J.N. Wilford, "The Iceman's Outfit, Cultural Clues", *International Herald Tribune*," Thursday June 23, 1994.
[4] L. Jaroff, "Iceman", *Time*, October 26, 1992, pp. 48–55.
[5] D. Lessem, *The Iceman*, Crown Publishers Inc., 1994.
[6] D.H. Fender, *General Physics and Sound To Advanced and Scholarship Level*, Cambridge University Press, 1957.
[7] J.D. Cutnell, K.W. Johnson, *Physics*, 3rd edition, Wiley, New York, 1995.
[8] G.R. Noakes, *A Textbook of Heat*, Macmillan, New York, 1965.
[9] A.B. Howard, *Chambers Biographies*, Chambers, London, 1950. Reprinted 1964.
[10] J.E. Gordon, *New Science of Strong Materials*, Penguin Books, 1968.
[11] The complexity of the calculations involved in studying a longbow are explored in W.C. Marlow, "Bow and Arrow Dynamics," *American Journal of Physics*, Volume 49, No. 4, April 1981, pp. 320–333.
[12] See discussion of vane mounting on an arrow given in *Coaches Manual Level 3*, of the Federation of Canadian Archers Incorporated published in 1980 by the Federation of Canadian Archers, 333 River Road, Vanier, Ontario, K1L 8B9.
[13] G. Rees, "The Longbow's Deadly Secrets", *New Scientist*, June 5, 1993, pp. 24,25.
[14] Homer's Odyssey XXI, in *The Illiad of Homer and The Odyssey*, rendered into English prose by Samuel Butler, William Benton, Encyclopaedia Britannica Inc., Chicago, 1952.
[15] *Brewer's Dictionary of Phrase and Fable*, first published 1870. Centenary edition, ed. by Ivor H. Evans, Cassell and Company Ltd., London, 1970.
[16] D. Bjerklie, "High Technology Olympians", *Technology Review*, January 1993, pp. 22–30.
[17] B.G. Shuster, "Ballistics of the Modern – Working and Recurve Bow and Arrow," *American Journal of Physics*, Vol. 37, No. 34, April 1969.
[18] L.B. Ackerman, "The Bow Machine", *Science*, July/August 1985, pp. 92,93.
[19] See information on pulleys in *Machines*, ed. by Robert O'Brien, a volume in the Life Science Library, Time Inc., New York, 1964.
[20] F. Bueche, *Principles of Physics*, 3rd edition, McGraw-Hill, New York, 1977, pp. 108,109

Chapter 3

Racketeering Missiles

Chapter 3

Racketeering Missiles

Section 3.1 Love and Tennis

My etymological dictionary of the English language [1] indicates that *tennis* is

an ancient game for two to four persons played with ball and rackets within a building specially constructed for the purpose.

Etymology means

the science or investigation of the derivation and original meaning of words; the source and history of a word. This word comes from Greek rootwords "etymos" meaning true and "logos" a discourse.

Tennis is sometimes used as a short form word for describing *lawn tennis*. Lawn tennis is described as

a ball and racket game, a variety of tennis played on an open lawn or other unenclosed space.

Today's athlete would be rather surprised at some of these definitions because many tennis games are now played on hard or gravelly type surfaces. The playing area is usually enclosed with wire mesh for the safety of the spectators and to conserve the energy of the players who would otherwise run around endlessly to retrieve out-of-court balls.

The word tennis itself comes from the French verb "tenir" meaning to take" or "receive". Apparently in the early days of tennis, the person serving the ball would yell out "tenez!" meaning "receive the ball" or perhaps, "take what is coming!" However, there are other scholars who suggest that the word tennis comes from the name of a city of the Egyptian delta, Tennis, noted in the Middle Ages for its linen from which the best original tennis balls were made [1, 2].

Whatever the origin of the word tennis, we can be sure that the game was very popular with the French as early as the 1500s since the word *racket*, variously spelled in modern English as *racquet* and racket, comes from a French word which meant "the palm of the hand." Thus, handball appears to have developed before tennis. The racket is a large substitute hand used to increase the reach of the player and reduce the wear and tear on the hand. Another remnant of the French origin of tennis is the rather odd scoring scheme which can cause confusion to the beginner. If one player has no points, the players score is said to be "*love.*" The umpire calls out the score as fifteen-love. The word love comes from the fact that the French when scoring a game, used to refer to 0 as "the egg," drawing a comparison between the shape of the 0 and that of an egg. The French for egg is l'oeuf. Just as cadet became caddy in golf, l'oeuf became "love" in tennis. It is interesting to note that if a cricket player is bowled out before he has made any score he is said to have acquired a *duck*. This is said to be short for duck egg. The idea is that the duck egg is a rather large egg and matches the large 0 put on the score card for the unfortunate cricketer.

The word *umpire* also has an interesting origin. It comes from an old French word "nompair" meaning "not a peer" or "not an equal." The umpire is the third person called in to decide a matter on which other people disagree. The idea is that the umpire is not involved as an equal in the progress of the game. The umpire is an impartial person chosen to enforce the rules and decide disputes. The umpire's role is the same as that of the referee in a football match. Just as people doubt the visual capability and the objectivity of the soccer referees, people have been known to doubt the impartiality of a tennis or cricket umpire. As we shall discover later in this chapter, because of the speeds involved in modern tennis, the umpire may be replaced by robots on the tennis courts of the future.

Commenting on the increased speed of modern tennis Brody [3] has pointed out that in modern rackets the sweet spot is higher up the head of the racket. As a result players can hit the ball higher when they serve, opening up more of the opponents court. Brody has also made tests on the way in which a larger ball would slow the game down. He has shown that increasing the ball's diameter by 20 per cent would slow a 200 kilometer per hour (113 miles per hour) serve to 170 kph (175 mph) which is much more playable by the receiver. Brody has suggested a change in the size of balls to the United States Tennis Association and envisages a time when the organizers of every tournament will choose a ball size appropriate to the court surface and the players' level of skill [4].

The title of this section hints at the fact that the racketeer is not someone who makes rackets for tennis. The word racket can also mean a noise or a distraction. When used with this meaning, the word is probably an imitative word mimicking the noise of a racket. The word racket can also mean a fraudulent activity. A criminal *racketeer* extorts money by threat or makes profits by illegal action. This meaning developed from the fact that, in the 1600s, pickpockets and other thieves used to start a disturbance before a robbery by throwing firecrackers into the street to distract the attention of their victims whilst robbing them. Thieves have not changed their strategies. In 1992 when walking through a courtyard in Rome I was accosted by two young ladies pretending to sell me a newspaper. Suddenly one of them started to clutch at my arm and screamed, creating a racket. In a very short period of time while my attention was diverted by the screaming, the other girl took my wallet out of my pocket. However my reflexes were very quick and I was able to grab the young lady making a racket whilst my colleague apprehended the wallet snatcher. That incident made me understand why thieves used to make a racket and became racketeers. Because of this strategy a law was passed by the British parliament in 1697 forbidding the throwing of "squibs, rockets and other distracting devices into the streets." A racketeering missile could well be interpreted by an unthinking computer as meaning bullets fired by a criminal!

The game of *court tennis*, also known as *royal tennis* in Great Britain, is still played in a rectangular space 33.5 meters (110 feet) long by 11.5 meters (38 feet) wide. In this game, the ball is made of tightly bound cloth and is slightly smaller than a lawn tennis ball. The floor of the court is made of concrete. The court is divided by a net into a service side and an opposite or hazard side. The rules of court tennis are fairly complicated and the player may make the ball bounce off the walls or even off the roof. This original form of the game has more in common with the modern game of squash or racketball than lawn tennis [1]. In this chapter, we will focus our attention on lawn tennis and its close relative, hard-court tennis.

Lawn tennis was invented in 1873 by Major Walter C. Winfield, a British Army officer. His intention was to devise a game that could be played on English lawns at garden parties. The game very quickly became popular. It was introduced into the United States by Miss Otterbridge of Staton Island in New York in the late 1870s. The game was then modified to be played on hard courts in parts of the world where it was difficult to maintain a good grass surface. Then the game of hard-court tennis became popular in its own right. Since the bounce of the ball is different on hard courts, a player has to learn to adapt when playing on different types of courts.

Section 3.2 Tennis Balls — Flannel Wrap and Gas!

About the time that lawn tennis was invented, the players started to use a rubber ball which was covered with cloth to give it the required dynamics when hitting the surface of the court. These balls contained pressurized gas which improved the bounce. The stages in the manufacturing of the modern tennis ball are as follows: A small plug of rubber, which has been hardened with the appropriate chemical additives, is placed in a heated press and made into a hollow hemisphere. Two hemispheres are then put in a curing press. Next, a small amount of a powdered mixture of sodium nitrate and ammonium chloride is placed inside the lower hemisphere. Under the action of heat, these two chemicals react to give ordinary table salt, water, and nitrogen. The nitrogen gas is at a sufficient pressure to make the ball more resistant to deformation than an unpressurized hollow ball and this increases the bounce of the ball. The hollow rubber ball is then covered with a special cloth made out of wool and nylon. The covered ball is then placed in a heated press to fuse the cloth covers to the rubber ball.

There is a significant transformation of kinetic energy into heat each time the ball bounces or is hit. For this reason, the ball becomes noticeably warmer during play. As the gas inside the ball warms up, the internal pressure of the ball rises and, as a result, the bounce of the ball increases significantly during the course of a game and, particularly, during a long exchange of shots. If one attempts to play tennis at a Mountain top resort such as Lake Tahoe in California, the atmospheric pressure is less than in a location near sea level such as the Netherlands. Therefore, in locations such as Lake Tahoe, the pressure inside the ball is effectively higher and the balls bounce in a more lively manner. I am told that some players try to overcome this effect by piercing the wall of their tennis balls to reduce the gas pressure. Others use slightly larger balls but since there is only a small market for mountain top tennis balls it is sometimes difficult to get hold of them [5].

High speed studies of tennis balls show that a ball loses energy after being served because of air resistance and the fact that the ball itself vibrates as a result of the deformation it experiences during the impact of racket and ball. An analysis reported by Lynne Ebert of the Department of Metallurgy and Material Science at Case Western Reserve University in Cleveland has shown that a tennis ball served at 240 kph (150 mph) has decelerated to less than 160 kph (100 mph) by the time it crosses the net. Typically, after it bounces up towards the player who must return the ball, it is moving at about 72 kph (45 mph) [5].

Currently the International Tennis Federation rules set the limits on the weight of the ball allowing variation of 3% with a specified diameter which can vary by 5%. The coefficient of restitution of the ball is fixed by stipulating that a ball bounce between 135 and 140 centimeters high when dropped from a height of 254 centimeters (coefficient of restitution 0.729 to 0.742). Some people prefer to play with bright yellow balls because these are more visible. Some other amateurs have been attempting to use two tone balls which are even more visible. The number of tennis balls used around the world is enormous: One manufacturer alone makes 56 million balls a year. 144 thousand are used in the Wimbledon Championships every year [6]. Because of the tremendous speeds involved in the game of tennis, there are often acrimonious debates about whether a ball is in or out of the playing area. In a study of the arguments about calls made by umpires it was found that in every ten line calls at a big tennis game one of them was wrong and that when players disputed calls made by the umpires they were right more often that not. Recently a robot umpire has been developed to judge whether balls are in or out of a certain playing area [7, 8].

The need for robot umpires can be appreciated from the following fact: Tennis balls can typically be served with a speed of 180 kilometers per hour and can reach 104 kilometers per hour during play. The ball is in contact with the court surface for as little as 5 milliseconds and the human eye has difficulty distinguishing events that last for under 20 milliseconds. (*milli-* is a prefix meaning a one-thousandth part). A particular robot umpire system which is receiving widespread testing in the tennis world is known as TEL from the phrase *tennis electronic lines*. The system was developed in South Australia by Baxter and Candy [7]. It consists of an array of strip antennas buried beneath the lines of the tennis court. They are buried at about 2.5 millimeters below the surface. The antenna gives out radio-type waves which vary in frequency depending on the direction of the line. Special tennis balls impregnated with 5 grams of iron powder can be detected by these wires buried beneath the lines of the court. As a ball moves into the field of the antenna it disturbs it and the system picks up a signal characteristic of the ball's trajectory. Each of the 13 lines on the court is connected by a system which feeds the signals from the wires to a voice synthesizer and to a hand-held computer on the umpire's chair. The system can announce electronically whether a ball is in or out. The system has been designed to ignore feet and metal objects and so human umpires are still needed to detect foot faults. In 1992 the system cost approximately US$ 70 000. Experiments have shown that this system is five times more reliable than human line judges [8]. In 1993 it was stated in a press story that as many as 12 TEL systems were being considered by Atlanta for the 1996 Olympic Games and that they would be cheaper than hiring and accommodating the necessary line judges [8].

Section 3.3 Vibrations, Sweet Spots, and Space Age Rackets

A sketch of a typical tennis racket is shown in Figure 3.1. Older rackets had strings made out of the guts of animals. Kate Charlesworth in a cartoon commentary on various aspects of tennis tells us that the guts from between six and eight sheep were required to string the classical wooden racket [6]. The reason for using a stringed rather than a solid racket is that if a racket of the size used in tennis was covered with a solid surface one would experience high air resistance when swinging it, making it difficult to play the game. (You can prove this for yourself by cutting a disc of cardboard to cover the strung area of a racket and try swinging it to play tennis!). People used to think that tight stringing of the racket enabled the player to hit the ball with more force. However scientific studies have shown that in fact loose stringing of the racket can improve the amount of energy transferred from the racket to the ball. Thus C. Arthurs [5] tells us

> *A racket is two inter-related systems, the strings and the frame. The strings form a trampoline for the ball. On impact the strings deflect absorbing some of the kinetic energy but then spring back to transmit almost all that energy back to the ball which leaves the racket as the strings rebound. Thus looser strings give more power than a tight one because they are more springy just as you can jump higher on a trampoline than on a concrete floor.*

The next question that science has to address in the design of a tennis racket is whether the frame should be flexible or rigid. It used to be thought that a flexible racket would

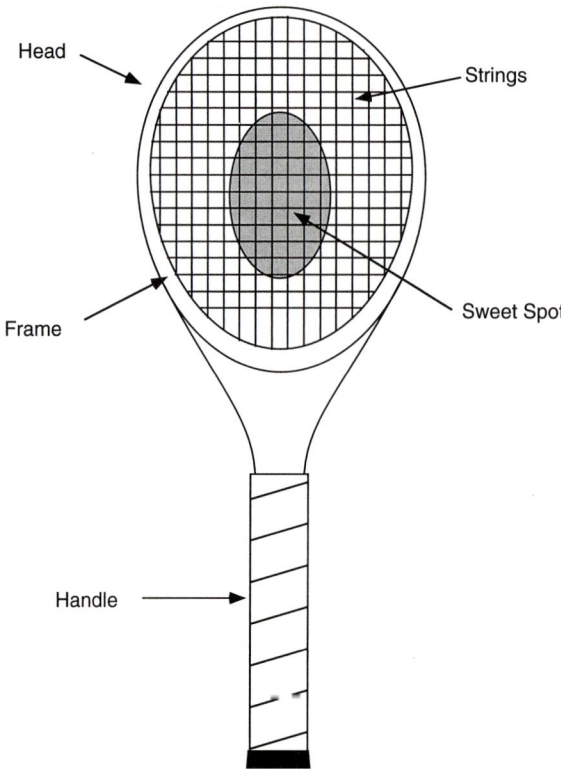

Figure 3.1 The basic structure of a tennis racket.

spring back when it hits the ball thus giving more energy to the ball. However if the racket bends when it hits the ball the impact takes place so quickly (about 5 miliseconds) that the racket cannot flex back in time to contribute to the energy of the ball. Arthurs [5] states that:

> *By deflecting less a stiff racket absorbs less of the balls energy than the flexible racket just as you can jump higher if your trampoline is standing on concrete instead of quicksand.*

Mark Powell product manager for Head Rackets tells us that if the rackets are too stiff they tend to cause tendonitis. *Tendonitis* is more widely known as *tennis elbow*. Tennis elbow is in fact the inflammation of the tendons and blood vessels in the elbow. It is caused by the high frequency vibrations arising from the balls' contact with the strings being transmitted by the racket to the players arm. More flexible rackets absorb more vibration so the player must compromise between the stiff racket to give energy to the ball and the effect on the elbow when playing a game of tennis.

3.3 *Vibrations, Sweet Spots, and Space Age Rackets*　　71

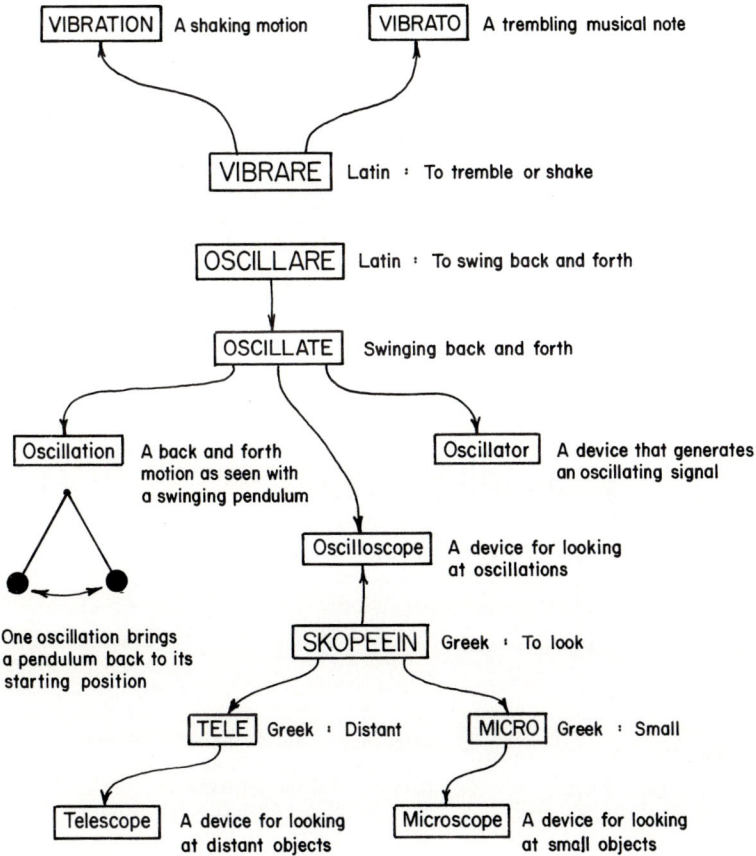

Figure 3.2 Roots of the words "oscillation" and "vibration".

In the racket of Figure 3.1 the shaded area in the middle of the racket is called the *sweet spot*. To understand what is meant by the sweet spot of a racket we need to explore in detail the scientific concepts involved in the descriptions of vibrations and oscillations of an object.

The origins of "to oscillate" and related words are shown in Figure 3.2. The Romans used to use the word "oscillarum" to describe a child's swing that moves back and forth. In scientific terminology, the term *vibration* is often used to describe a general shaking rather than a regular motion. When scientists wish to study something that is vibrating, they use a device to change the movement of the object into an electrical signal which can then be viewed on a television screen. Any device which changes the nature of a physical quantity into a different kind of physical signal is called a *transducer*. Thus, a *photocell*, a device which turns light energy into electrical energy is referred to as a light-to-electricity transducer. The device that turns sound into electrical energy is a *microphone*. A microphone is a sound-to-electricity transducer. One way to study the vibra-

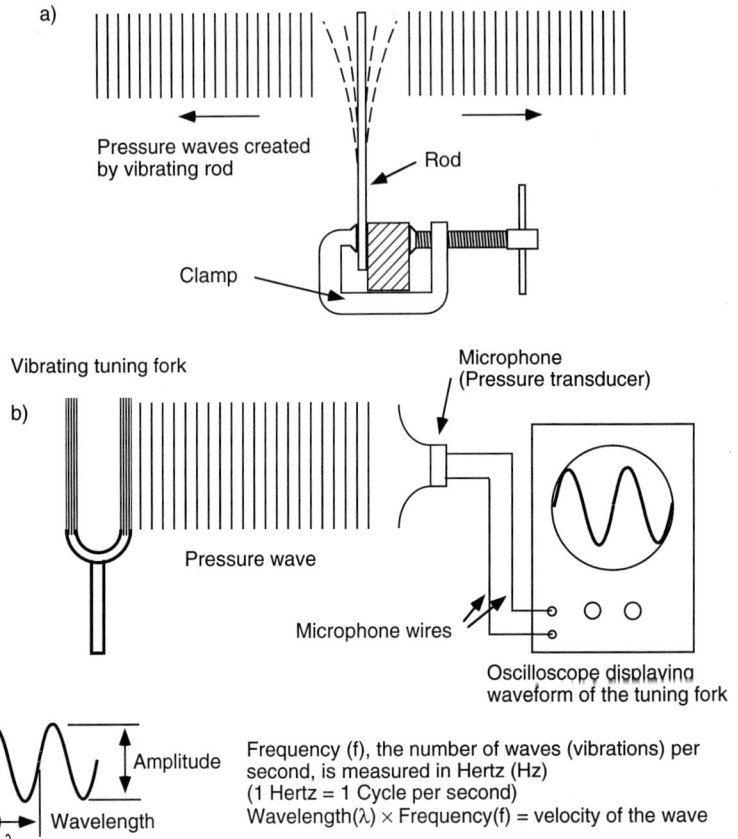

Figure 3.3 The motions of a vibrating body can be studied using a microphone and oscilloscope [3]. a) A clamped rod can be heard vibrating because the movement of the unclamped end of the rod sets up a series of pressure waves which the ear interprets as a particular sound or musical note. b) Sounds can be made visible by using a transducer to convert the sound to electrical energy which can then be viewed using an oscilloscope.

tion of the end of a clamped bar of Figure 3.3 would be to pick up the sound waves generated by the vibrations with a microphone and display the variations in air pressure reaching the microphone as a wavy line on the television screen. The instrument used to make oscillations visible on a screen is called an *oscilloscope*. The Greek word "skopeein" meaning "to look at" or "to see" has given us the familiar words *telescope*, *microscope*, and *periscope* from the Greek rootwords "tele" (distant), "micro" (small), and "peri" (around).

Figure 3.3(b) illustrates the way in which pressure waves leaving the end of a *tuning fork* can be transduced and displayed on an oscilloscope. (A tuning fork is just a bar bent into a horseshoe shape.) The build-up and drop-off of pressure in the air which constitutes the sound looks like a wavy rope when displayed as a pressure/time graph on the oscilloscope. For this reason, sound is referred to as a wave motion. The distance between repeating pressure values in one complete cycle is known as the *wavelength* of the motion. To calculate the wavelength from Figure 3.3(b), you have to know the velocity of travel since this graph is displaying how the pressure varies with time.

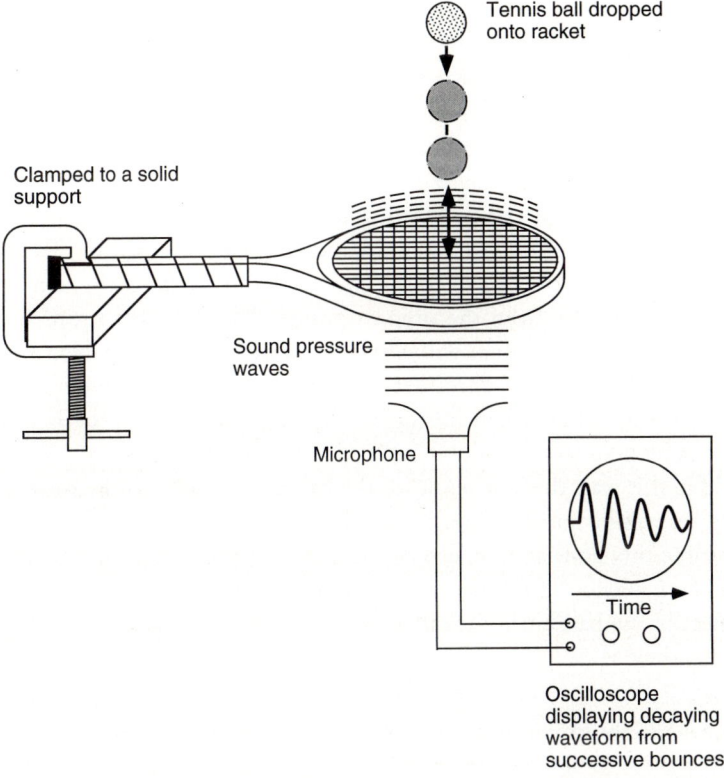

Figure 3.4 We can study the design of tennis rackets, and the resulting sweetspots (places where the collision between racket and ball feels good), by looking at the vibrations produced in the racket when a ball is dropped from a fixed height onto various parts of the racket face [10, 11].

One can modify the equipment shown to study the vibration of a rod in Figure 3.3 to study the vibration of a tennis racket when a ball is allowed to drop onto the racket clamped in the position as shown in Figure 3.4. Professor Howard Brody of the University of Pennsylvania has used such a system to study the vibrations in the racket caused by balls of different energy (dropped from different heights) falling onto different positions on the racket. He has also taken stroboscopic pictures of the balls to see how much energy is left in the ball after a rebound (i.e., he measures the coefficient of restitution of the ball). Using such equipment Professor Brody can outline what is known as the sweet spot on the tennis racket. This is the area of the racket which gives maximum energy back to the ball and also feels good to the tennis player, hence the name. If one were to compare a modern racket with an older model one would find that it is lighter and also has a much larger strung area. The strung area can be up to 20% larger which results in a much larger sweet spot facilitating the players return of the ball [3, 9, 10].

To appreciate the physical significance of the sweet spot it is necessary to realize that a uniform bar can vibrate in many ways. Two of the basic ways a bar can oscillate are shown in Figure 3.5(a). As the wooden bar vibrates there are positions of no motion known as *nodes* and between the nodes there are what are known as *antinodes*. In a *xylophone* the wooden bars which produce the musical notes are fastened to the frame at the nodes of the basic vibrational mode shown in Figure 3.5(a)(i). Note that the word xylophone comes from the Greek word "xylon" meaning wood and "phone" meaning a sound [1]. Note that in the sketch of a basic xylophone shown in Figure 3.5(b) the lengths of the piece of wood between the two mounting points is the same as the combined lengths pretruding from the support points. When the bar is hit in the middle with the player's hammer the bar will vibrate up and down with the fixed points joining the bar to a support beam at a position equal to the minimum vibration. In the same way it can be shown that when hitting a ball with a simple bat, the latter should be held at the same node as that where the support point of a xylophone is nailed down. When held in this way, the best point for the ball to hit the bat is the node at the other end of the bar as illustrated in Figure 3.6. However we must also take note of the fact that if possible this node should also be the center of gravity of the bar since this will minimize the tendency of the bat to rotate in the hand of the player. Thus in Figure 3.6 the effect of the ball hitting at different points of the bat with respect to the center of gravity is shown.

Before 1970 the most popular material for making tennis rackets was Canadian ashwood. This wood was cut into long strips steamed, glued, and pressed together. In the older rackets the strung area was 450 square centimeters (70 square inches). An average racket weighed 355 grams and had a head frame about 18 millimeters deep measured perpendicular to the strings. This racket design was actually dictated by the natural limitations of wood. In the older rackets if one attempted to increase the area of the strings the wooden frame was either too weak to sustain the repeated impacts of balls or, if reinforced, too heavy for play. In the 1960s, players began to use bigger and bigger rackets until the international tennis federation ruled, in 1980, that the racket should have certain dimensions with a maximum area of 478 square centimeters. Scientific experiments showed that a lower racket weight was an advantage, since increasing the mass of the racket by 33% produces only a 5% rise in the speed of the ball, whereas a 33% increase in racket head speed increased the ball speed by 31%. Therefore since one can swing a light racket faster, light rackets have come to dominate the game. Readers interested in reviewing the development of modern rackets should consult the article by C. Arthurs [5].

If you are a player with money to invest in your game you can have your own customized racket designed to fit around the way in which you play tennis. In 1987 such a tai-

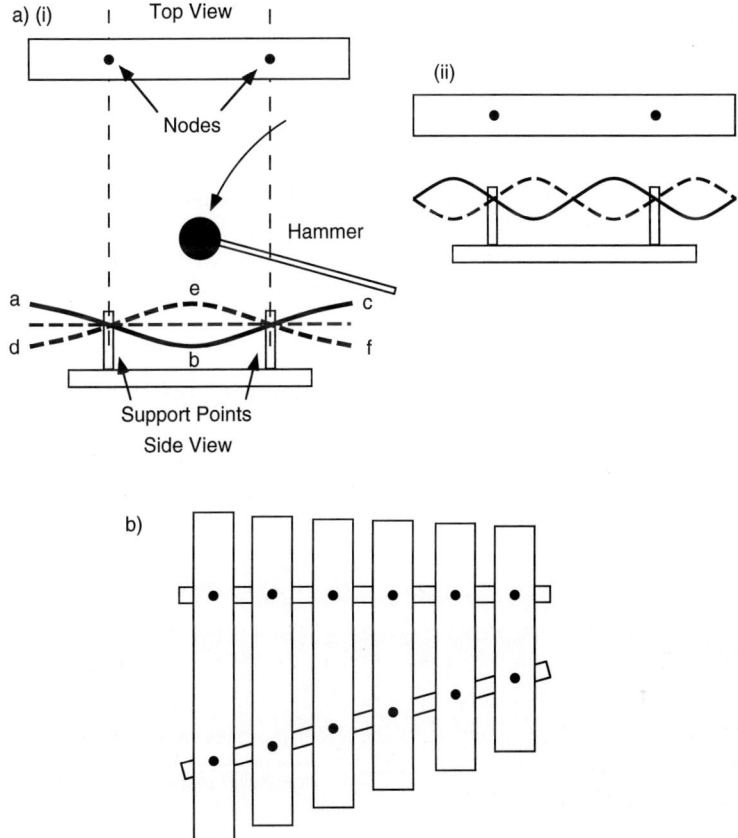

Figure 3.5 A uniform bar of wood can vibrate in many ways. Positions where there is no motion are known as nodes. Positions of maximum motion are called antinodes. a) Two of the many different ways a wooden bar can vibrate when supported by two pins. b) The bars of a xylophone are supported at the nodes of the natural vibration mode illustrated in (a).

lor-made racket cost a thousand dollars; you could get a second racket for only $500 dollars more!

In the early 1970s experiments were carried out to investigate the effect of various structures of the strings in the racket. Thus in the so-called spaghetti racket double strings fitted with plastic tubes were used. This racket surface imparted such a tremendous top spin to the ball in play that relatively unknown players using this racket started to beat top tennis professionals. In a tennis match in the late 1970s, a player by the name of Ilie Nastase used a spaghetti racket to defeat Guillermo Vilas, a top Argentinian professional. Vilas stormed off the court. As a consequence of this incident the International Tennis Federation devised a new rule which said that [11]:

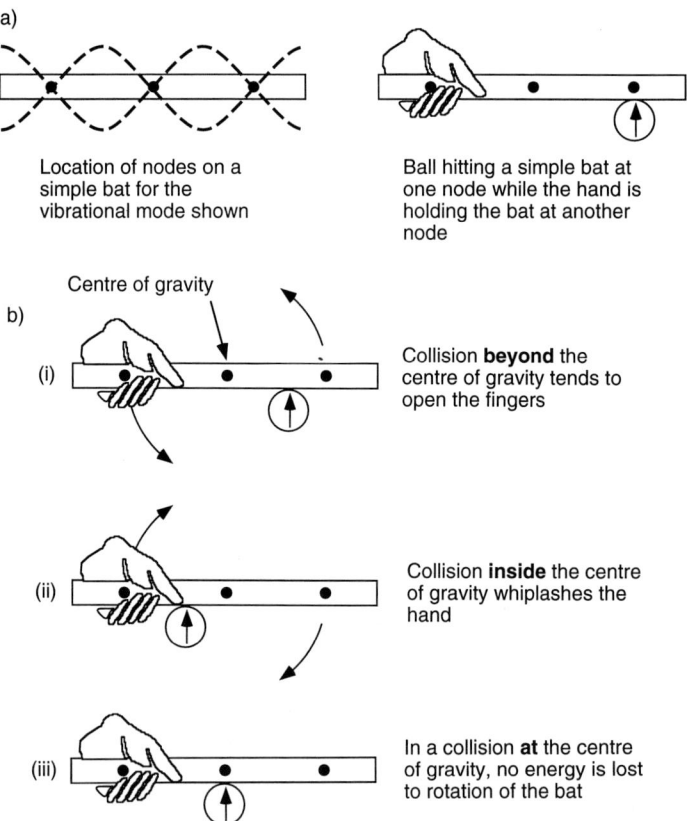

Figure 3.6 When hitting a ball with a simple bat, a player soon becomes aware of "sweetspots" that result in the hitting action feeling good. Collisions at these sweetspots also maximize the kinetic energy in the returned ball. a) One can make use of the nodes of the natural vibration of the bat to obtain a comfortable hit. b) The effect of hitting a ball at three different locations along a bat with respect to the nodes of the natural vibration.

Rackets could be any material, weight, size, or shape but attachments to the racket must not alter the flight of the ball and the strings must be evenly spaced.

Another innovation in the design of rackets developed by sport ergonomists is to provide a racket with a head which can be rotated to various angles with respect to the straight line of the racket handle. This ergonomic interfaced racket has been developed by a group of scientists and engineers from six American research institutes. They state that the conventional tennis racket only suits about 15% of the population, which is why aspiring tennis players have to spend many hours learning how to hit balls in the right direction. The spokesman for the group Andrew Brown states that the new racket, known as the *Index Handle Racket*, cures the problem [12]. Apparently the optimal alignment of the tennis racket face is determined by the arrangements of bones in the players wrists and arm. In the general population natural differences in these arrangements produces a 43° range of different optimal alignments. It is claimed that within ten minutes the player is able to find an alignment of the head with the racket handle that suits his anatomy. Another innovation in racket design by the group headed by Brown is that they have changed the shape of the handle to six-sided rather than the eight sides found in conventional rackets. Rick Robertson, director of the Biomechanics Laboratory at California State University, Sacremento, tells us that the six-sided shape fits more naturally in the surface of the closed hand. The International Tennis Federation has ruled that the six-sided handle and the adjustable racket face are allowed in International Tennis so long as the racket face is not adjusted during a game.

Section 3.4   Serving the Ball

In tennis, the ball is initially thrown into the air and hit with the racket to send it over the net into the opponent's court. This is known as *serving the ball*. During the act of serving, the ball is hit by a player standing at full height with arm plus racket raised to the maximum comfortable height. A ball hit in this manner by a good professional player can leave the racket at 250 kph (150 mph). Ball speeds of the order of 128–160 kph(80–100 mph) are usual in top international players' services. In modern television broadcasts of international matches, the speed of a delivery is often shown on an electronic display. The speed is measured in a fraction of a second using the Doppler effect. This technique for measuring speed is discussed in the next section.

In order to improve one's tennis service, it helps to understand the physics of a falling ball. In his hints on how to improve their game, Fontana advises tennis players as follows:

> *The trick (behind a good service) is to throw the ball up to the right spot for hitting in a consistent manner.*

The right spot according to Fontana is the point where the ball hovers for a very short time before beginning its descent after the force of gravity has robbed the ball of upward motion. The reason for hitting the ball at the hover point is that, at this point in the trajectory, the ball momentarily has no velocity in any direction. Therefore the tennis player is more likely to be able to give it the desired direction with the racket drive. Fontana advises players to practise throwing the ball until they know just how much energy to give it so that it hovers at the right point in the air when being hit. A ball should be tossed slightly to the right by a right-handed player and about a foot and a half in front of the body. Fontana states that many players serve the ball too much to their left shoulder resulting in a cramped and awkward service motion. Fontana also tells the player not to squeeze the ball as it is thrown into the air because this puts some spin on the ball and reduces control. Many beginners hold the ball in the palm of the hand to launch it into the air; this apparently does not lead to a good trajectory for the flight of the ball [13].

Section 3.5   How do they Measure the Velocity of Those Speeding Balls?

Modern technology makes it possible to measure the speed of a ball in flight using what is known as the *Doppler effect*. This effect is named after the Austrian scientist Christian Johan Doppler (1803–1853). Doppler was interested in the well-known fact that a sound emitted by an object moving towards a listener has a different pitch to the sound heard when the source is moving away from the listener. Anyone who has heard the siren of an ambulance approaching, passing by, and traveling away has heard this phenomenon. The change in frequency caused by motion is now known as the *Doppler shift*. Earlier, we discussed how sound waves are a series of compressed air packets moving through the air, created by the vibrating object generating the sound. If we imagine a vibrating tuning fork on a trolley moving toward us, then the time interval between the compressions received would be less than if the tuning fork had remained stationary. This is because the movement of the trolley enables each compression to "catch up" slightly with its predecessor. Since the time interval between compressions is shorter, the ear hears a note of a higher frequency. This effect of the movement of the source of the sound is illustrated in Figure 3.7. The opposite effect is created when the source of sound is receding from the listener. As the sound source recedes, the time interval between the compressions received from the vibrating tuning fork would be larger than if the fork had remained stationary. In other words, the arrival of successive compressions is delayed because of the movement away of the source of sound. The ear hears this as a sound of lower frequency. After working out his theory to explain the changes in pitch as the source of the sound moves past the listener, Doppler carried out an experiment. This was in 1842, long before there were precise instruments for measuring the frequencies of sound. To test his theory, Doppler arranged for a locomotive to pull a flatcar back and forth at different speeds. On the flatcar were trumpeters sounding this note or that. On the ground, musicians with a sense of absolute pitch recorded the note they heard as the train approached them and as it receded. Doppler's theories correctly predicted the shift in the notes heard by the musicians and their dependence on the speed of the locomotive. The Doppler effect also occurs when sound is reflected off a moving object.

3.5 How do they Measure the Velocity of Those Speeding Balls?

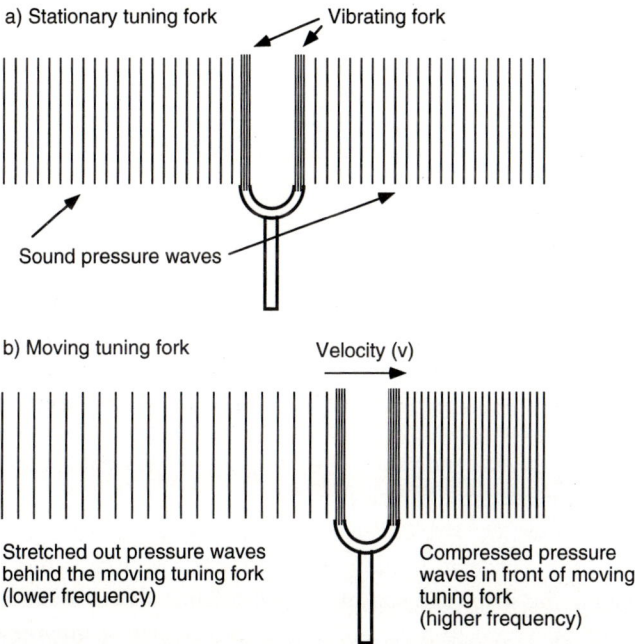

Figure 3.7 The Doppler effect for a moving source of sound occurs because sound waves ahead of the moving source are squeezed together so that the sound received at the ear has a higher pitch than the original sound. After the source has passed, the distance between the pressure waves is stretched out causing the sound arriving at the ear to be lower pitched.

The Doppler effect is also observed with light waves. Modern police equipment for measuring the speed of cars utilizes this effect. A laser light of known frequency is directed toward the moving object. The reflected signal is frequency shifted by an amount that depends on the speed of the object. (The physical nature of laser light will be discussed in greater detail in Chapter 6.) With modern electronics, the frequency shift can be displayed and interpreted immediately to give the speed of the car or other object. These devices are sometimes referred to as *radar guns*, but this is a misleading name stemming from early police use of an older technology based on the reflection of radio waves off an object. With this old equipment, the time for a reflected pulse to return to the detector for two radio pulses separated by a known time interval was measured. This information evaluated the position of the car at the beginning and end of the time interval between the pulses. This in turn enabled the velocity of the car during the time interval to be measured. The term radar means RAdio Detection And Ranging, (to get a *range* on an object means to determine its distance). The word radar was first coined in 1935 and was used as a secret code name for the equipment being developed to track and follow aircraft during World War II. Modern Doppler-based equipment can be used to measure the speed of a ball in a tiny fraction of a second.

Section 3.6 Slow down you move too Fast

The title of this section is taken from a song that was popular in the 1960s urging you to slow down and watch the flowers grow. Sports are not only activities for amateurs but are big business in the professional arena. In recent years the technology of tennis has developed to the point that some specialists feel that the professional game of tennis is too fast and is becoming boring for the spectators. This is because with the very high speeds of deliveries of service (as noted up to 250 kph) all that the spectators see in some professional games is the flash of a sizzling ball which the opposing player is unable to return. Thus Lendl in 1992 said that he wouldn't bother to watch a match between the Dutchman Richard Krajieck and the Croatian player Goran Ivanisevic since both players were known to serve the ball at over 190 kph. Lendl said "I wouldn't watch it I would just read the papers the next day to see what speeds their serves were and how many untouchable serves they hit."

To try and improve the attractiveness of professional games for paying audiences the International Tennis Federation has considered many options. Thus in 1992 they considered changing the shape and size of the court. They also considered altering the area where the ball must land to score a point. They decided against these changes. Other ideas were to ban rackets above a certain stiffness and to ban players from jumping when they served. However, there was strong opposition to both ideas. Another suggestion was to make the structure of balls softer or lighter. Making the ball softer would limit its speed because it would deform when hit hard and lose some of their energy. Making it lighter would cause it to decelerate faster because the drag force of the air would be more effective. However, this idea did not go down well because it would require a major shift in manufacturing technology and very strict tests to find out what kind of balls were being used. Another suggestion was to make the wool/nylon felt covering the ball more fluffy. When the balls are new the felt lies flat on the surface of the ball. After a few encounters with the racket and the court surface the felt begins to stand up from the surface increasing the ball's cross section. This dramatically increases air resistance and makes precise ball control possible. However the air resistance then falls off again as the felt gradually rubs off the ball during further play.

At present, there does not seem to be any consensus as to how to slow down tennis games for the spectator, but we can expect this to be an area of continuing debate in the coming years [7].

Section 3.7 Carbon Feathers and Plastic Skirts for Battered Birds

I had always wondered where the name badminton came from for the game in which one battered a simulated bird back and forth across a net. I shudder to think what the original players used in their game and hope it was not a de-winged dead bird. One theory of the origin of badminton was that British army officers brought it back to England from India. Cooke and Mullins tell us that a game similar to the modern game was first played in the 1860s by the Duke of Beaufort's family of Badminton House in Somerset, hence the name. Apparently *badminton* is now played in over 123 countries, and by an estimated 4 million people in England alone. Until I began to research for this chapter, I had no idea how sophisticated the technology of badminton has become in modern times. Apparently experts say that

> Badminton is a sport of cunning. Success depends on deceiving your opponent with changes in racket angle and wrist movement in the last fraction of a second before playing the shot.

The so called "*bird*" used in this game, which is also known as a *shuttle cock*, is made from a hemispherical cork or plastic nose with 16 feathers attached to it and weighs around 5 grams. The original type of shuttle cock is made with goose feathers. The spines of the feathers are glued into holes in the cork and the feathers fan out behind the nose to form a cone. There is also a synthetic version of the bird in which the feathers have been replaced with a plastic skirt.

The name of this section anticipates the fact that we shall discover that some modern shuttle cocks have synthetic feathers made from carbon fibers and will perform better than the ones with simple plastic skirts. In 1994 synthetic shuttles were banned from top level badminton competition.

In a game of badminton shuttles fly at speeds in excess of 240 kph (150 mph). Professional players feel that the plastic skirted shuttles reduce the variety of shots available to the player thus leading to a less tactical game with more emphasis on smash attacks from an aggressive player. Wind tunnel studies of the two types of shuttles have shown that they behave quite differently at high speed. When the feathers are attached to the cork nose they should form part of a cone of 58–68 mm base diameter. Manufactures produce several speeds by varying the weight of the shuttle between 4.74 and 5.50 grams. The feathers are arranged like the blades of a turbine so that air flowing through them produces an anticlockwise rotation as viewed by the player as the shuttle leaves the racket. This spin compensates for the fact that the feathers differ slightly in size, weight and shape. The difference between the plastic skirted shuttle cocks and the goose feathered standard shuttle cocks is due to the stiffness of the feathers. High speed photographs of shuttles striking walls show that the plastic skirts distort more than feather designs. Again high speed studies have shown that plastic shuttles tend to rotate at only half the speed of the feather versions. Wind tunnel tests show that both the feather and synthetic designs experience the same forces for speeds up to 83 kph but at higher speeds the difference in dynamic behavior between them becomes pronounced. Manufacturers of shuttle cocks are constantly trying to make better synthetic shuttlecocks because the feathered versions cost twice as much. They claim that the plastic skirted birds last three or four times as long as the goose feathered versions. Currently the manufactures are hoping that synthetic shuttle design will be revolutionized by carbon fiber synthetic feathers which are much stronger than plastic. New artificial feathers will use a central core of carbon fiber surrounded by plastic. Manufacturers claim that prototype shuttles made with these carbon fiber feathers look, feel and behave like the real thing and last longer [14].

Section 3.8 Are you Being Served? (The Robot is here)

We have already mentioned how robots are invading the game of tennis because in some circumstances they can perform better than human umpires. More robots for the tennis player are on the horizon. Machines have long been used to enable tennis players to practise shots, but up until 1995 these machines had limited capability. They could fire balls at a constant height and speed allowing a player to practise routine ground strokes, but a new robot called *SAM (Sport Action Machine)* will enable the players of tomorrow to practise against the type of shots hit by professional players [15]. It is claimed by the inventors of SAM that the robot can play every shot in the book at speeds of up to 160 kph. The robot was developed in Australia by Daniel Elbalm, working with Neale Fraser, former Captain of an international Australian tennis team. As a result of this collaboration, a robot has been designed and built that can produce top spin, back spin, lobs, drop shots, flat serves, kick serves, slices, and even mimic shots from both left- and right-handed players. The shots can be directed to land consistently within an area of 15 square centimeters, which is more accurate than any tennis coach could achieve. SAM can operate at eight levels of skill from novice to top professional. The key feature of SAM is a pair of rollers which can move independently at different speeds. These control the speed at which the ball is fired, its trajectory, and spin. Serves are fired at intervals of as little as six seconds. Future development of the machine will enable the player to specify the type of tennis game to match that of any international player. All that will be required is a computer card which will be inserted into the machine to make it play shots in the style of the selected player.

References

[1] J.T. Shipley, *Dictionary of Word Origins*, Littlefield, Adams and Company, Trotowa, New Jersey, 1970.
[2] *Chambers Etymological English Dictionary*, ed. by A.M. MacDonald, W. and R. Chambers Ltd., ll, Thistle Street, Edinburgh.
[3] H. Brody, *Tennis; Science for Tennis Players*, University of Pennsylvania Press, Philadelphia, 1987.
[4] B. Holmes, "New Balls Please and Make Them Bigger," *New Scientist,* February 25, 1995, p. 9.
[5] C. Arthurs, "Anyone for Slower Tennis", *New Scientist,* May 2, 1992, pp. 24–28.
[6] See Kate Charlesworth's cartoon entitled "Wimbledon Special," *New Scientist,* 1989.
[7] I. Anderson, "Metal Balls Settle Out of Court Disputes," *New Scientist,* September 5, 1992, p. 19.
[8] I. Anderson, "Line Calls to Swear By From Electronic Judge," *New Scientist,* February 6, 1993, p. 7.
[9] M. Weiner, "Playing Tennis in the Dark", *Technology Review,* January 1992, p. 12.
[10] S. Bernardo, "Physics of the Sweet Spot", *Science Digest,* May 1984, pp. 60, 64, 65, 95.
[11] J. Randall, "All Strings Attached. A Study of the Design of Tennis Rackets", *Science,* May 1981, pp. 91–93.
[12] V. Kerinan, "Adjustable Racket Heads", *New Scientist,* July 17, 1993, pp. 22.
[13] D. Fontana, "Development of Good Ball Toss Essential to Effective Tennis Serve", news item in *Toronto Globe and Mail,* Wednesday July 15, 1981.
[14] A. Cooke, J. Mullins, "The Flight of the Shuttle Cock", *New Scientist,* 12th March 1994, pp. 40–42.
[15] I. Anderson, "Play it Again SAM, and Again and Again", *New Scientist,* June 24, 1995, p. 5.

Chapter 4

Bolas, Boomerangs and Bouncing Bombs

Chapter 4

Bolas, Boomerangs and Bouncing Bombs

Section 4.1 Gauchos, Bolas, and Spinning Tops

Gauchos are the cowboys of South America. They have become familiar to western audiences through movies depicting them herding cattle in such areas as Argentina and Chile. These movies have also familiarized the viewers with an interesting missile with which the Gauchos capture the cattle they are chasing. The instrument is known as a *bola* or *bolas*.

When I visited Argentina in October 1993 one of the first items I purchased at the tourist store was my own personal bola. It consists of three balls attached to strings which are joined together as shown in Figure 4.1(a). To use the bola the gaucho first invests energy in the three ball system by whirling it above his head. When the system is spinning rapidly the balls are spread out as shown in the sketch. The gaucho then throws the system at the errant animal. The spinning ball system has both a linear velocity above the ground (called the *translation velocity*) and a velocity of rotation. When the bola reaches the animal, the balls wrap themselves around its leg using the energy of rotation as illustrated in Figure 4.1(b). (For the sake of clarity only one ball is shown wrapping itself around a rather straight leg!) As the balls wrap themselves around the leg, they go faster and faster as the radius of rotation of the ball decreases. The bola is an ingenious device which makes use of the fact that maximum energy can be stored in a given amount of rotating material if the material is concentrated at a distance from the axis of rotation. Also, the rotating energy of the ball/rope system stabilizes the flight of the device and makes it easier to predict its trajectory. In this chapter we are going to study the general use of rotational energy to stabilize missiles. To understand the basic technology of rotating missiles, we need to explore the ideas of angular motion and the storage of energy in a rotating system. The basic concepts involved in describing stored energy in a rotating object can be appreciated from the system shown in Figure 4.1(c). If you take a ball and attach it to a thin string which can be passed down the center of a cardboard tube, you can store energy in the ball by rotating it around the top of the tube. If you hold the thread at the point p, you become aware of the fact that the faster the ball rotates the more you have to pull on the string to keep it from moving the tube. The larger the distance between the axis of rotation and the ball (shown as r in Figure 4.1(c)), the harder you have to pull on the string for a given speed

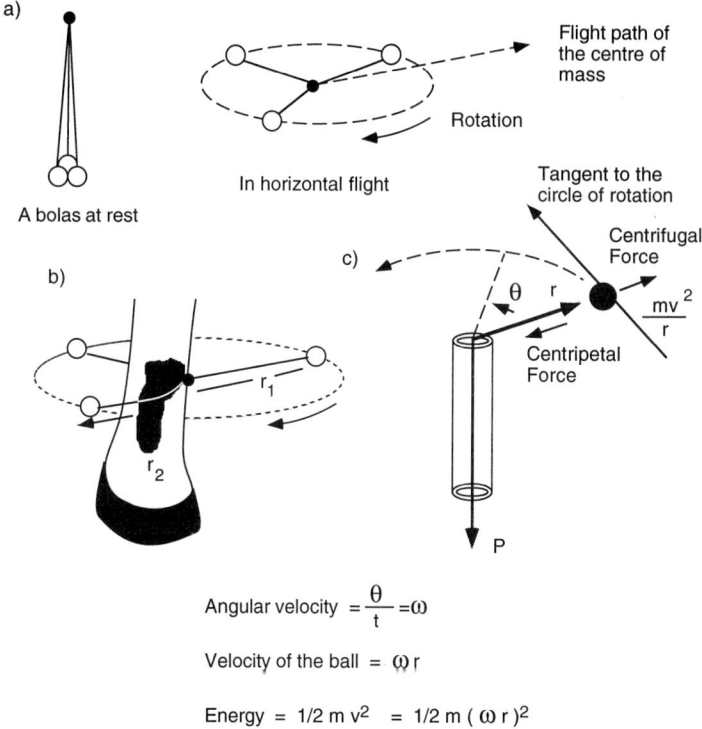

Figure 4.1 The bolas thrown by a Gaucho wraps itself around the legs of an errant animal and tumbles it to the ground. a) In a bolas, energy is stored in the rotating balls. b) The rotational energy of the bolas is used to wrap the ropes connecting the balls around the legs of the animal being chased. c) A simple toy made with the help of a cardboard tube can be used to illustrate the basic concepts used to describe the energies of rotating bodies.

of rotation. Engineers call the force that has to be exerted by the link between the axis (the cardboard tube) and the ball a *centripetal force*. The name comes from two Latin root words: "centrum" meaning "center " and "petere" meaning "to move towards an object." Thus the centripetal force is the force needed to keep the rotating object from flying away from the center of rotation. If the string should break during the experiment, the ball will appear to move off from the circle along a line just touching the circle of rotation at the point the ball had reached when the support string broke. The word *tangent* comes from the Latin word "tangere" meaning "to touch." A tangent line just touches the circle. The same root word also appears in the terms *tangible* for objects which can be touched and *intangible* for ideas which do not have a physical existence and therefore cannot be touched. The way the ball rotating on a string flies away from the circle of rotation when the string breaks has given us the everyday figure of speech "to go off on a tangent." This expression is used to describe someone changing the subject to another topic from that being discussed or failing to concentrate on the problem or discussion at hand. Engineers say that there is a force created by rotation which causes the ball to try to move away from the center of rotation. This is called a *centrifugal force*. The term centrifugal comes from another Latin root word "fugere" which means to "run away." This same word has given us the word *fugitive* for someone who runs away. The centrifugal force is the force that causes objects to try to run away from the center of rotation. In modern air pollution studies, *fugitive emissions* are traces of gases and dust which escape during the cleaning procedures used in an industrial process.

It can be shown that the centrifugal force experienced by a ball of mass m rotating at distance r from the center of rotation is equal to

mv^2/r

where v is the linear velocity of the ball around the circle. Engineers use the quantity *angular velocity* to describe the rotation of objects. The quantity angular velocity is equal to the angle moved through, θ, in time t. Thus for the bob of Figure 4.1(c),

angular velocity = θ/t

The kinetic energy of the bob is $\frac{1}{2}mv^2$. As shown by the formula at the bottom of Figure 4.1, this can be written in the form

$\frac{1}{2}m(\omega r)^2$

where ω is the symbol used to denote angular velocity.

Now, if we pull down on the string so that r decreases, because the energy in the ball has to remain the same, the angular velocity around the smaller circle must increase. This fact is well known and used by skaters and ballet dancers. They store energy of rotation in as they spin with their outstretched arms. Then, as they pivot on one toe, or the point of one skate, they can increase their speed of their rotation by pulling their arms inwards. This increase in speed of a rotating body as its diameter is decreased is a result of the *conservation of angular momentum*. Thus when spectators applaud a skater who spins very fast on a point by pulling in her arms, they are applauding the conservation of angular momentum! [1, 2]

In order to further develop our understanding of the mechanics of spinning bodies, it is necessary to have an understanding of the vocabulary given in Figure 4.2. The line about which the rotation occurs is called *the axis*. If you push with a force F at a distance R from the axis, scientists say that you are applying a *torque* which has the magnitude $R \times F$. The word torque comes from the root word meaning to twist. As shown by the word web of Figure 4.2, the word has given us the name for many tangible objects and intangible concepts. It is not always realized that British and American English differ in the use of certain words. An example of this difference is that in North American English, a *torch* is used almost exclusively to describe a device used for welding in which gases coming through a pipe are ignited at the end to form a welding torch. In Great Britain, a torch means an illumination system which is known in North America as a flashlight.

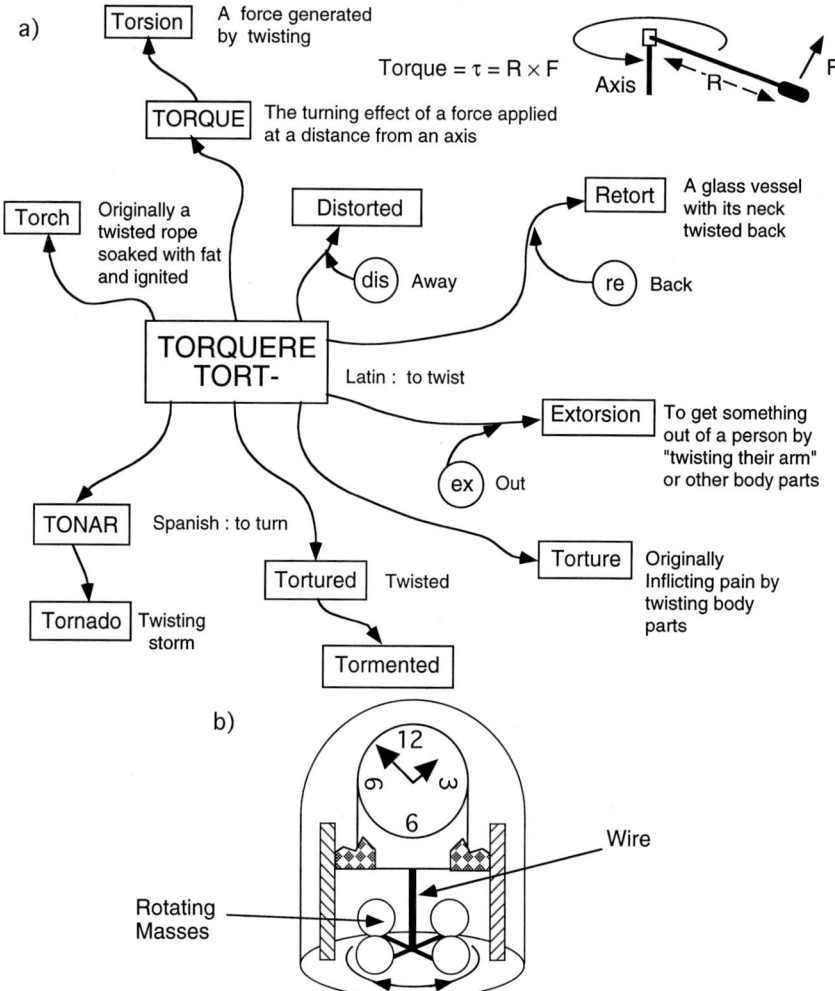

Figure 4.2 To understand the mechanics of a rotating body it is necessary to master the concepts of torque and torsion. a) Twisted words. b) Clock equipped with a torsional pendulum.

Sometimes a clock is equipped with a *torsional pendulum* instead of a swinging pendulum. In a torsional pendulum, energy is stored in a rotating system of objects. These objects are often made into fancy decorations such as angels or exotic animals. They are placed at the end of thin bars to store the maximum energy for a given mass. In the torsional pendulum clock, the mass at the end of the wire rotates until the torsion force created in the twisted wire is sufficient to bring the pendulum to rest. The energy stored in the wire now twists the pendulum back in the next phase of the oscillation. It is usual to cover such a pendulum with a glass cap since dust settling on it can increase its mass and thereby slow down its torsional oscillation so that the clock loses time.

With the more familiar swinging pendulum, the mass of the pendulum does not enter the equation for the period, so it does not require protection from settling dust.

In engineering science, energy is often stored in a wheel known as a *flywheel* (short for flying wheel) [3, 4]. One of the ways to save energy and cut down on pollution in a city would be to equip buses and cars with sophisticated flywheels. Instead of slowing a bus with standard brakes one would slow the bus down by letting it drive a flywheel up to a certain speed. The energy stored in this flywheel could be used to set the bus in motion again after it has stopped to pick up passengers [3, 4].

In Figure 4.3 the amount of energy that can be stored in rings of a given mass, but different dimensions, for a given speed of rotation is illustrated. The major design limitation on such a flywheel is that as mass is moved outwards to create the efficient energy storage system, the material and structure of the spars must be strong enough to resist the centrifugal force created as the wheel spins. Because a flywheel can occasionally disintegrate under the effect of centrifugal forces, flywheels should always be covered by protective screens strong enough to capture the fragments of the wheel should it disintegrate.

In modern archery the bow is often equipped with a thin rod that extends from the front of the bow with a relatively large cylinder attached to its end. Cutnel and Johnson [2] explain how this addition to the bow increases the moment of inertia of the bow and helps stabilize it when the archer is firing. They point out that the relatively massive cylinder placed at some distance from the bow increases the moment of inertia about the axis of the system located approximately near the armpit of the arm holding out the bow. This increase in moment of inertia enables the archer to steady the system when holding the bow fully drawn [2].

If an aircraft moves over the top of a loop, the pilot experiences a centrifugal force just as he would if he were sitting on the bob of the system of Figure 4.1(c) or on the outer edge of a flywheel. If the centrifugal force equals his weight, mg, a situation expressed by the equation

$$mv^2/r = mg \text{ so that } v = \sqrt{rg}$$

the pilot will momentarily feel that gravity does not exist, a situation known as *weightlessness*.

In very high speed turns, the centrifugal force experienced by the pilot can cause loss of consciousness. The force in such turns is expressed as a multiple of the weight of the pilot. Thus, it is said that in such and such a turn, the pilot was subjected to $5g$. This means that the centrifugal force is 5 times the weight of the pilot (or the equivalent of his being acted upon by five times the gravity of the Earth, g). The maximum centrifugal force that a human can tolerate is about $16g$. (See discussion of roller coasters in Chapter 8).

When astronauts spend time in orbit around the earth, they are subjected to very weak gravitational forces and they begin to suffer from problems such as bone weakening. It has therefore been suggested that spacecraft intended for long journeys in outer

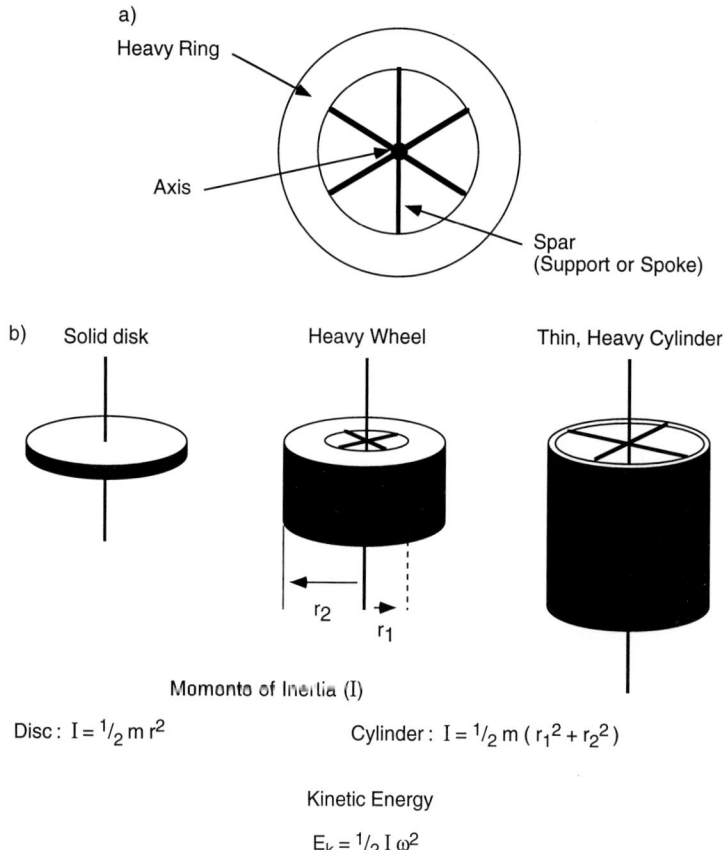

Figure 4.3 Flywheels are used to store energy. a) In a flywheel, most of the mass is moved outwards but the thin spars supporting the rim of material must be strong enough to resist the centrifugal force created when the wheels spin. b) The energy stored at a given speed of rotation for a disc of mass M depends on how the mass is distributed about the axis. This is quantified by a property known as the moment of inertia.

space would have to be in the form of a wheel spinning to create an *artificial gravity* at the rim of the spacecraft. Thus spacecraft of tomorrow may look more like flywheels than aircraft. They will have to spin at just the right speed, given by the formula above, so that people will feel they are walking around as on the earth. In such spacecraft "out toward the rim" would be "down". People would have the same sensations experienced by the motorcyclist at the fair ground who is able to ride around a vertical circle if his speed is sufficient to press him against the wall of the enclosure.

When I was writing this chapter, I was trying to remember the dynamics of the 'whip and top' toy that used to be very popular with children in Europe before the Second World War. During this time, on a trip to Germany, in the toy area of the craft market that is part of the walls of Nürnberg, I discovered a store selling wooden toys. There

4.1 Gauchos, Bolas, and Spinning Tops

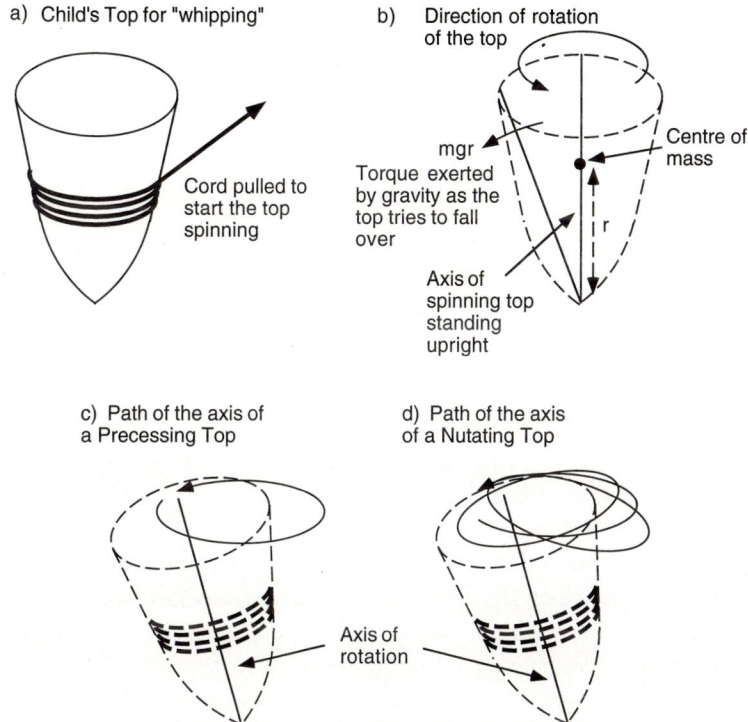

Figure 4.4 A spinning top will precess and nutate as it slows down.

at the front of the store was the "whipping top" shown in Figure 4.4. My German was not sufficient to be able to ask how it worked and so I gestured to the lady at the store to give a demonstration. The good lady, who was well into her seventies, with great gusto wrapped the cord of the whip around the grooves of the top and quickly set the top spinning on the sidewalk by pulling the whip away. She then proceeded to whip the top as it was spinning to increase speed and maintain its spinning action. The cord of the whip as it pulled away from the top exerted the torque necessary to set the top spinning. The angular momentum of the top kept it spinning upright. A top spinning on its point represents a position of unstable equilibrium. We call it unstable because the slightest disturbance will cause it to try to fall over. In the sketch of Figure 4.4(b) it is shown that as the top tries to fall over, the weight of the top exerts a torque to try to make the top spin about an axis into the plane of the drawing. However, the energy of spinning around the vertical axis cannot disappear immediately. The net effect of the torque due to the weight of the top trying to disturb the spinning action is to cause the tip of the top to trace out a circle around the vertical axis. This movement is called *precession* of the top. We will discover in later chapters that this type of precession is important in the dynamics of a boomerang. Other forces acting on the top cause the top to nod as it precesses. This nodding is described in scientific terms as *nutation*, a word derived from the Latin word for nodding. The nutation and precession of a top are phenomena that can be observed very easily if you whip the top and watch its behavior as it slows down or if you spin a simple top. The phenomena behind these complex patterns of a spinning top are very important in the study of long range weather variations on the Earth because the earth in space is a spinning top which both precesses and nutates. Some scientists believe that the nutation rate of the Earth is a major reason for the cycle of ice ages which the Earth experiences[5]. (A complete discussion of the forces acting on a top is beyond the scope of this chapter; the dynamics of a top are usually only discussed in studies undertaken in the latter part of a physics degree.)

The French scientist Jean *Foucault* (1819–1868) was interested in demonstrating to the world the reality of the Earth's rotation. In 1851 he carried out a series of experiments in which he studied the behavior of a massive pendulum with an extremely long suspension system. Foucault realized that if a large pendulum was set in motion it would maintain its plane of oscillation while the earth twisted under it. Thus if one had a very long pendulum at the North Pole the earth would make a complete twist beneath it in 24 hours. Foucault also realized that if he hung such a pendulum at the equator there would be no twist. Napoleon III, the emperor of France, arranged to have a large church in Paris used for the most famous test carried out by Foucault. Foucault suspended a large iron ball about 2 feet in diameter from a steel wire more than 200 feet long hung in the dome of the church. The sphere was equipped with a spike which, as it swung backwards and forwards, just cleared the floor. This spike could make a mark in sand sprinkled on the floor of the church. To carry out the experiment, the iron ball was drawn far to one side and tied to the wall by a cord. All the doors of the church were kept closed during the experiment. When all was quiet, the cord holding the pendulum was set on fire (if he had used knives or scissors to cut the cord this would have interfered with the experiment by generating vibrations). As the audience watched, the cord broke and the pendulum began to swing. As time passed the mark made by the pendulum spike visibly changed its orientation. It twisted in the direction and at just the rate that was expected for the location of Paris: one rotation in 31 hours 47 minutes. The spectators were watching the Earth rotate under the pendulum [6]. In the words of Asimov [7]:

> *Foucault's experiment demonstrated the rotation of the Earth where other experiments normally only deduced the rotation of the earth.*

For such an experiment a massive pendulum is needed to overcome the fact that as the Earth rotates under the pendulum there is a very small torsional force present in the support of the pendulum. The reason the pendulum is able to maintain its position relative to the Earth is that there are no forces operating on it sufficient to twist it around with the rotating earth. Foucault's original experiment is often demonstrated in the scientific museums of the world with the large ball pendulum being suspended in the entrance hall of the museum.

In 1852 Foucault again demonstrated the rotation of the Earth by setting a wheel with a heavy rim into rapid rotation. The wheel maintained its original direction, demonstrating the rotation of the Earth underneath. In this second experiment in 1852, Foucault had invented what we now call the *gyroscope*. As illustrated by the word web of Figure 4.5, gyroscope means "something which enables one to look at rotation."

The support system for the gyroscope is known as a *gimbal*. This word came into use in the 13th century as sailors developed devices to keep the compass and clocks used on the ship steady as the ship rolled. The type of gimbal used on a ship is a little more complicated than the one shown in Figure 4.5, but the idea is that a gimbal was composed of two rings, one inside the other with the rings able to pivot in such a way that the compass or clock stayed steady. The twin rings were called gimbals from the Latin

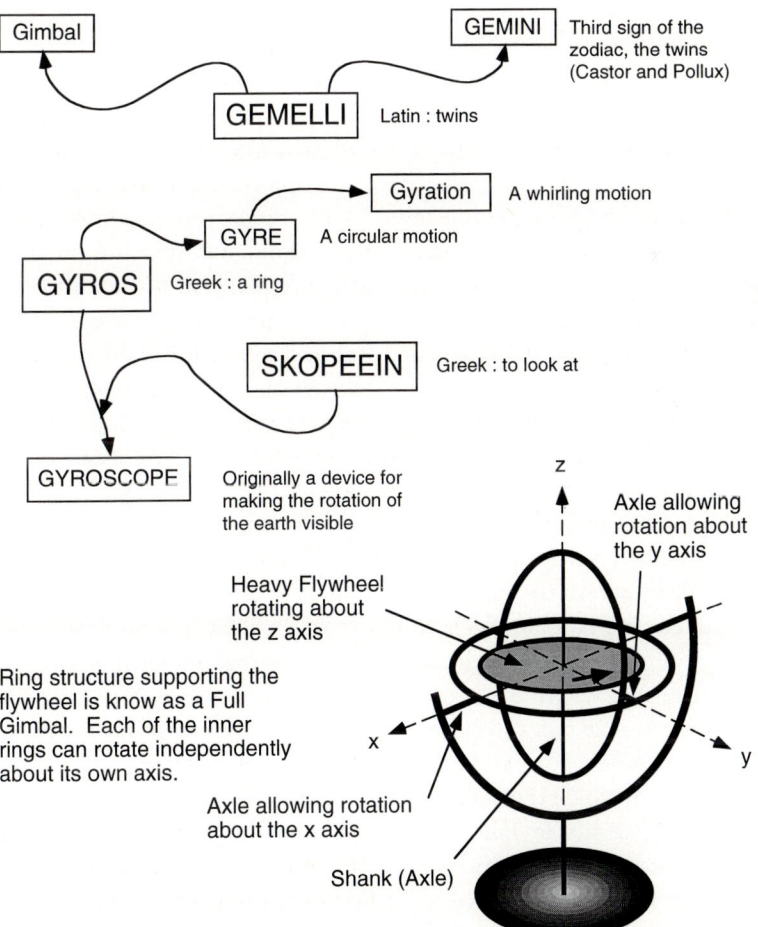

Figure 4.5 Gyroscope: A fascinating toy and a component of the guidance systems of many deadly weapons.

word for twins. This word has given us *Gemini* as the name for the third sign of the zodiac. In Greek mythology, we are told about the twins Castor and Pollux. These twin brothers loved each other dearly. Because of this love, we are told, they were placed together forever as a constellation in the sky by the god Jupiter [8].

The toy gyroscope known to many from their youth, has a *universal joint* at one end of the support axis of the flywheel. It is a *ball-in-socket joint*. It is called a universal joint because a body supported in this way can turn in many directions. To set the gyroscope rotating, a thread is wrapped around the shank of the flywheel and pulled. When the thread is pulled away from the shank, it exerts a driving torque which sets the wheel of the gyroscope in rapid rotation. Note that the flywheel of the gyroscope is made with the mass concentrated toward the outside of the wheel. When such a wheel is spinning fast, the support stand of the gyroscope underneath the spinning wheel can be moved about without the gyroscope changing its orientation in space. This fact is used to create a reference point for an object traveling through water or air, the support system carrying the gyroscope being adjusted to keep the moving object in a fixed direction of travel with respect to the direction of the gyroscope in its original position. When used in this way the spinning wheel is called a *gyroscopic compass*.

A missile widely used in sea battles is the *torpedo*. The missile is named after a fish which has the ability to give its victims an electric shock. The word "torpidus" in Latin meant "numb, without feeling, and sluggish." This original meaning lingers in the word *torpid* used to describe something which is inactive, dull and sluggish. Presumably the torpedo fish was so called because its electric shock rendered its victims torpid.

The first *torpedo missile* intended to destroy ships was invented in 1866 by the English engineer Robert *Whitehead*. It had a cylindrical steel case and was driven by an engine powered by compressed air. The early torpedo was an inefficient weapon, which had a tendency to wander off target. In 1896 Whitehead added a gyroscope to control the steering of the torpedo. The gyroscope made the torpedo a highly accurate and deadly weapon.

Originally, the compressed air used to drive a torpedo revealed the line of travel of the torpedo to sailors on vessels being attacked. They could then try to intercept the torpedo with shells and/or depth charges or take evasive action. Modern torpedoes are equipped with electric motors which do not leave the tell-tale trail of bubbles and are therefore more difficult to detect. Moreover, modern torpedoes are equipped with intelligent nose cones that detect and move towards the sound of a ship's propellers.

It is interesting to note that the gyroscope, which was developed for the purely scientific purpose of demonstrating the rotation of the Earth, became the essential part of a control mechanism which created a deadly weapon of war. This example demonstrates how futile it is to attempt to split research in science into pure and applied research, or into peaceful and weapons-related areas. It is impossible to predict today how the discoveries of the scientist will be used by society tomorrow.

Scientists have long known that *dolphins* and *sharks* swim much faster than they should be able to based on calculations of the expected drag of fluid around their bodies and the available muscle power in their anatomies [9]. Recent research has indicated that the reason for their speed is similar to the reason why the dimpled golf ball goes farther than the smooth ball. The shark and the dolphin have rough skin. The shark's skin has fine ridges of scales that act to reduce the drag by suppressing turbulence. In the 1970s NASA scientists found that they could reduce drag on models of aircraft and boats by cutting small grooves into the surface of the bodies. They did not follow up this research because it was felt that the cutting of the grooves on the surface of the aircraft was too difficult. However, in the early 1980s scientists at the 3M Corporation developed an adhesive tape whose surface had many narrow grooves running down its length. This tape is called *riblet tape*. When riblets are placed parallel to the flow of the air or water, they reduce turbulence in the layer of fluid close to the surface. The 1987 America's Cup yacht race was won by the US boat "Stars and Stripes." The hull of this yacht was coated with riblet tape, a factor which contributed to the higher speed of the boat [10].

The Boeing Corporation is exploring the use of such tape on commercial aircraft, and in March 1990 it was estimated that the use of riblet film on certain parts of commercial aircraft could result in the saving of US$80 000 per year per aircraft. Tests by Boeing indicate that the riblet tape should reduce drag by 4% resulting in savings of 50 000 gallons of fuel per year on each 757 [11].

The navy is not as enthusiastic about the potential use of riblet tape since the bottoms of ships become quickly covered with barnacles and algae which clog the grooves and prevent the basic operation of the tape. However, underwater weapons such as torpedoes are not in the water until launched. Covered with riblet tape, the torpedo can travel faster and farther than if it had a smooth surface. One of the terms used by sailors to describe torpedoes is tin fish. It looks as though the tin fish of tomorrow will have their own synthetic shark skins to enable them to behave like golf balls!

Russian scientists adopted a different technique for increasing the range and speed of a torpedo. As pointed out by Kwing-So Choi, a mechanical engineer specializing in drag reduction at the University of Nottingham, the principle of the procedure used by the Russian scientists was well known. But Western scientists had not found a practical way of using it. Choi states that a layer of gas a fraction of a millimeter thick can reduce the friction at the surface of a torpedo by up to 40%. The drag reduction arises because the gas strips the torpedo of the thin layer of water clinging to its surface, a source of frictional drag. Russians applied the principle by pumping gas through a second skin that covers the torpedo. John Downing, a marine analyst at the International Institute of Strategic Studies in London, speculates that the gas is produced by a device similar to a rocket engine positioned near the head of the torpedo. The Russian torpedo is codenamed Shkval which means "squall" in Russian.

When a missile is moving through air the problem is not a clinging layer of gas but the turbulence of the layer of air that surrounds it, as we saw in the case of the golf ball. Aircraft designers have tried to reduce the drag over a wing surface of an aircraft by sucking the turbulent layer off of the wing surface to reduce the drag. In tests with aircraft flying at Mach 1.8, it has been shown that a titanium metal glove with many tiny holes in it can suck off the turbulent air and increase the range of the aircraft for a given amount of fuel. The suction surface was created from a sheet of titanium covering about 40% of the left wing of a F16FL fighter. This surface was smoothed into the rest of the wing with foam and fiberglass. It was perforated with millions of tiny holes cut by laser through which a pump sucked the turbulent air. We are told that this is a continuing area of research to improve the fuel efficiency of both military and commercial aircraft [12, 13].

The yo-yo is an ancient toy. A Greek bowl that dates from 450 BC depicts a boy playing with a yo-yo. A simple *yo-yo* consists of two discs joined at their center by an axle; it is basically two wheels joined by a very short axle. The string used to suspend and operate the yo-yo is wrapped around the axle. Technically, a yo-yo is a mechanical device that stores energy in a rotating mass. To a physicist, a yo-yo is a flywheel on a string. The behavior of a yo-yo depends upon the way in which the flywheel is attached to the suspending string and it is useful to discuss the behavior of the yo-yo in terms of systems shown in Figure 4.6. (The discussion of the physics of a yo-yo given in this chapter is based on a fascinating account written by Wolfgang Berger [14].) The construction of the ideal yo-yo is shown in Figure 4.6(a)(i). It spins as it falls and then, because the string is fixed to the axis, immediately rolls back up again as shown in the sketches Figure 4.6(a)(ii). The length of the unwound suspending string is equal to zero when the yo-yo is at its highest point. In this position, the yo-yo has gravitational potential energy due to its height. As the yo-yo is allowed to fall, it unrolls along the suspension string and some of this potential energy is converted into linear kinetic energy of the fall. Some of the energy is also converted into angular energy of spinning as the yo-yo unrolls along the string. The ideal yo-yo has an infinitely thin string so that the support axle of the yo-yo does not change in diameter during the winding and unwinding. With a thick or real string, the diameter of the axle is much larger at the start of the fall, as shown in Figure 4.6(b), and as the string unrolls the axle decreases. The way in which the velocity of the fall (not the speed of the rotation) changes with the falling yo-yo for both infinitely thin and real string is illustrated in Figure 4.6(d).

The people of the Philippines claim that the yo-yo was developed as a hunting weapon. They describe how a hunter would sit in a tree with a heavy flint tied to a long leather strap waiting for the prey. The stone is said to have unrolled down the strap to hit the prey and then if it missed the prey it would return back to the hunter. Berger thinks that this explanation of the history of the yo-yo is most improbable since letting the stone roll down the string, as with a yo-yo, would result in the missile hitting the prey with a much lower speed than if dropped freely onto the prey. Some of the energy of the fall would be stored in the spinning rock and would not contribute to the efficiency of the weapon. Berger points out that it would be much more efficient to simply drop the rock on a string onto the prey and haul the rock back up if the object of aggression were missed. It is said that yo-yo in the Philippine language means "come-come," the idea being that it is a missile that comes back.

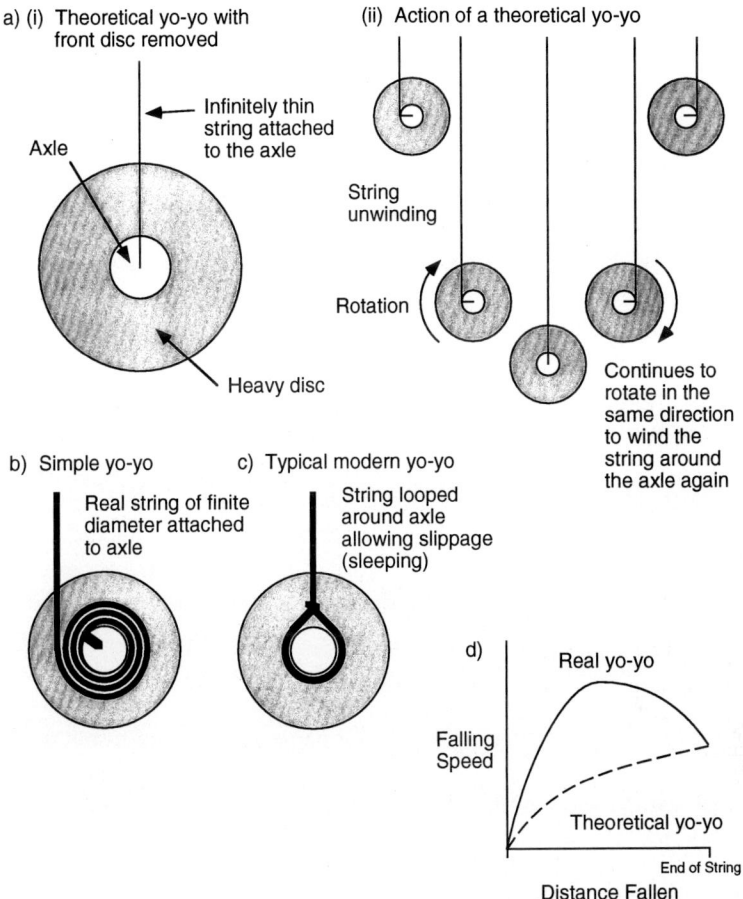

Figure 4.6 The difference between real and theoretical yo-yos is due to the effect of the real diameter of the string.

The modern popularity of the yo-yo in North America can be traced back to the commercial initiative of Donald Duncan who also invented the parking meter. In 1928 Duncan watched a Filipino named Pedro Flores playing with a yo-yo which was different from those available in the toy shops. Flores' yo-yo had the string looped around the axle (as shown in Figure 4.6(c)). This loop allowed the yo-yo to spin at the bottom of its fall instead of immediately climbing back up the string. In the language of yo-yo specialists, this spinning at the end of the string is known as *sleeping*. A yo-yo is often allowed to sleep before performing tricks. If dropped passively from the hand, the simple yo-yo will never roll back up to the hand. This is because part of the energy of the initial position is lost during the unrolling and rolling up on the suspension string as a result of friction in the system. The yo-yo player instinctively learns to adjust for this loss of energy by either throwing the yo-yo with extra energy as it begins its downward roll or by jerking it upwards on its way back to give it extra energy to roll back up to the hand. With a typical sleeping yo-yo, however, using a passive release (i.e. just dropping it) will result in a sleeping period of approximately two seconds before most of the energy is lost by friction. For several yo-yo tricks it is essential to keep the yo-yo sleeping for more than four seconds and the specialist player must throw the yo-yo vigorously towards the ground to give it additional energy. Berger reports that experienced players can increase the rotational energy of the yo-yo more than twenty-fold and push the rotation to 140 cycles per second. This corresponds to the energy it would have from a vertical drop of 25 meters. For such a throw, the rim of the yo-yo can be moving at nearly 100 kph when it initially starts to spin at the end of its fall. Berger describes some very interesting tricks that can be carried out with such high energy yo-yos and also describes some special yo-yos designed to facilitate difficult tricks [14, 15].

Modern yo-yos sometimes have a Teflon-coated axle and Teflon coating on the inner surfaces of the two flywheels constituting the yo-yo. Teflon reduces the friction between the moving parts of the yo-yo and the string. *Teflon* is a special plastic made out of carbon and the chemical fluorine. Fluorine is a very chemically active and poisonous gas, but when combined with carbon in the correct proportions it becomes a plastic in which the molecules are so tightly bound to one another that the material has only a very low ability to bond with other substances which it contacts. That is why it is very difficult for substances to stick to Teflon surfaces. For this reason Teflon is used to coat frying pans. Teflon-coated bullets are often used by terrorists. These bullets are banned in North America because they are especially dangerous as they lose less speed when traveling through the air, or the body, and can cause particularly nasty wounds.

The use of angular momentum to stabilize missiles is much older than the development of the gyroscope and the torpedo. Today many missiles used in sports are given spin to stabilize their trajectories. Thus in American football, in which the ball is essentially an egg-shaped object, the football will experience the least air resistance if it travels with a pointed end forward. If this type of football is thrown without any spin it will tumble chaotically in the air and travel only a relatively short distance. The tumbling is caused by the fact that the force of the air at the front is greater than that at the back. If this type of football is thrown with a moderate amount of spin, its path will be less erratic even though it may show some wobble as it moves forward. We now know that this wobble is precession about the spin axis and that it increases as the ball slows down. The more spin is given to the football, the more stable its flight characteristics. American football players sometimes refer to a well-thrown short pass in which the ball is spinning about its axis as a *bullet pass*. The idea is that the football is spinning like a bullet (see Chapter 6).

Section 4.2 Boxcar Integrators and Lasers

One of the challenges faced by a scientist who develops new ideas is the coining of a suitable vocabulary which can help to convey the relevant ideas and discoveries to colleagues and to the public at large. All too often the scientist does not pay much attention to, or show enough concern for, the communication of ideas to ordinary people and, as a result, manages to develop very obscure vocabulary. For example, let us look at some strange vocabulary in the electronics industry. In modern electronics jargon two electronic devices go by the names *bucket brigade detector* and *boxcar integrator.* I can remember being puzzled by both terms when I first read them in a scientific paper. I could not find either of them in a dictionary. After considerable effort, I established that both terms are metaphors. The bucket brigade detector turned out to be an electronic device for capturing and passing on a signal through a chain of electronic components to its final destination in the same way that buckets of water used to be passed from one person to another in a chain to take water from a pond to fight a fire. The boxcar integrator repeatedly records a set of signals which are basically the same each time but distorted by noise. The idea is that by integrating (summing) the signals from successive recordings, the noise tends to be self-canceling whereas the signal of interest grows stronger. The inventors of this device said it was like looking at the whole series of identical boxcars on a railroad under very feeble illumination and taking a snap shot on the same film frame every time the box car was in the same position relative to the camera. The cumulative photograph of many boxcars would look like a good photograph of a single boxcar under strong illumination. Unfortunately when this explanation was given to me, I did not know what a boxcar was. In British railroad terminology, an American *boxcar* is a *railway wagon,* or wagon for short. Thus the British would have called such an integrator, if they had used the same kind of metaphor, a wagon integrator. Such metaphors may be vivid to the people developing the equipment but can be very obscure for others.

My experience with dictionaries in the attempt to discover the meaning of boxcar integrator, and similar terms, illustrates a major difficulty experienced by students studying a new science. New scientific terms can take up to twenty years to make it into the standard dictionaries. In some situations the problem of gaining an understanding of new scientific vocabulary is aggravated by the existence of an older meaning for a word given new meaning by scientists who did not check existing dictionaries when creating what they believe to be new terms. This type of problem is illustrated by the difficulties one of my students encountered with the word holograph. Laurentian University has a school of translators. From time to time students ask me for help in their translation projects. On one occasion, when *holography,* a system for making three-dimensional images of an object, was developing, a student came for help very puzzled by the fact that a dictionary defined a *holographic manuscript* as a "document written in one handwriting." I was able to explain this older meaning of the word holographic by a reference to a current news story that reported the recent discovery of a long-lost poem written by Keats. The poem was an original manuscript written out by Keats himself and was described in the news story as a holographic manuscript. The word comes from Greek root words "holos" meaning whole, and "graphein" to write. The root word "holos" is also to be found in *holocaust* where it combines with the Greek root word "kaustos" (burnt) to create a word meaning wholesale destruction. (*Caustic soda* is sodium hydroxide, a chemical which will burn holes in many things.) *Photography* is a word meaning "writing with light." Ordinary photographs are two-dimensional. When scientists developed a system of photography capable of generating three-dimensional images they described the system as holography because it generated the whole three-dimensional reality. The piece of photograph film which stored the information to create a hologram was called a holograph. The scientist who created the process never checked a dictionary to see if the word holograph was already in use. In our discussion of rockets used to explore space, we will take a brief look at how holography is being used to image the unburnt fuel in rocket exhaust.

In modern English there has been a major change in the way in which scientists make their words as compared to fifty years ago. Today's scientists rarely use Greek and Latin root words; instead, they use acronyms and root words from living languages. Thus fifty years ago scientists talked about the apparatus that they assembled to carry out their experiments. The word *apparatus* is defined in a dictionary as

> things prepared or provided, material; sets of tools, equipment, from "ad" — to, and "paratus" — prepared.

In modern scientific reports of experimental investigation, apparatus is often described as "the *set-up* used in the experiments." The word set-up is obviously created from the living language to replace the word based on classical language word roots. Today a popular technique for word making in science is to create an acronym from the tips, that is, initial letters of words. The word *acronym* itself is formed from the Greek root word "akron" meaning "the tip." The *Acropolis* in Athens, was a town built on the tip, or the highest point, of the city. "polis" is Greek for "city." *Politicians* were originally people who governed a city, and an *acrobat* walks on the tips of his toes. Nym comes from the Greek word "onoma", a name.

One of the most famous acronyms in modern English is the word *laser*, a word fabricated from the initial letters of the phrase:

> *Light Amplification by Stimulated Emission of Radiation.*

Most people are unaware of the origin of the word laser and also adapt it for use as a verb. Thus the statement "a laser lases to give out light" illustrates both usages. Lancelot *Hogben* who has written a book on the vocabulary of science despised acronyms. In his book [16] he said,

> *The acronym operates with initial letters of vernacular phrases such as in operation PLUTO (PipeLine Under The Ocean) of World War II. The separate elements of the*

acronym are not even suggested to a person whose native language is English and the interpretation is meaningless to a person unfamiliar with English. Of late, monstrosities of this sort have penetrated the laboratory with the creation of such words as laser.

Such anger in scholars, and my grumbling against bucket brigade detectors, is not very fruitful. In condemning such words as laser, Hogben is a modern day Canute. (The Danish king who is said to have told the sea tide to retreat but who got his feet wet in the process.)

The move to create words using such techniques as metaphors, living language roots, and acronyms instead of Greek and Latin root words began at the end of World War II. At that time the requirement in England that scientists should study Latin and Greek in their pre-university training was abandoned. One of the reasons, (there are several others) that I was not educated at Oxford or Cambridge is that I did not take Latin or Greek in my high school days. I preferred to study French and German. Scientists who had a basic knowledge of Latin and Greek, would naturally draw upon these languages to create their new vocabulary. However, even this classical method of creating scientific words is open to criticisms, and artistic grumbles about scientific vocabulary are not new. We are told [16] that *Lucretius*, a Roman writer, said,

> *I know how hard it is in Latin verse to tell the dark discoveries of the Greeks. This is chiefly because our poor speech finds strange terms to fit the strangeness of the thing.*

The Romans just did not have enough Latin root words to create new words to describe the Greek ideas and technologies that they assimilated when they conquered the Greeks. Hopefully as scientists start to cope with the information that their research has discovered, they will pay more attention to the way in which they coin phrases. When English scientists developed words for the telephone and television they used Greek root words: "tele" meaning "distant," "phone" meaning "sound." *Phonetics* is the study of the sounds of words. On the other hand, German scientists described telephones as *Fernsprecher*, literally German for "distant speaker," and *Fernsehen* "distant seeing" for television. It should be noted that German scientists have often turned to living languages to create new words.

When the word laser started to make its way into the dictionaries in the 1960s, I came across one definition in a British dictionary that said that lasers stood for the phrase "Light Amplification By **Simulated** Emission of Radiation" which just goes to show how meaningless things become if one has to go without one's "t" –especially if one is an Englishman [17]!

The strange mixture of vocabulary that can develop in a change-over period from an old to a new custom for creating scientific words occurred in the period when the first rockets for exploring space were being developed. The first rockets were called *Titan rockets*. In the Greek stories about the gods, the Titans were the children of the goddess of Earth, *Gaia*, and the god of Heaven, *Uranus*, names which have given us such words as *geography* and *uranium*. Thus the Titan rockets left the earth to search the heavens.

On the other hand, a cord linking the rockets to the control station before the rocket was fired was called an *umbilical cord,* a metaphor taken from medical science. A particular type of rocket with three legs was called a *milk stool rocket* which is a living language metaphor [8, 17].

When an object is named after a person, the term is described as an *eponym*. This comes from the Greek words "onoma" "a name" and "epi" "upon." An eponym is a personal name placed upon another object. Thus the term Titan rocket is an eponym. Eponyms can have their problems. The unit of force in the new scientific system of units is the newton. I had one student whose native language was not English ask me how the "new ton" differed from the old ton. It took me a long time to explain that the name Newton came from an English word meaning "a new town" and that some ancestors of the famous physicist Newton (after whom the new unit of force was named) must have come from a new town. I have always felt that it would be useful to have another name for terms which are based on myths. A *mythonym* would be an eponym when the person involved in the imagery is not a real person. Thus a newton would be an eponym and a Titan rocket would be a mythonym.

Section 4.3   Some Circumspect Vocabulary

When one graduates from university one is awarded a degree. This word comes from the Latin "de" meaning "down" and "gradus" meaning "step." Although we usually regard a degree as being a step up in the world, it literally means to step down from student status to become a knowledgeable person. We have already met degrees of temperature and now we will discuss what we mean by *degrees of angle*. The word *angle* in Latin comes from a word meaning bent. If you look at the two sides of an angle you can see that they could be formed from a bent stick. When early engineers started to mark out circles on the ground in their efforts to build granaries for storing grain and also to build such things as *Stonehenge,* the famous ancient circle of stones in Great Britain, they probably used a string tied to a pole and paced out the circle as suggested in the sketch of Figure 4.7. Probably they had a standard string and a standard number of steps to make a good reference circle. In any case, they came to measure the angle swept out by a point P moving completely around the circle in terms of 360 steps which they called degrees; a standard symbol used for an angle is the Greek letter theta, θ, and the degree is represented by the symbol, "°". The basic measure of 360° for a complete circle can be divided into four quadrants, each of 90°.

Number systems can be set up using any number as a basic unit. The western world is now heavily involved in metric measurements in which the basic unit of counting is 10. When numbers are expressed in terms of the number 10, the system is described as *decimal* from the Latin word "decima" for "a tenth." The ancient Babylonians, how-

4.3 Some Circumspect Vocabulary

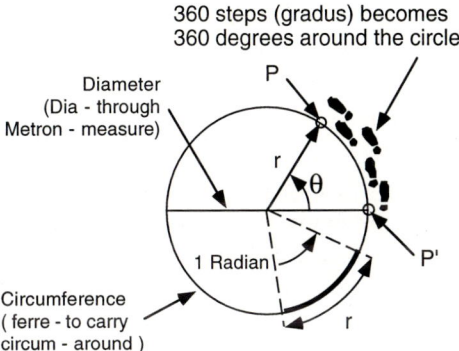

Figure 4.7 Quantifying the concepts of angular rotation.

ever, preferred to work in units of 60. Sixty is quite a good system to use since it has more factors, numbers which it can be divided by, than ten has. Thus 60 can be divided by 30, 20, 15, 12, 10, 6, 5, 4, 3, 2, and 1. Ten can only be divided without remainder by 5, 2 and 1. The counting based on 60 also naturally leads to units of twelve since 5 times 12 is 60. Until very recently the grocer used to work in terms of a *dozen* (12) and a *gross* which was twelve dozen (12 × 12=144). The ancient Babylonians, who developed astronomy before the Romans existed, were also responsible for our concept of angular measure and of time measurement. Therefore we still measure hours in terms of 60 minutes, each of which can be broken into 60 seconds. The mathematician uses exactly the same measures to divide a degree into 60 minutes, each of which is equal to 60 seconds.

Two different words are used to describe the boundary of the circle. One is the word *circumference* which comes from the Latin "ferre" "to carry"; presumably the circumference was the line traced out by carrying the string around a complete circle. The word "circum" for "around" has also given us the word circumspect which is in the title of this section. To be circumspect means "to look all around" before doing something.

Another word for the boundary of a circle and other geometric figures is *periphery*. This is exactly the same idea as circumference, but from Greek root words. In Greek "to carry" is "pherein" and "peri" means "around." In everyday speech when we say something is peripheral to our concerns we mean that it is outside of the main circle of our concerns.

The word diameter comes from the Greek root "dia" meaning "through"; thus, originally, the diameter of a circle was the through measure, that is, the length of the line through the center of the circle. The Greek root "metron" combined with "peri" has given us *perimeter* which originally meant "the measure around the circle" but now means the boundary itself. When we use the word *radius*, for the line drawn from the center to the circumference of the circle, we are using the Latin word for the spoke of a wheel. The word radius has also given us radiation, the idea being that the first radiation observed by the scientists were the rays of the sun coming down to the earth. In the evening, the rays of the sun often look like the spokes of a wheel The word *radiation* has obviously broadened its meaning to cover all sorts of transmission of energy by means of waves. When something travels around a complete pathway like a circle it is said to have completed the *circuit*. This word also comes from circle and the Latin word "ire", to go. Thus to do a circuit means to travel around a circle. An *electrical circuit* is the pathway followed by the electrons moving around the system.

The symbol π (pi) is used for the ratio of the circumference to the diameter of a circle because it is the Greek form of the letter P which starts the word perimeter. If we time the movement of a point on a circle (such as point P in the circle of Figure 4.7) as it goes all around the circle back to its starting position, this time interval is said to be the *period* of the rotation. This word is coined by linking "peri" with "hodos", the Greek word for "a road" that we met when discussing the origin of the word "method" in Chapter 2.

In recent years many scientists have switched from using degrees to measure angle and rotation to using what is known as *radian measure*. The radian measure of the magnitude of an angle is the number of radii (radii is the plural of radius) of the circle spanned on the perimeter between the two radii defining the angle. Thus, as shown in Figure 4.7, *one radian* is generated by a length of arc of the circumference equal to the radius of the circle. Since the circumference of the circle is given by the formula 2π, it can be seen that there are 2π radians of arc comprising the circumference of the complete circle. Two popular measures of rotational speed are *"revolutions per minute"* abbreviated to *rpm.* and *"cycles per second"* abbreviated to *cps*. In scientific units 2π radians per minute is equal to 1 rpm.

To describe motion in a circle we often make use of what are known as *trigonometric relationships*. You will find it helpful to refer to Figure 4.8 throughout the following discussion. If we look at the point P rotating on a disk fixed to the wall by an axle at the point O and if we have a pendulum or bob hanging from the point P, then as P rotates around the circle, the line carrying the bob moves back and forth along the diameter with its distance from O being the length b. In the same way, if we were to be looking at the projected shadow of a pin at P onto the line AB, a point would appear to move up and down the distance a from the center O. Then if we look at how a varies with the angle of rotation, θ, of P around the circle, we would find that it traces out the solid line shown in Figure 4.8(b). Such a pattern of movement is called a *sine wave*. The word "sine" comes from a Latin word meaning "a bend." (The *sinuses* are curved ducts which when congested or inflamed can give a bending headache!) The plotted value of b varies in the same pattern as the sine wave but is displaced and is called the *cosine wave* as shown by the dotted line of Figure 4.8(b). It is said to be complementary to the sine wave. At any time in the movement of the point P you can specify the angle, θ, of P with the reference line, CD, using any one of the three ratios shown in the diagram. Engineers and scientists use these values so frequently that almost all numerical tables and pocket calculators include them. The three ways of describing the magnitude of the

4.3 Some Circumspect Vocabulary

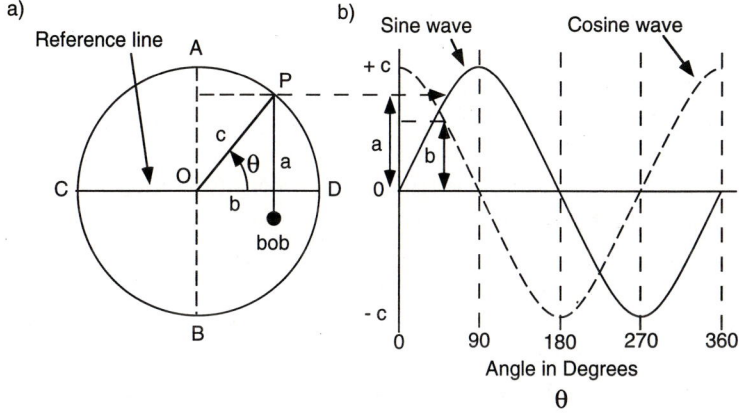

a = distance of the pin (P) from the horizontal line (CD)
follows the solid line on the graph ———

b = distance of the bob line from the vertical line (AB)
follows the dashed line on the graph – – – –

$$\sin \theta = a / c$$

$$\cos \theta = b / c$$

$$\tan \theta = a / b$$

Figure 4.8 The sine and cosine of an angle follow the co-ordinates of a point moving around the circumference of a circle as shown above.

angle with respect to circular motion are called *trigonometric functions* because the study of triangles is called *trigonometry*. In the study of trigonometry you find yourself moving back and forth between Greek and Latin words. Thus *triangle* is a Latin word meaning "three angles", but the Greek word for an angle is "gonia" and so the study of triangles is Greek — trigonometry (three-angled shapes). We will now digress a little to explore some of the mysteries of scientific vocabulary and the way in which David the shepherd boy used high speed missiles and psychology to defeat Goliath the Philistine Giant [18].

Section 4.4 Killing Giants and Catching Fish

In the western world, one of the most famous stories of heroism is about a shepherd boy called David who kills a heavily armed giant of a man named Goliath [18]. David was from the tribe of Benjamin, one of the twelve tribes of the Israelites; Goliath was a Philistine. In modern English a *philistine* is defined as a person

> *Indifferent to culture whose interests are material and whose ideas are ordinary and conventional.*

This view of Philistine culture has arisen because they lost their independence when the Israelites conquered them. The people who write the history books often provide a very poor view of the people against whom they have struggled. For instance, King John, who was the English king at the time of Robin Hood, made major efforts to break the power of the monasteries in Great Britain. Since the monks wrote the history books, King John emerges as one of the major villains of history instead of a reasonable man (as judged by the standards of his time) who tried to weaken the power of a group with vested interests. It is wise to be very cautious when interpreting the social aspects of history and it is equally important to assess the perspective of the historian. The Philistines, in fact, were ethnically and culturally related to the Greeks who spread a high level of culture around the Mediterranean sea. Many scholars would judge their culture to have been far superior to that of the Israelites whom the Philistines probably regarded as primitive Judean hillbillies! For reasons which are not fully understood, there was a large migration of people who were culturally Greek out of Greece proper and down the eastern Mediterranean to Egypt around the 12th Century BC. These migrants were called "the sea people" by the Egyptians. Major battles were fought against the sea people by Rameses III (1197–1165 BC). The Egyptian victories against the sea people are recorded in their histories [19]. About the year 1200 BC some of the sea people settled in what is now known as the Gaza Strip and became known to history as Philistines. This area, shown in the map of Figure 4.9, is a fertile strip of land bordering the desert and Judean hills. The part of the Judean hills bordering on the Philistine territory was inhabited by the Israelite tribe of Judah which was always closely allied with the small tribe of Benjamin. It is thought that the election of Saul, crowned in the year 1010, as King of Israel was actually a defensive move by the league of the twelve tribes of Israel to protect themselves from the Philistines. In *The First Book of Samuel*, Chapter 7, a major confrontation between the Israelites and the Philistines is described. It is interesting to note that it is believed that one of the reasons the Greeks and the sea people were generally successful in their campaigns against others in the area is that they had access to iron weapons which were superior to the bronze equipment and armor of the people they conquered. The gods of the Indo-European peoples, such as the Vikings of the North and the Dorians of Greek history, always included a powerful blacksmith god. To the Greeks this god was known as Hephaestus. The art of making iron weapons

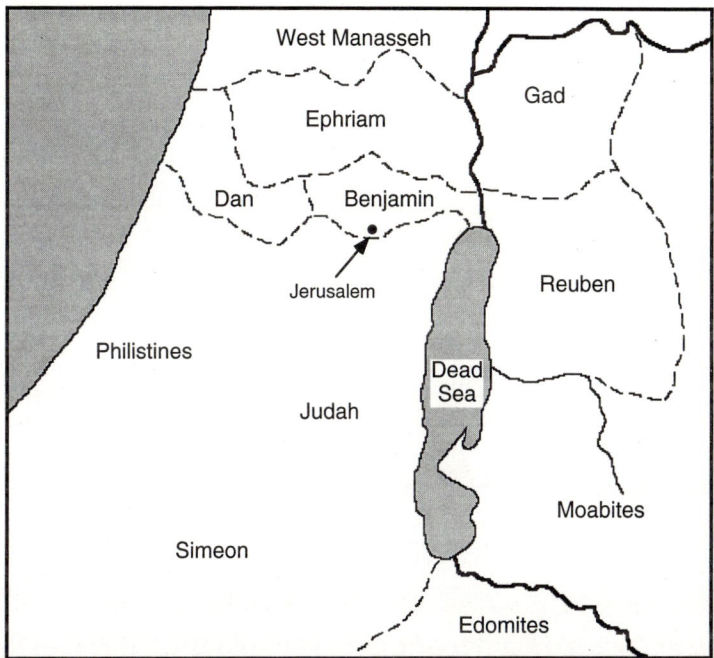

Figure 4.9 The Philistines were essentially Greek in culture and ethnically Indo-European. They arrived in what is now known as the Gaza Strip as part of a migration, which in Egyptian historical records is called the invasion of the "People of the Sea." The figure shows the locations of some of the ancient tribes in the middle east.

was a closely kept secret. We are told in the Bible in the thirteenth chapter of The First Book of Samuel that

> There were no blacksmiths in Israel because the Philistines were determined to keep the Hebrews from making swords and spears. The Israelites had to go to the Philistines to get their plows, hoes, axes, and sickles sharpened. The charge was one small coin for sharpening axes, and for fixing goads, or two coins for sharpening the plows or hoes. And so on the day of battle none of the Israelite soldiers except Saul and his son Jonathan had swords or spears.

Sometimes in the battle practices of the time, each side would select a champion to fight the battle between the two armies in one-on-one combat. The side of the loser would then assume that their god was inferior to the god of the opposition and would run away rather than fight. Therefore when the Israelites and the Philistines confronted each other we are told that

> The champion of the Philistines, a man called Goliath, from the city of Gath, came out from the Philistine camp to challenge the Israelites. He was over nine feet tall and

wore bronze armor that weighed about 125 pounds, and had a bronze helmet. His legs were also protected by bronze armor and he carried a bronze javelin slung over his shoulder. His spear was as thick as the bar on a weaver's loom, and the spear head weighed about 15 pounds. A soldier walked in front of him carrying his shield. Goliath stood and shouted at the Israelites, 'What are you doing there lined up for battle? I am a Philistine. You slaves of Saul choose one of your men to fight me. If he wins and kills me we will be your slaves, but if I win and kill him you will be our slaves. Here and now I challenge the Israelite army. I dare you to pick someone to fight me.'

We are told that Saul and his men when they heard this were terrified. Probably the statement that Goliath was nine feet tall was a little bit of an exaggeration. (See, however, the views of Dr. Greenblat [20] at the end of this section.) As the story of David and Goliath was retold over the centuries, Goliath probably grew a little bit each time, in the same way that the captured prize fish of the local angler grows with each bottle of beer. However, the blonde and blue-eyed Greek people could grow to heights well over six feet and, to a "five foot nothing" warrior on the other side, a six-and-a-half foot high Greek with a feather-crested helmet dressed in bronze armor must have looked like an invincible giant. The typical appearance of a Greek warrior as he went into battle is shown in Figure 4.10(a). We have many statues showing that Greek warriors often wore their helmet tilted on the top of their head as illustrated in the picture of Athena, the warrior goddess of the Greeks, in Figure 4.10(b). Like foolish hockey players, some warriors liked to show they did not need a helmet for protection. This is an important point that we will return to later in our study of the saga of David versus Goliath.

In The First Book of Samuel, we are told that David the shepherd boy had brought supplies to the Israelite camp for his older brothers who were in the army of Saul. When he discovered that no one would accept the challenge of Goliath, David volunteered to fight him.

When Saul heard of the offer by David, the King offered to loan his own armor to David. We are given the following account in verse 37 of the 17th Chapter of The First Book of Samuel.

> 'Go then', said Saul; ' go and the Lord be with you.' He gave his own armor to him to wear, a bronze helmet which he put on David's head and a coat of armor. David strapped Saul's sword over the armor and tried to walk, but he couldn't because he wasn't used to wearing them. 'I cannot fight with all this,' he said to Saul, 'I'm not used to it.' So he took it all off and then he took his shepherd stick and picked up five smooth stones from the stream and put them in his bag. With his sling ready he went out to meet Goliath

Commentators on this story often make play of the fact that David, out of his great faith in God, went out with the sling to meet Goliath. They often overlook the fact that he would have gone with armor if he had been able to walk under its weight. Furthermore, if he had been so confident in victory, why did he pick up five stones instead of just one?

Figure 4.10 Goliath the Philistine was probably killed as much by his arrogance as by the stone from David's sling. a) A Greek warrior in full battle armor. b) A typical depiction of the warrior goddess Athena.

I always had difficulty understanding the details of the story of David and Goliath because I did not know what was meant by a sling. To me, when I first heard the story, I visualized David using the instrument shown in Figure 4.11(a). This device, properly called a *catapult*, is also widely known in North America as a slingshot. In a catapult of the type shown, the missile is launched by means of energy stored in a stretched piece of elastic. The missile flies between the Y-shaped arms of the stick when the catapulter releases the stretched elastic. Thus the missile is launched by converting the potential energy of the stretched elastic into kinetic energy, in the same way that an arrow is launched from a bow. In England small boys usally call such a catapult a *"cattie."* The sling used by David is quite different and it is illustrated in Figure 4.11(b). It consisted of a string in the middle of which is a piece of leather or cloth which holds the missile when the string is folded in two. To launch the missile, the string is usually whirled around the head, in the same way that the gauchos whirled the bola around. When the top speed of the sling is achieved, one end of the string is released and the missile flies off along a tangent. The horizontal circle launching technique illustrated in Figure 4.11(b) is probably used at relatively short range. An alternate method of using the missile is to swirl the support strings in a vertical plane. This method would be used to lob the missile into a parabolic path so that it would fall down onto the enemy. From the historical account of the battle, we can deduce that Goliath was insulted that the best warrior the Israelites could send out to meet him was a shepherd boy with a stick and a sling. As the biblical account states, the Philistine started walking towards David with his shield bearer walking before him. He kept coming closer and when he got a good look at David,

> He was filled with scorn for him because he was just a nice good-looking boy. He said to David, 'What is that stick for; do you think I'm a dog?' and he called down curses from his God on David. 'Come on', he challenged him, 'and I will give your body to the birds and the animals to eat'. Goliath started walking towards David again and David ran quickly towards the Philistines battle line to fight him. He reached into his bag and took out a stone which he slung at Goliath. It hit him on the forehead and broke his skull and Goliath fell down.

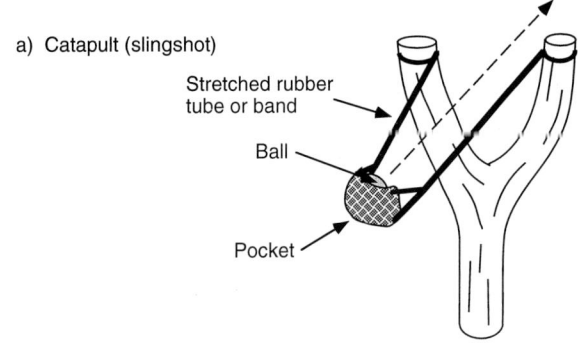

a) Catapult (slingshot)

b) Sling for launching missiles

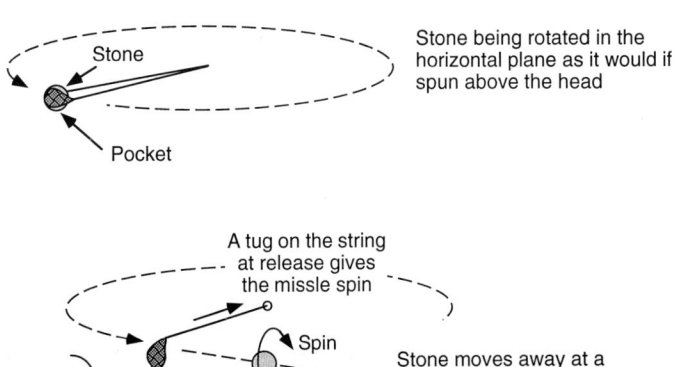

Figure 4.11

4.4 Killing Giants and Catching Fish 113

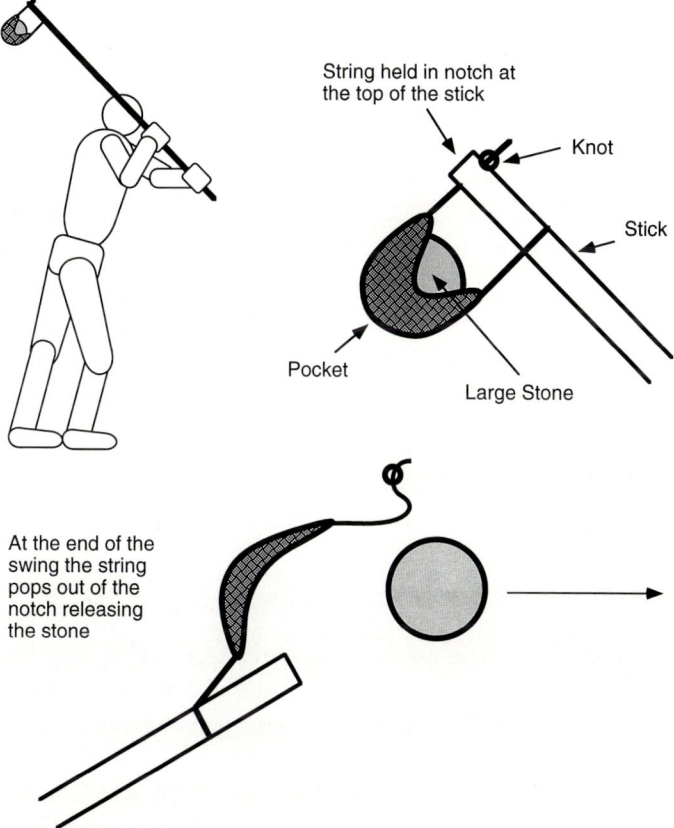

Figure 4.11 Catapults and slings can be used to lob stones at a target. a) A typical slingshot, more properly known as a catapult. b) The sling was used to lob missiles by accelerating them in a circular motion. c) In close range fighting the stick sling was used to launch more massive missiles.

The reason that David was able to achieve such a quick victory was probably due to the fact that, while pouring scorn on the little shepherd boy, Goliath kept his helmet above his head in the position shown in Figure 4.10(b); otherwise the stone would not have been able to inflict the fatal wound. We can also assume that Goliath was not well versed in the skills of slingers; otherwise, he would not have had to ask David why he was carrying the stick. When I was originally told the story of David and Goliath, I also wondered why David would take a stick when he obviously needed the maximum maneuverability and speed to avoid the giant. Korfmann who has written a very interesting article on the sling as a weapon tells us that when going into battle in biblical times, a slinger would use three different lengths of sling depending on the range he wished to achieve [21]. The kinetic energy of the missile rotating at the radius r is proportional to r^2; longer slings can store more energy in the missile and therefore have a longer range. Korfmann also tells us that when fighting heavily armed men, the slinger, if failing to injure the soldier dressed in the way illustrated in Figure 4.10(a) with a high speed missile, would use what was known as a *stick sling* to launch a much larger missile at the soldier from close quarters. Blyth who wrote an interesting article on ancient armor notes that a surprising feature of Greek style equipment was the lightness of the javelins and arrows used by such warriors [22] Blyth points out that arrows and javelins could not be too strong otherwise they could be picked up and used by the opposing forces. There is a suggestion in Blyth's account that the arrows and javelins were intended to be used only once, having a probable breaking strength very close to the expected impact strength. He also points out that the armed warrior needs to find the best combination of armor and offensive weapons to restrict the overall weight he must carry into battle. The large spear carried by such warriors was part of the defensive equipment used by the array of soldiers when facing cavalry attacks.

When the elite Greek warriors fought against cavalry the formation that they adopted was called a *phalanx* from a Greek word meaning battle array. In medical science, the whole row of finger joints is called a phalanx because of its similarity to the line of soldiers ready to fight.

Historical documents of battles between the Greeks and the Persians record that at the battle of Thermopylae the Greek spears broke during the battle and in another battle the Persians were said to have caught hold of the spears with their hands and had broken them at the tip [22]. In hand-to-hand combat the warrior would use the large spear in an attempt to pierce the armor of the opponent and hence the heavy weight of the spearhead given in details of Goliath's equipment.

In a discussion of the famous encounter between David and Goliath, Korfmann points out that many people try to explain the victory of David over Goliath in terms of medical and or supernatural circumstances. One attempt to explain David's victory in medical terms has been made by Dr. Robert *Greenblat*, a professor of endocrinology [20]. He suggests that Goliath suffered from *tunnel vision*. This is an illness which involves an inability to see anything except objects directly in front of you as you have no side vision. Thus, Greenblat suggests that if Goliath suffered from tunnel vision, the agile David could have skipped around the supposedly 10-foot-tall Philistine who would never be sure where the young Israelite was. In an article based on an interview, Greenblat is quoted as speculating,

> *Did David's boldness border on youthful irresponsibility and impulsiveness in issuing the challenge to Goliath? It is far more likely that his keen powers of observation disclosed Goliath's peculiar movements.*

Greenblat [20] goes on to comment that,

> *David probably realized his opponent was forced to turn his entire head to focus his full gaze on an object. Perhaps David suspected what we know today: Human giants are prone to suffer from lateral blindness. Giantism is frequently caused by a tumor*

on the pituitary gland, the so-called master gland of the body in the brain. This can affect nerves of the visual process to produce tunnel vision in which sight is clear only in a straight line. David therefore would step agilely aside when he had drawn close enough to Goliath. Then, as his adversary hesitated, clumsily turning his head to bring the youth within his limited field of vision, David took deadly aim with the slingshot and struck the long forehead spot unprotected by heavy armor.

Dr. Greenblat concludes his article with the statement,

David won this victory by superior knowledge, skill and agility rather than brute force.

Dr. Greenblat's charming little book entitled *Search The Scriptures: The Physician Examines Medicine in the Bible* [20] examines twenty biblical stories — ranging from the love affair of David and Bethsheba to Esau's sale of his birthright for a bowl of soup — from the perspective of what might have been happening to the biochemistry of the individuals involved. Although the Associated Press account of an interview with Dr. Greenblat focuses on the story of David and Goliath, it is not one of the stories that is actually in the book.

Korfmann suggests that attempts such as those of Dr. Greenblat to explain the victory of David in terms of medical problems is an area of scholarly speculation fuelled by the fact that many people in the western world are unaware of the lethal potential of the sling as a weapon. Korfmann tells us that this neglect of the sling as a weapon extends to professional historians. He notes [21] that,

The British pre-historian V. Gordon Childe more than once toward the end of his life attempted to bring the significance of the sling as a weapon to the attention of his colleagues, he met with little success.

Korfmann also explores the possible reasons why the sling is almost ignored in the stories of the Greek wars. He suggests it is because the warriors felt that the sling was not a gentlemen's weapon and it was only used by inferior people. This attitude is very similar to that which brought about the Papal condemnation of the crossbow, mentioned earlier.

Although the elite Greek fighters considered their slingers less than respectable warriors, the heavily armed soldier could be in real danger without their support. This is illustrated by what happened to a group of Greek soldiers who had been recruited to help overthrow the King of Persia in the year 401 BC. At the battle of Cunaxa in Persia, the rest of the rebel leader's army left when the leader of the revolution was killed. The Greeks then had to fight their way back to Greece. They were led by an Athenian called Xenophon. Xenophon led the 10 000 soldiers on a 1500 mile march to the Greek colony of Trapezus (modern name Trebizond) on the Black Sea. The march took five months through unknown country against obstacles of weather, difficult terrain, and enemies who attacked them continuously. Xenophon told the story of this famous march in a book called *Anabasis*. On the first day of the march, the infantry were attacked by small numbers of enemy cavalrymen, archers, and slingers. Because of these enemy tactics, the Greeks in the first day of retreat moved less than three miles! That night Xenophon called his captains together and said,

> We need slingers ourselves at once, and also horsemen.

Xenophon records that

> There are Rhodians (citizens of the island of Rhodes) in our army, most of them understand the use of the sling and their missiles carry no less than twice as far as those from the Persian slings. For the latter have only a short range because the stones that they are using are as large as the hand can hold; the Rhodians, however, are versed also in the art of slinging leaden missiles.

Xenophon noted that the two hundred Rhodian slingers that he recruited from amongst the elite troops,

> fired missiles that carried further than the Persians, further even than the Persian bowman's arrow.

In these interesting details given by Xenophon, the difference between the missiles fired by the Persians and the Greeks is noted. Archaeologists in their excavations sometimes found piles of what looked like eggs. Korfmann tells us that some *archaeologists* wondered what purpose such odd clay eggs had served. He believes these archaeologists were unfamiliar with the sling as a weapon. Studies have now established that as early as 4000 BC, slingers started to use egg-like missiles. Korfmann notes that such missiles were more dangerous than spherical ones because of the way in which they were able to damage the human body on impact. He does not mention in his article, and there is no physical evidence or historic records that I know of to support my claim, but in fact it is probable that the early slingers had learned the knack of stabilizing the trajectory of egg-shaped missiles by pulling the string, at the time of release, inwards towards themselves as indicated in Figure 4.9(b). This would spin the missile and give it stability in its trajectory. It would also keep the nose of the missile up and stop it from tumbling end over end so that the pointed end would hit the target, thus exerting greater force at the point of impact. In fact, unless the slinger did indeed use spin to stabilize his missile there really is no advantage to the ovoid shape. Korfmann records the interesting detail that clay missiles for use in a sling were sun dried and not baked in the fire. If these pure clay missiles had been baked, they would have cracked or would have contained internal cracks. For this reason they were dried slowly in the sun. By the time of the Greek epics and the later Roman wars, the slingers were using missiles made of lead. The Romans called them "glandis" because of their supposed resemblance to acorns. In his article, Korfmann shows pictures of several of the lead missiles found at various battle sites. He notes that the soldiers scratched messages on their leaden missiles, one of which has been translated (politely) as, *"for Pompei's backside "*.

Each slinger would tend to use uniform missiles, manufactured either from clay or lead, so that he would be able to judge the trajectory of his missile from experience and practice. Lead was the heaviest metal readily available in Greek and Roman times; it was suitable as a missile because of its high density, meaning that, for a given speed, more energy was invested in the missile. Therefore its impact would be more forceful and its potential range greater. Calculations show that slings of the type seen in statues are of such a length that missiles could easily achieve a release speed of 100 kph. Korfmann points out that a 25 g missile released at this speed could still reach the target with an energy equal to that of a golf ball falling from the top of a seven story building. These missiles were obviously able to embed themselves in an enemy's body and Celsus, a medical author of Greek and Roman times, wrote in his book *De Medicina* instructions for extracting sling missiles of lead and stone from the bodies of wounded soldiers. If a soldier had been the victim of a more massive stone launched with a stick sling, the injuries would be in the form of internal injuries, damaged organs, and internal bleeding.

Korfmann, in his discussion of the sling, points out that slingers generally out-ranged archers and that, in a battle formation, they would stand behind the archers to help protect them. It appears that most slingers could fire lead missiles up to 400 meters.

Korfmann presents a map showing the distribution of records of sling use throughout the world. He notes that sling users tend to cluster around the equator. He wonders why there is this preference for the sling in the warmer climates, and writes

> *When the explanation is found it may transcend the weapons themselves and thus lie outside the realm of archaeology proper.*

We have already found the answer to Korfmann's problems in Blyth's explanation that the problem of building bows in the Mediterranean was the absence of a suitable springy wood that could be used to store energy under the temperature and humidity conditions of the Mediterranean. Blythe pointed out that

> *A composite bow is an expensive weapon and simple wooden bows did not function in the heat of the equatorial regions. Therefore for a combination of economic and practical considerations the sling was used in the Mediterranean. Furthermore, it must be remembered that most of Northern Europe was covered with forest. The use of a bow for ambush was probably much more practical than a sling which is an open terrain weapon.*

As mentioned at the beginning of this section, the modern catapult makes use of elastic as a device for storing energy by stretching the elastic. In a technical article describing the effect of the use of natural rubber products on the sales of catapults, White, of the Malaysia Rubber Producers Research Association, points out [23] that

> *The arrival of rubber as a material for storing energy made design of small catapults of the type used by errant school boys very simple. The accuracy of such devices in expert hands is legendary.*

In the article, White goes on to describe the use of catapults for catching fish! Apparently, fishermen use catapults to place ground bait accurately in position in a particular stretch of water. Then the fisherman settles back with his rod and line and waits for the fish to bite. White tells us that the fishing catapults use natural rubber latex tubing to propel the bait accurately into place. The angler selects the type of water he suspects a particular species inhabits, slow moving deep water for some types of fish and fast moving shallow waters for others. To concentrate the fish in the right area, the angler will throw some ground bait into the area close to his hook. The consistency of ground bait is important. For still or slow water it can be light, but in fast moving or deep water a heavy lump will be necessary to take the bait to the bottom where the fish are lurking. Aiming the bait in the right place is a tricky task and here the catapult comes into its own. "Tackletime," a British company, produces 50 000 catapults a year for fishermen [23].

Another missile used by fishermen is the weight at the end of the line that enables it to sink into the water at the right place. In the 1980s, the British government banned lead-based fishing weights because of the effect of lead on fish, birds, and other wildlife — particularly the swans of the River Thames. "Tackletime" now produces a heavy plastic-covered weight which has enough tungsten (a very heavy metal) inside it to make it sink in the same way as a lead weight. In the next section we will explore the way in which ancient and modern people throw stones without either catapults or slings

Section 4.5 Dancing Aborigines and Skipping Stones

In an interesting article entitled, "One, Two, Three Strikes You're Dead is an Old Ball Game," Sarah Boxer tells us that paleoanthropologists (specialists in the study of very ancient man) have long wondered why early tool making sites like Olduvai Gorge in Tanzania are littered with smooth roundish stones not suitable for making into tools.

Barbara Isaac of the Peabody Museum at Harvard examined these stones and suspected that in fact they were throwing stones used in hunting. She then searched historical accounts of primitive peoples to see if there were records of the use of special throwing stones. One record she discovered [24], by a French explorer of the South Pacific, a Count De La Perouse, recorded that he lost 12 of the 61 men he took with him to what is now the Island of Samoa in 1787 to obtain supplies of fresh water. His men were subjected to

> *a fusillade of rocks thrown so hard that they produced almost the same effect as our bullets and had the advantage of succeeding one another with greater rapidity.*

Pictures of these stones reproduced in the article by Sarah Boxer show them to be egg-shaped. Although it is not discussed in this article the accompanying picture also

4.5 Dancing Aborigines and Skipping Stones

appears to show the natives using slings. In 1870 the British historian J.G. Wood wrote in *The Natural History of Man* about the Aborigines in Australia. (The word aborigine comes from two Latin words: "ab" meaning "from" and "origo" meaning "a beginning." Therefore, aboriginal means someone who was in a land from the very beginning of recorded time. The word is used in Canada to describe the Indians who originally inhabited the country. In Australia, the word aborigine is the name used to describe the native inhabitants of the country that European explorers discovered when they first went to Australia.) Wood's report states:

> *Aborigines in Australia had occasionally killed gun totting British soldiers by dancing crazily from side to side, to prevent the soldiers from taking aim, and unleashing a shower of stones with a force and precision that must be seen to be believed.*

Miss Isaac located nine Pacific Island war handstones in the collection of the Pitt Rivers Museum at Oxford. Subsequent searches turned up more handstones in the Museum of Mankind in London than at the Peabody itself. According to Isaac, most of them were lemon-shaped which suggests that they were thrown with a spin rather like tiny footballs. From these accounts it can be seen that the use of spin to stabilize a hand-thrown missiles is a very ancient art [24].

Sooner or later in their stone-throwing careers, most small boys who live near ponds or larger stretches of water learn to throw thin, flat stones that skim over the surface of the water, dipping down and bouncing off the water in a series of leaps. In England, the throwing of stones to make them skip in a series of leaps is known as "playing ducks and drakes." The origin of this term is very obscure. I learned to play ducks and drakes on the shores of the North Sea in England. I still have in my possession two or three limestone pebbles ideal for playing ducks and drakes that I picked up on my favorite beach. The gravel of the local beaches near my home town include stones from many different geological origins. But the best stones for ducks and drakes are thin limestone pebbles washed down the coast from the chalk cliffs farther north. This chalk is often layered and apparently when chalk boulders break up they form thin, flat stones which become rounded by the tumbling action of the waves on the beaches. One of the stones I have in my possession is an inch and a half across by an inch at right angles and about a quarter of an inch thick. This type of stone can be made to take seven or eight leaps even by inexperienced throwers. It can be made to do more by those who know how to spin the stone very fast on launching. Again, here is an example of learning to use sophisticated physics by instinct or experience without needing to study the conservation of angular momentum. Alan Green has written a short article on the science of skipping stones [25]. In it he tells us that an unofficial world record for a skipping stone is between 30 and 40 skips. It is unofficial because the organizers of the annual stone-skipping tournament at Mackinac Island, Michigan, recognized by the *Guinness Book of World Records*, has sanctioned this event for the world's best skipping stones but will not permit artificial stones (man-made) in the competition. The official world record for ducks and drakes using a natural stone is 24 skips, a record equalled by three different contestants. The unofficial record is held by John Zehr who set out to make the perfect skipping stone by trying different materials and different shapes. As Alan Green tells us, the stone that holds the record was generated by a mistake by a machinist making a stone out of a mixture of plaster and sand. Zehr tells us that the machinist accidentally made a dimple a few thousands of an inch deep and half an inch across in the center of the skipping surface. This unofficial champion's stone is two inches wide and half an inch thick with the half inch dimple. Apparently, the dimple decreases the surface area in contact with the water so there is less loss of energy through friction so that the stone travels farther and makes more skips.

The science of throwing spinning stones is not as simple as it looks. Green however, points out that the speed at which a stone is thrown does not necessarily mean it will skip more. This is because of the balance of forces involved in the perpetuation of the skipping action. You would think increasing the flat surface area of a stone should produce more skips. But because the volume of any solid increases faster than its surface area, a rock with twice the surface area of a smaller, similarly shaped stone has more than twice the weight. This means a larger skipping stone must be proportionately thinner if advantage is to be gained from the higher surface. The stone skipper can also increase the number of skips by waxing the underside of the stone to reduce friction when the stone is in contact with the water.

During the Second World War, the physical principles employed in the skipping stone were built into a special type of bomb intended for use against water dams and to circumvent defences placed near ships in harbors which were used to protect them against torpedoes. These so-called bouncing bombs, developed by Sir Barnes Wallis, were actually a refinement of ricocheting cannonballs [26, 27]. The use of ricocheting cannonballs is reported in military literature as early as the middle of the 16th century. They were used to great effect by Admiral Nelson in his wars against the French. It is probable that the effectiveness of cannonballs which ricocheted off the surface of the sea was originally an accidental discovery which the observer was able to turn into a regular practice. A ricocheting cannonball would make better use of the energy vested in the cannonball to inflict damage near the waterline increasing the probability that the ship being attacked would sink.

During the Second World War (1939–1945), one of the aims of the allied bombing of Germany was to stop the production of war weapons. The destruction of the large hydro-electric dams generating electricity for the German factories would obviously help to stop weapons production. The problem faced by the would-be designers of special weapons to attack the dams was that such dams would be protected against torpedoes launched from aircraft by booms suspended in the water as indicated in the sketches given in Figure 4.12. The two sketches of this figure are based on sketches drawn in the notebooks of Sir Barnes Wallis. In his original design, Wallis proposed giving backspin to a spherical bomb in order to give it aerodynamic lift by the Magnus effect as used in tennis. This would also generate lift while the bomb is in contact with the water, reducing frictional losses and substantially increasing the range and number of

4.5 Dancing Aborigines and Skipping Stones

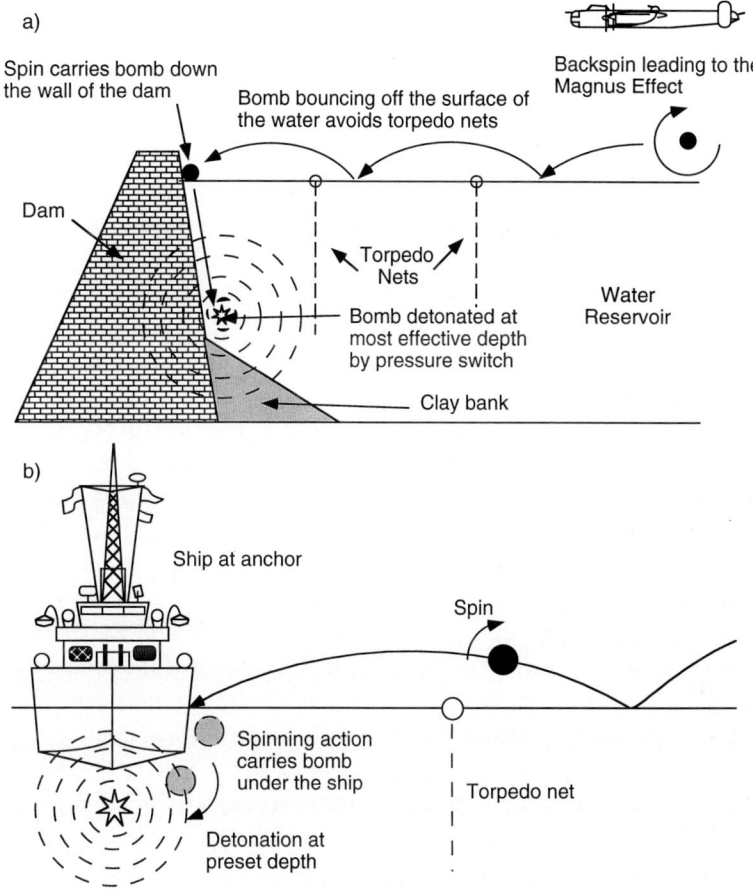

Figure 4.12 The principles of operation of the "Bouncing Bomb."

bounces obtainable. In the first plan, Wallis proposed a gigantic spinning sphere containing 6500 pounds of explosive to be dropped from a low flying aircraft onto the surface of the reservoir. The bomb would then bounce across the water, over any protective netting, and strike the top of the dam. Then it would sink, pressed tightly against the dam by the forces generated by its spin. It would finally explode at the correct depth, triggered by a pressure sensitive device measuring the water depth. In fact, attempts to use spherical bombs proved unsuccessful because the bombs broke up when they hit the surface. However, the tests carried out with spherical bombs were sufficiently encouraging for the design to be modified to a cylindrical bomb which was easier to spin, easier to carry, and just as effective at bouncing over the torpedo nets. The development of the spinning bombs proceeded in parallel with the development of a similar weapon intended to be used against battle ships sheltering in the Norwegian fiords and protected on the seaward side by torpedo nets. This proposed operation is illustrated in Figure 4.12(b). From the beginning of the project to develop such spinning bombs, the British Royal Air Force worried that, should the bombs prove to be successful, the design would be copied immediately by German scientists. Therefore attempts were made to ensure that any bomb that did not reach its target would self-destruct before it could be captured.

When the bouncing bomb had been perfected, it was used in a famous air raid. This was an event which was later made into a movie. In the raid, nineteen Lancaster bombers were used to deliver the bombs which destroyed the Mohne Dam. However, to successfully deliver bombs, the specially trained air crews had to fly very low. There were severe losses of bomber crews — sufficient to prevent any further use of the device. Barnes Wallis himself was so deeply affected by the crew losses in the dam raid that he gave his £10 000 award ($24 000 Canadian) for inventing the bomb to help educate the children of the Royal Air Force personnel. At a 50 year anniversary of the actual raid on the dams carried out on May 17, 1943, Royal Canadian Airforce Navigator Danny Walker recalled how 30 of the dambusters were Canadian. He described how during the raid you had to be only 60 feet above the water and going at 240 mph (384 kph) and the bomb had to be rotating at 500 revolutions per minute. Only 16 of the Canadians survived the raid.

Unfortunately, one of the spinning bombs failed to explode and was captured by the German forces. They immediately recognized the importance of the concept and began to develop similar weapons. The development of such spinning missiles was pursued by both the British and the Germans but remained incomplete when the war ended. Such missiles have now been superseded by guided missiles, but the fact that spinning bombs were used provides another interesting example of how simple child's play can be turned into the basic idea behind deadly missiles.

Section 4.6   The Deadly Missiles of Cricket and Baseball

Baseball and *cricket* are two closely related games. There is some argument as to which game was developed first. Hostetter [28] tells us that the games appear to develop in parallel until the late 1800s; however, one British writer to the London Times in 1874 said "Some American athletes are trying to introduce us to their game of baseball, as if it were a novelty, whereas the fact is that it is an ancient English game long discarded in favor of cricket". In both games, the ball is a high speed missile which has to be intercepted by a player with a bat. The bat in baseball is round. The cricket bat is made of willow and has a broad flat surface on one side. In this section we will look at the structure of the balls used in both games and the dynamics of launching those balls. We will also discuss the ways in which the pitcher in baseball and the bowler in cricket use physical phenomena to modify the trajectory of the ball as it moves toward the batter [28, 29].

If we consult the official baseball rules, we find from rule 1.09 that

The ball should be a sphere formed by yarn wound around a small sphere of cork, rubber or similar material covered with two strips of white horsehide, or cowhide, tightly stitched together. It shall weigh not less than 5 and no more than 5.25 ounces avoirdupois, and measure no less than nine nor more than 9.25 inches in circumference.

Avoirdupois is defined in the dictionary as, "the system of weights in which one pound equals 16 ounces. The phrase comes from an old French phrase meaning, 'to have weight.'" The detailed construction of the ball used in baseball is shown in Figure 4.13. At the center of every baseball is a small ball of cork [29]. Natural *cork* is the outer bark of a type of an oak tree that grows in Portugal, Algeria, and Spain. These trees can grow to a height of 100 feet, although the average tree is about 30 feet. The word cork comes from the Latin word "cortex" meaning "the bark of a tree." When doctors started to explore the structure of the human brain, they thought that the outer part of the brain looked like the bark of a tree and hence the term cortex for the outer layer of the brain.

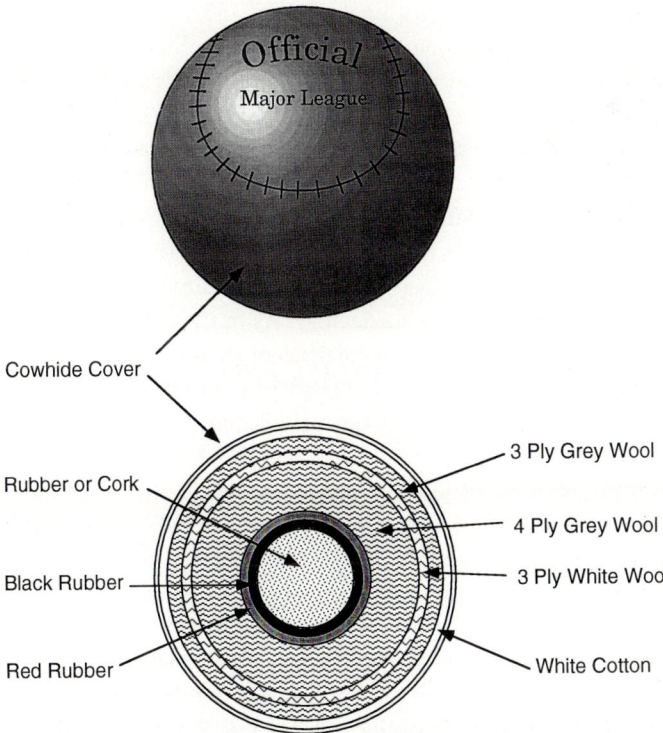

Figure 4.13 The basic design of the missile (ball) used in baseball. The cover is made from two hourglass-shaped pieces of horsehide or cowhide stitched over layers of various types of wool threads which are wrapped around a rubber or cork core.

When a cork tree is about 15–20 years old, the bark is stripped from the tree — the harvester being careful to leave the inner living bark undamaged. It is possible to remove fresh layers of cork from the tree every ten years. The bark is boiled to remove tannic acid and to make it pliable. It contains natural fatty substances that make it almost impermeable to water and gases. The beginning of the 1500s saw the first widespread use for cork as stoppers in glass bottles. This led to the use of the name "cork" for the stopper in a bottle. Corks can become missiles when a bottle of champagne is opened. In such cases the compressed carbon dioxide in the bottle is described as the propellant driving the missile made out of cork. Generations of young boys have also been equipped with "pop guns" which fire tablets of cork by means of compressed air.

Most of the baseballs used in competitive baseball are made by the Rawlings Company, which in 1974 moved its factory to Puerto Rico to avail itself of the lower labor costs. In the first stage of the manufacture of a baseball, a cork ball is compressed and coated with rubber to produce what is known in the industry as a *pill*. At this stage of manufacturing it looks like a red golf ball. The pill is enlarged by wrapping woollen yarn around the pill. The first layer is 219 yards of a blue-grey yarn which is wrapped around the ball. Next, white yarn is wrapped around the ball followed by a third layer of blue-grey yarn. The ball receives a final wind of 150 yards of cotton string and it is then dipped into a barrel of rubber cement. Traditionally, baseballs were covered by horsehide but because of the shortage of horsehide, the balls are now covered with cowhide [29–31]. (The official move to cowhide-covered balls in major league play was permitted in the early 1970s.) An official baseball has 104 stitches each made by hand. An expert seamstress at the factory in Puerto Rico can complete the stitching of one ball in 10 minutes and has an output of 36 balls a day.

In recent years there has been some controversy over the possible variation in the quality of the balls made by Rawlings. Some people maintain that the "bounceability" is not controlled closely enough by the factory. As in the case of golf balls, the bounce of the ball is measured in terms of the coefficient of restitution. In the officially approved test for measuring the coefficient of restitution of a baseball, the ball is fired from an air-driven cannon, at a speed of 85 feet per second, at a wall eight feet away. The rules say that when this test is carried out, the ball should bounce back at slightly more than half speed to give a coefficient of restitution of 0.514–0.578. Professor Brandt, a physicist of New York University, employed by a competitor of the Rawlings company who wished to break their monopoly, has carried out measurements on the coefficient of restitution which he maintains vary markedly from ball to ball. Brandt claims that some balls may have a coefficient of restitution as high as 0.607 which is about 5% over the limit and results in a livelier, that is, bouncier ball than is permitted by the rules. Professor Brandt also claims to have studied the change in the performance of baseballs over a period of time and has discovered a statement that in 1938 the U.S. Bureau of National Standards established that the average baseball had a coefficient of restitution of 0.46. Professor Brandt calculates that the increase from 0.46 to 0.607 in coefficient of restitution means that a baseball player hitting the ball of 1938 with a force sufficient to drive the ball 300 feet would in 1983 be able to drive the ball a distance of 400 feet using the same force.

A comparison of the playing actions of baseball and cricket turns up some differences between British and American English. In baseball the word *pitch* is a verb and a noun. "Pitch" denotes both the act of throwing a baseball and the throw itself. In cricket, the pitch is the area of the field in which the game is played. In the game of baseball, the player throwing the balls at the batter is described as the *pitcher*; in cricket, the player launching the balls is described as a *bowler*. In order to be able to understand the different ways in which the pitcher and bowler launch their missiles, it is necessary to have a basic understanding of the way in which the two games are played. An article on the game of cricket in an encyclopedia starts with the following statement [32]:

> *Cricket is an outdoor game played between two teams of eleven players each. The cricket field is from about 450 by 500 feet to about 525 by 550 feet. In the center of the field parallel to its short ends two wickets are thrust into the ground, 66 feet apart. Each wicket consists of three wooden stumps (slender sticks) between 27 and 28 inches high, placed equidistant in a straight line so that the distance between the first and third stumps is between 8 and 9 inches. On the top of the stump two strips of wood between 4 and 4.5 inches long and known as bales are placed end to end . . . . The wicket is centered lengthwise in a white line 8'8" long known as the bowling crease. Four feet in front of each bowling crease and parallel to it is drawn another white line called the popping crease or simply the crease.*

The central action of the game of cricket takes place between the batter who stands behind the popping crease and in front of the wicket and the bowler who delivers the ball from behind the opposite bowling crease. The bowler pitches the ball either underhand or overhand and without bending his arm. The ball must hit the ground in front of the batter. A batter may be out in various ways. He is said to be *bowled out* if the ball hits the wicket. He is *caught out* if the ball he hits is caught before it reaches the ground. If the wicket keeper knocks down the wicket when the batsman is not between the bowling and popping crease, the batsman is said to be out *stumped*. This fact has led to the everyday English phrase of being "stumped by a question," which means being unable to respond to it. A person is considered to be defeated by such a question in the same way that a batsman is defeated by the wicket keeper.

In the course of the game, a bowler delivers a series of six balls from one end. A sequence of six balls is known as an *"over"* (if the bowler delivers six balls without the batsman scoring any runs the batsman is said to have delivered a *maiden over*). At the end of an over, another bowler bowls from the opposite end of the pair of wickets and all of the fielders change their positions relative to the batsman. Some of the terms used to describe the fielding positions in a game of cricket are shown in Figure 4.14. During the course of a game of cricket, the ball is gradually damaged by the many hits it usually receives, and after 85 overs the bowler can take a new ball. In the course of a long cricket match, the bowlers used in the early stages of the game are usually *fast bowlers*. They

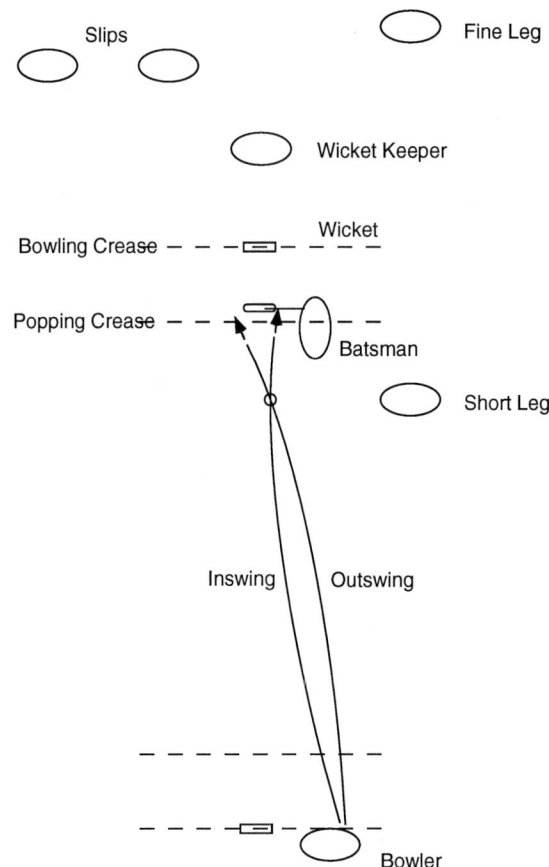

Figure 4.14 Some essential names for the positions of players and components of the pitch used in the game of cricket.

send the shiny new ball through the air at very high speed. As the ball becomes roughened and slightly deformed, it is not so easy to control at high speed and the team will usually switch to *spin bowlers*. Spin bowlers deliver the ball more slowly but they have greater skill in spinning the ball with the seam. They also exploit the Magnus effect we met earlier to make the ball curve in its flight towards the batsman. If, during a cricket match, a ball is lost due to a massive hit from a batsman or for any other reason, it must be replaced by a ball in the same state as the one that was lost. In cricket, batsmen usually have greater difficulty when facing a fast bowler with a new ball, so sometimes in a game of cricket, the fielding side will try to bowl defensively until the time comes for taking a new ball. Sometimes when cricket is played on a very damp ground (after rain), the ball will break on the bounce in a very complex manner and one says that the game is being played on a *sticky wicket*. This has become an everyday phrase for

4.6 The Deadly Missiles of Cricket and Baseball

carrying out any kind of negotiations under difficult conditions (the negotiators are said to be playing on a sticky wicket) [33–35].

North American viewers of cricket games sometimes find the whole procedure rather mystifying. In a humorous speech to an American audience Tom Alton a British Counsel in America described the game as follows [28]:

> Cricket is played with two sides, one out in the field and one in. Each man that's in the side that's in goes in and when he's out he comes in and the next man goes in until he's out. When they are all out, the side that's out comes in and the side that's been in tries to get those coming in out. Sometimes you get men in and not out. When both sides have been in and out, including the not outs, that is the end of the game.

In Figure 4.15, the basic layout of the playing area of baseball is shown. In baseball each team has nine players. The catcher occupies a position directly behind the position of the batter described as *home plate*. The pitcher delivers the ball to the catcher and attempts in so doing to prevent the batter who stands beside the home base from safely hitting the ball. Three basemen are situated at or near the bases of the baseball diamond. Players also include a shortstop situated behind the second to third baseline (about midway between the bases) and three outfielders who occupy positions in left, center, and right field. In our summary we are basically concerned with the conflict between the launcher of the ball and the batter. Those who would like further background on the games of cricket and baseball will find more information in Refs. [30, 32].

The way in which the pitcher and the bowler launch their missiles differs radically. A pitcher is not allowed to run up to the launching point and during the delivery must keep one foot in contact with a rubber slab two feet long and six inches wide. This means that the pitcher's range is severely limited in comparison with that of the bowler who can release the ball from any point of the 8'8" long bowling crease. A fast bowler may run a distance of 20 yards to build up speed before launching the cricket ball. The pitcher must throw the ball without a bounce. Furthermore, a successful pitch must pass through the *strike zone*: an imaginary vertical rectangular area 18 inches wide in front of the batter and extending from the batter's knees to the armpit. A skillful pitcher will aim for the corners of this strike zone. The professional baseball pitcher can throw the ball at 135 kph (85 mph) or more.

To make the ball curve in space a good baseball pitcher can actually spin the ball at rates of 1800 revolutions per minute. Cricket balls have been measured at speeds of up to 160 kph (100 mph) leaving the bowler but, of course, since the ball bounces once, there is some loss in speed before the it reaches the batter. The ways in which the cricket ball is spun by the bowler to achieve complex trajectories is discussed in Refs. [33–35].

Brancanzio has discussed the length of time the batter in baseball has to decide how to hit the pitch thrown at him [36]. The ball leaves the pitcher's hand 17 meters from the batter. It usually travels at a speed of about 135 kph (85 mph) and reaches a batter in 0.45 seconds. If the pitcher is a top line pitcher throwing a ball at 150 kph (95 mph), the flight time is only 0.4 seconds. It takes a good hitter about 0.25 seconds to accelerate

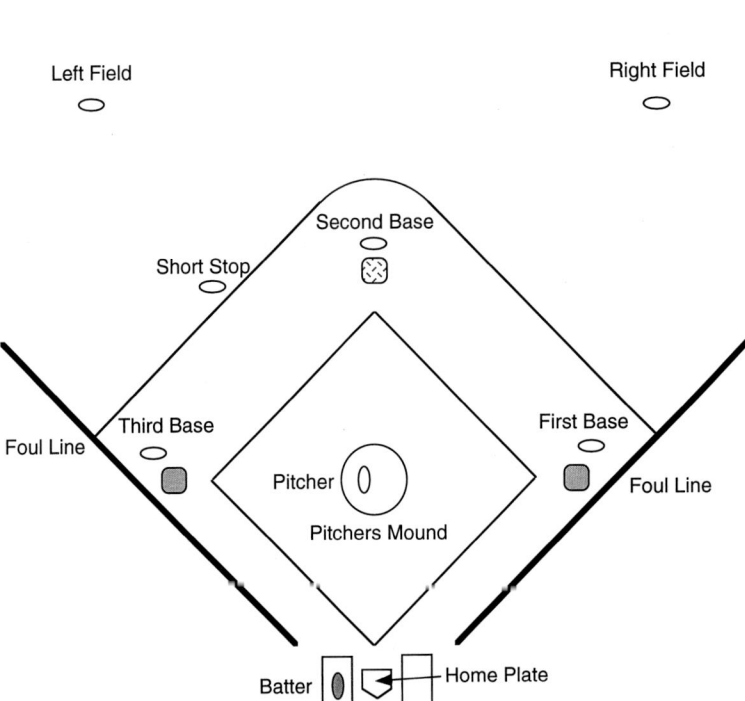

Figure 4.15 Basic baseball diamond showing player positions and names.

the bat from rest to the point where it meets the ball. This means that a batter has about 0.15 seconds to decide whether or not to swing at the pitch. During this interval the batter must also estimate the speed, the rate, and the direction of the curvature of the pitch and calculate the path of the ball as well as decide whether to swing or not to swing.

Tudge tells us that in cricket it is a most complex task to play against good fast bowling. Thus in an article on the conflict between the bowler and batsman [37] he states,

> It takes at least as much time for the nerve impulses to travel from eye to limb and for muscles to contract appropriately as it takes for the ball to travel the statutory length of the cricket pitch. In practice, batsmen don't react to bowling, they anticipate it; they begin their stroke at the moment the ball leaves the bowler's hand, or even before, judging its angle of approach, the position of bounce from secondary clues such as the height of the bowler's arm, the angle of his wrist, and the slope of his shoulder. Somewhere

in the cerebellum (the brain) mathematical calculations of awesome complexity are carried out which predict the trajectory and the height and the angle and the bounce of the pitch. If these calculations are done correctly, if the batman's brain is wired appropriately to his body, and his technique is sound, then the arc of the bat does indeed meet that of the ball and the impossible is achieved.

These two brief discussions of the difficulties of intercepting a ball, in baseball and in cricket, show that in fact the reason players have to practise is that the skill required to intercept the ball requires the highest level of mental alertness and experience.

An important difference between the rules of cricket and baseball with respect to the ball is that in cricket the bowler is allowed to shine one side of the ball on his trousers to exploit the potential of the ball for swerving in space. Such deliberate polishing of the ball is absolutely forbidden in baseball.

As in the case of tennis discussed earlier, some people feel that modern cricket playing styles have spoiled the game for spectators and some experts are calling for the game to be slowed down. This aspect of cricket has been discussed by Tudge [37].

Players playing baseball and cricket found that they could alter the trajectory of the ball in a unpredictable manner by adding small amounts of substances to the surface of a ball. Such adulteration of a ball is strictly forbidden but is still practised surreptitiously by some players. Regulation number 3.02 of the baseball rule book states:

No player shall intently discolor or damage the ball by rubbing it with soil, resin, paraffin, licorice, sand paper, emery paper, or other foreign substance.

Guttman tells us that up to 1988 no one had been caught tossing a licorice ball but pitchers have used everything from belt buckles and wedding rings to bent eyelets on their gloves to scar the skin of the baseball and change the aerodynamics [38]. One of the most blatant cases in baseball history was that of Honeycott of the 1980 Seattle Mariners. He was caught on the baseball mound with a thumb tack sticking through a bandaid on his finger. He became the first pitcher to be suspended for throwing the scuff balls since that type of pitch was banned in 1920. When discussing illegal forms of pitching Brancanzio [36] comments

Around the turn of the century some pitchers discovered they could produce unevenness in the flight of the ball by rubbing certain substances into the ball to produce a smooth spot. A few pitchers mastered the ability to throw a spit ball (I need not explain the details) which behaved so unpredictably in flight and gave the pitcher such an advantage that it was banned. A few pitchers have kept this in their repertoire occasionally sneaking it past an unsuspecting batter (and umpire). It is rather difficult to expectorate surreptitiously on a baseball in the presence of 10 000 spectators. The trick is to hide various equivalent substances (petroleum jelly is a favorite) on parts on one's uniform to be skillfully transferred to the fingers and ball as the need arises.

The advantage to be gained by such secret doctoring of balls tempts even the top players to cheat. Thus in an international cricket match against South Africa played in 1994 the English captain Mike Atherton appeared to pull something from his pocket and rub it into the ball. Atherton said he had merely been wiping sweat from his hand. The following day however he acknowledged that he had dried his hand with dust that he kept in his pocket. For using an illegal substance that could give the ball an unpredictable trajectory Atherton was fined $3000. His critics said this was not enough and a poll conducted by the readers of the British Daily Mirror newspaper found that 100% of respondents wanted Atherton to step down as captain of England [39]. For more information about the ball dynamics in the two sports see Refs. [40–43].

In the title of this section, the balls in baseball and cricket were described as deadly missiles and indeed in both baseball and cricket there are many injuries and several deaths a year as a result of players being hit by balls. When I played cricket, nobody wore protective gear, other than leg pads and an athletic support to protect "the family jewels". However, in recent years, cricketers have started to wear protective helmets with face guards. In the next section we will discuss the problem of personal injury from balls in games such as cricket and baseball and in other games using missiles such as the badminton shuttlecock.

The aerodynamics of a baseball can be influenced by many factors including the airflow around the arena in which the ball game is going to be played. Thus studies carried out at MIT by Paul Lagase demonstrated using wind tunnel experiments that a new observation box built at the Boston ground of the Red Sox Baseball team deflects the wind in such a way that a deep flyball will travel 8–12 feet less than it would have done before the observation box was built.

Section 4.7 Keep your Eye away from the Ball!

The coach usually tells the player of ball games, "Keep your eye on the ball!" When giving this advice he should say, "Keep watching the ball." The last thing a player wants to do is to bring the eye and the ball together. Unfortunately, all too often, the ball does meet the eye, or another part of the body, causing injury [44–48]. Indeed history may have been very different were it not for an historic death in a cricket game. During the reign of George II, the Royal Family of Great Britain were criticized for their love of cricket. As a result, we have the comment recorded by an English woman that members of the King's family were "diverting themselves with Cricket, a play all who are or who have been school boys are well acquainted with." The Prince of Wales (the title given to the eldest son of the king), Frederick Louis, son of George II, was an ardent cricket fan. When, in the opinion of his father, he should have been at the palace attending to affairs of the kingdom, he was often away playing cricket. Eventually the Prince of

Wales was killed by a cricket ball and his son George became King George III. King George III was thought to be mentally unstable, and it may have been his early succession to the throne because of a cricket death that caused political difficulties in the then colonies of North America that led to the American Revolution! [28].

The numbers of injuries in high-school sports in America are summarized in Figure 4.17. According to a national consumer product safety division there were 51 baseball-related deaths amongst 5–14 year olds between 1973 and 1983. An average of five deaths a year. Of these, 85% died after being hit in the chest. Fred Engh of the National Youth Sport Coaches' Association observed when these figures were released that the deaths were avoidable because every child had been hit by a regulation hard ball, the same kind used in the Major Leagues. Parents fail to understand that many youngsters below the age of nine lack basic motor skills. Engh advocated the use of safer balls until children are nine or older. In Figure 4.16 the internal structures of two of the safer baseballs are shown. There is some resistance to the use of the safer balls because they are not as lively as the regulation balls and make a different sound when hit by the bat, so that the children feel they are not playing real baseball. People concerned with safety in sports recommend that younger children should also wear a protective padded jacket. One victim of a baseball hit, ten-year-old Ryan Wojick of Citrus Park, Florida was killed in March

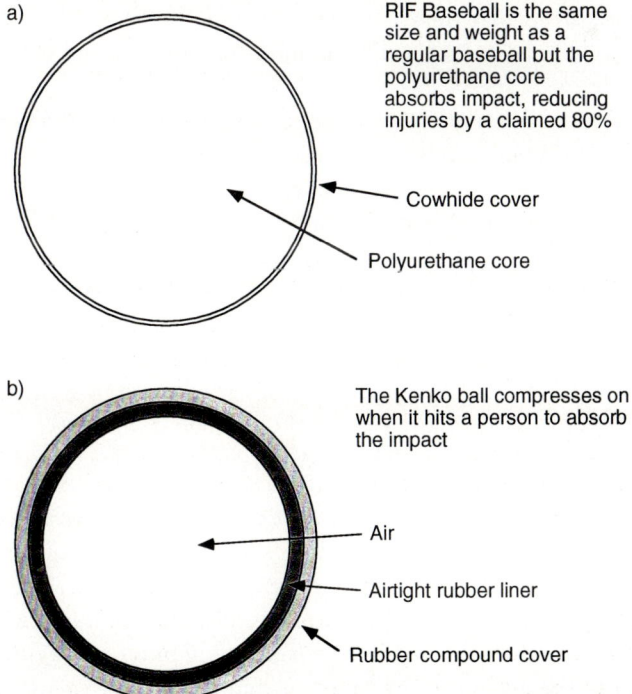

Figure 4.16 The inner design of two safer baseballs for use by younger players which are available commercially.

of 1990 after being hit in the chest by a ball. The Hillsborough County coroner who performed the autopsy said Wojick would have survived had he been wearing a protective vest. The padded jacket would have increased the mass around Ryan's heart, diffusing the impact. After the accident, free protective vests were donated to the Citrus Park Little League, but many players did not use them. In the words of Jane Wojick, Ryan's mother [49],

> *My son died because I didn't know this equipment existed. How many dead children will it take to make others wear the equipment?*

When we look at protective clothing in general in a later chapter, we will find that the same resistance to other safety equipment has been overcome in the long run. A death in junior baseball is considered by society at large to be an acceptable lifestyle risk, yet is relatively easy to protect children against sports injury. In an article discussing safety in Little League baseball, safety advocates have said that another major improvement in Little League baseball would be to do away with fixed bases and provide breakaway bases that come apart on impact. Stationary bases, in the words of Carolyn White, "give rise to a myriad of ankle and knee injuries." A study conducted by the University of Michigan says breakaway bases, could reduce sliding injuries in junior baseball by 96%.

Another sports area where there is a rising concern over the accident rate is in racket sports such as squash racket ball and badminton. Squash balls and badminton shuttlecocks can hit the eye at speeds of over 160 kph (100 mph), causing severe injury. The Squash Rackets' Association of Great Britain estimates that 3.5 million games of squash are played in Great Britain every year and that there are approximately 5 serious injuries per 100 000 games. Since the wearing of seatbelts in cars became mandatory in Great Britain, sports have overtaken road accidents as the primary cause of serious eye injuries. Medical experts are pressing for rules to make eye protection mandatory in some racket sports. The wild swing of the racket can also be dangerous to the eyes. One squash player was hit by a racket; the blow ruptured supporting structures of her eye lense. Because of a dense hemorrhage, persistent glaucoma, and the development of a cataract the individual lost a major percentage of her vision. In one British study it was reported that 11% of squash injuries to the eye resulted in blindness. To prevent such serious injuries in racket sports, medical experts give the following advice [50]:

> *Players must first toss away old misconceptions. Many people believe that plastic lenses won't break; others mistakenly think that prescription lenses with hardened glass are safe. These beliefs leave a racket sport player with a false sense of security. When a 21-year-old male player wearing hardened lenses was struck by a squash racket, the lense of his glasses shattered upon impact into hundreds of pieces causing subsequent severe bleeding in the eye and a torn retina. Do not use an open eye guard. This type has no protection from a direct hit by a small ball. In one study, five out of 67 squash players and nearly half of the racketball players suffered eye injuries wearing open guards. The*

closed eye guard provides better protection. Prescription and non-prescription lenses should be either CR39 plastic or polycarbonate plastic with a minimum 3mm thickness at the center of the lense. The frame should be strong enough to withstand the force of a ball traveling at 265 kph (140 mph).

Doctors also advise players never to slam a ball in anger or frustration after losing a point because such wild balls have caused some severe and unnecessary injuries. A Canadian expert, Michael Easterbrook, has written two excellent articles on the type of protective equipment available and on the inadequacies of some of the popular open guard equipment [51, 52]. The Easterbrook article also shows some graphic pictures of injuries to the eye and to the skull from both balls and rackets during sports events. Anyone who plays squash or badminton should be given copies of these pictures before they decide whether or not to wear protective equipment.

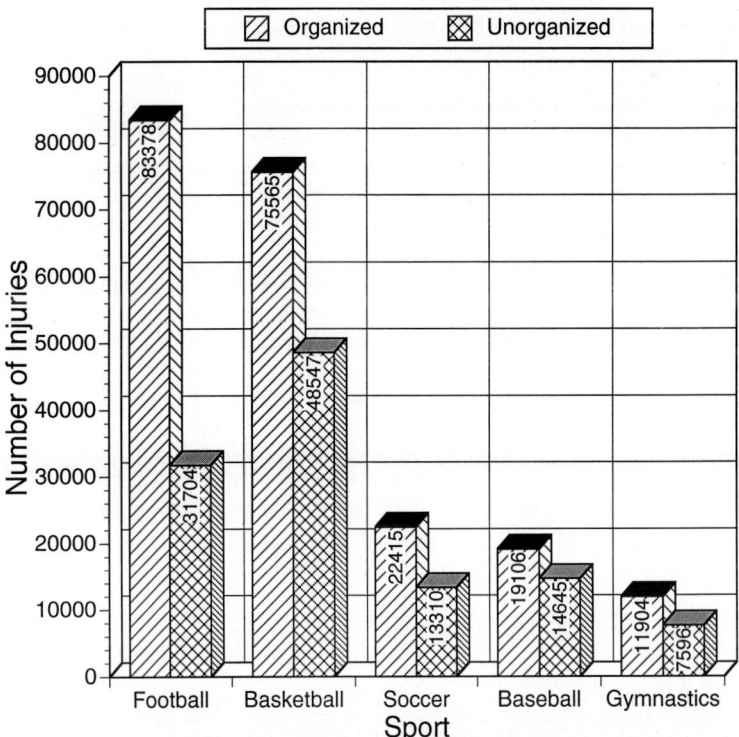

Figure 4.17 Sports injuries are more common than many people appreciate. Shown above are the statistics for sports injuries which occurred in schools for the academic year of 1988/1989.

References

[1] D.H. Fender, *General Physics and Sound To Advanced and Scholarship Level*, published by the English Universities Press, 1957.
[2] J.D. Cutnel and K.W. Johnson, *Physics*, 3rd Edition, John Wiley & Sons Inc., New York, 1995.
[3] R.F. Post, S.F. Post, "Flywheels" *Scientific American*, Vol. 229, No. 6, December 1973.
[4] A.R. Milner, "Flywheels for Energy Storage", *Technology Review*, November 1979, pp. 32-40.
[5] V.H. Booth, *Physical Science: A Study of Matter and Energy*, 2nd Edition, Macmillan, New York, 1967, pp. 25, 26.
[6] J.L. Synge, B.A. Griffith, *Principles of Mechanics*, McGraw-Hill, New York, 1949, pp. 408-411
[7] I. Asimov, *Biographical Encyclopedia of Science and Technology*, Doubleday, Garden City, New York, 1948.
[8] Edith Hamilton, *Details on myths quoted from Mythology*, A Mentor Book published by the New American Library, New York, original edition 1942, many reprints.
[9] See discussion of Reynolds number and drag in B.H. Kaye, *Chaos & Complexity; Discovering the Surprising Patterns of Science and Technology*, VCH, Weinheim, 1993.
[10] B. Rosenberg,."Speed in the Groove" (A Discussion of Riblets), *Technology Review*, November/December 1987.
[11] F. Tarada, "Ribbed Planes Fly With Greater Ease." *New Scientist*, March 10, 1990, p. 38.
[12] J. Mullins, "Gas Skin Makes Torpedo Go Like a Bomb, *New Scientist*, June 3, 1995, p. 20.
[13] J. Beard, "Metal Glove Speeds Super Sonic Planes", *New Scientist*, December 7, 1991.
[14] W. Berger, "The Yoyo: A Toy Flywheel," *American Scientist*, Vol. 72, No. 2, March/April 1984, Volume 72, pp. 137-142.
[15] W.F. Allman, "Physics on a String, A Discussion of Yoyo Dynamics," *Science 84*, Vol. 5, 1984, pp. 92-94.
[16] L. Hogben, *The Vocabulary of Science*, Heineman, London, 1969
[17] R.P. Brennan *Dictionary of Scientific Literacy*, J. Wiley and Sons Inc., New York, 1992.
[18] The version of the Story of David & Goliath that we shall use in this discussion in to be found in the Old Testament of the Bible in the First Book of Samuel starting at Chapter 17.
[19] See entries in Encyclopaedia Britannica on the Sea People (Vol. 21, p. 129) and on the Philistines (Vol. 17, p. 858).
[20] R. Greenblat, "Search The Scriptures – The Physician Examines Medicine in the Bible".
[21] M. Korfmann, "The Sling as a Weapon", Scientific American, Vol. 229, No. 4, October 1973, p. 34-42.
[22] H. Blyth, "The Structure of a Hoplite Sheild in the Museo Gregoriano Etrusco". Bollettino of the Monumenti Musei E Gallerie Pontificie, Tipografia Poliglotta Vaticana, Vol. 3, 1982.
[23] L. White, "Caught by a Catapult", Rubber Developments, Vol. 40, No. 3, 1987, pp. 82-84
[24] S. Boxer (ed.) "One, two, three Strikes You're Dead in this Old Ball Game", *Discover*, Vol. 7, No. 6, June 1986, pp. 6,7.
[25] A. Green, "Skipping Science", *Science 85*, Vol. 6, No. 8, October 1985, pp. 86,87
[26] I. Hutchings, "Bouncing Bombs of the Second World War". *New Scientist*, March 2, 1978, pp. 563-566.
[27] G. Jones, "Sir Barnes Wallis, Inventor without a Monument". *New Scientist*, November 8, 1979, pp. 434-435.
[28] H.C. Hostetter, "Sticky Wickets: Did Cricket Spawn Baseball?" *American Way* (the House Journal of American Airlines), October 1983, pp. 187-191.
[29] C. Frohlich, Resource Letter, PS. 1, "Physics of Sports", *American Journal of Physics*, Volume 54, Number 7, July 1986, pp. 5902-5903.
[30] R.K. Adair, *The Physics of Baseball*, 2nd Edition, Harper Collins, New York, 1994
[31] R.K. Adair, "The Physics of Baseball", *Physics Today*, Vol. 48, No. 5, May 1995, pp. 26-31.
[32] See entry on Cricket in *Funk & Wagnalls Standard Reference Encyclopedia*, Ed. by H.R.S. Phillips, Standard Reference Works Publishing Co. Inc., New York, 1979.

[33] W. Brown and R. Mehta, "The Seamy Side of Swing Bowling," *New Scientist,* August 21, 1993, pp. 21–24.
[34] R. Mehta and D. Wood, "Aerodynamics of the Cricket Ball," *New Scientist,*. August 7, 1980, pp. 442–446.
[35] N.G. Barton, "On the Swing of a Cricket Ball in Flight," *Proceedings of the Royal Society of London,* A379, pp. 109–131 (1982).
[36] P.J. Brancanzio, "The Hardest Blow of All" (A discussion of the difficulties of hitting balls in baseball and cricket). *New Scientist,* December 22/29, 1983, pp. 880–883.
[37] C. Tudge, "Why Cricket is no Longer Cricket," *New Scientist,* May 29, 1986, pp. 58,59.
[38] D. Guttman, "The Physics of Foul Play" (A study of the way in which some players attempt to meddle with both bat and ball in baseball), *Discover,* April 1988, pp. 70–77.
[39] News Story Entitled "A Handful of Dirt is Threatening to Soil the Image of English Cricket." *Time,* August 8, 1994, p. 10.
[40] C. Frohlich, "Aerodynamic Drag Crisis and Its Possible Effect on the Flight of Baseballs." *American Journal of Physics,* 52(4), April 1984, pp. 325–337.
[41] W.F. Allman, "Twisting Slowly in the Wind," (A study of baseball pitching using wind tunnels), *Science,* 1983, June pp. 92,93,l03.
[42] W.F. Allman, "Pitching Rainbows," The Untold Physics of the Curve Ball." *Science,* 1982, pp. 32–39.
[43] J. Kluger, "What's Behind the Homerun Boom?" (This article contains a good discussion on the manufacture and testing of baseballs) *Discover,* April 1988, pp. 78,79.
[44] R.G. Watts, and A.T. Bahill, *Keep Your Eye on the Ball: The Science and Folklore of Baseball*. W.H. Freeman and Company, 1990.
[45] For a vivid account of the type of injuries you can sustain on a bicycle see "Rough Ride," by E. Rosenthal, *Discover,* April 1992, pp. 36,37.
[46] A.T. Bahill, T. LaRitz, "Why Can't Batters Keep Their Eyes On the Ball? *American Scientist,* Volume 72, May/June 1984, pp. 249–253.
[47] A useful source of information on sport injury is the National Centre for Catastrophic Sports Injury Research, Directed by Frederick Muller, University of North Carolina, Chapel Hill, N.C., 27514.
[48] The National Youth Sports Foundation for the Prevention of Athletic Injuries Incorporated, 10 Meredith Circle, Needham, Massachusetts, MA 02192.
[49] Ryan Wojick, private communication.
[50] For a discussion of protective eye care when playing ball sports in Canada see the question and answer session of the journal in *The Canadian Occupational Health and Safety Journal,* Volume 3, Number 4, July/August 1987.
[51] M. Easterbrook, "Ocular Injuries in Racket Sports," *International Ophthalmology Clinics,* Volume 28, Number 3 (1988) pp. 232–237.
[52] M. Easterbrook, "Eye Injuries in Racket Sports", *International Ophthalmology Clinics,* Ocular Sports Injuries, Vol. 21, No. 4 (1981) pp. 87–119.
[53] Readers interested in the dynamics of racketball should consult the article by J. Walker, "Success in racketball is enhanced by knowing the physics of the collision of the ball with the wall", *Scientific American,* September 1984, pp. 215–227.

Chapter 5

Darts, Stone Discs, and Boomerangs

Chapter 5

Darts, Stone Discs, and Boomerangs

Section 5.1   Javelins and Snow Snakes

In this chapter we will explore the design and performance of missiles launched by hand. One example is *quoit throwing*, a game in which heavy flat rings are thrown at, or onto, a vertical pin. A dictionary definition [1] of a *quoit* is

> *a heavy flat ring for throwing as near as possible to a hob or pin, the game played with such rings, origin of the term obscure.*

Quoit throwing was often played with rope rings on the decks of cruise ships and on sandy beaches. It is probable that horseshoe throwing, in which the player tries to lodge the horseshoe onto a pin, developed from the sport of quoit throwing.

The second missile that we will consider is the throwing spear or *javelin*. In the description of Goliath's equipment given in the bible David says to Goliath [2]

> *You are coming against me with sword, spear and javelin.*

The javelin was a short throwing spear thrown at the enemy prior to a close encounter. Blyth tells us that it was designed so that if it missed a soft target and hit a shield, or the ground, it would break preventing the enemy from picking it up and throwing it back [3]. It appears that short throwing spears were standard equipment for soldiers throughout history. The size and shape of the javelins varied from army to army. Thus among the Byzantines in the 6th century AD it was said that the soldiers carried three fighting darts which were approximately 18 inches long, heavily weighted with a barbed head and flighted like an arrow. They were carried clipped behind the soldier's shield. This type of javelin was used for many centuries, particularly in Eastern Europe, in close fighting [2].

In earlier chapters we have talked about the metamorphosis of weapons of war into peaceful tools and instruments. The javelin has metamorphosed into a peaceful pastime in several parts of the world. Thus among the North American Indians, the javelin has become a *snow snake* [4]. In recent years North American Indians have applied to have their ancient sport of snow snaking added to the winter Olympic games. The snakes are actually intricately crafted wooden poles that are thrown for long distances across the snow. Thomas Maracle lives on the Tyendinaga Indian Reserve near Kingston, Ontario in Canada and he describes the sport in the following way.

> *Snow snakes were actually used to end wars. Our people warred against each other for a long time. These (snow snakes) were a gift from the creator to our people to compete by sport rather than war*

To begin the competition a log is dragged through the snow to leave a trough. The snakes are thrown down the trough reaching speeds of up to 240 kph and can travel a distance of up to two-and-a-half kilometers. Maracle explains that the snakes actually float on a cushion of air. Maracle makes about six snow snakes each year and these then need to be seasoned for several years. They are soaked in a solution of animal and other fats for a few months and then dried. Before a competition they are heated on a wood stove and polished with the oils that ooze out during heating. They are then frozen again before being used in competition.

Although at one time Indians competed in snow snake competitions all across North America the sport waned before the second world war. A player only handles the snow snake with deerskin gloves because contact with bare hands can reduce the distance that the it travels by about 15 meters. An Onodaga Indian holds the world record for throwing snow snakes at the New York State Fair with a throw just over two-and-a-half kilometers. Maracle tells us that, to the untrained eye,

> *The snow snakes look like pool cues. The ends are carved with the name and the story of the owner. Some of the snow snakes have ends tooled in silvers and others are zinc tipped.*

In Britain, Javelins have metamorphosed into small darts which are thrown at a target. Groups of people are often found playing the game of darts in a British *pub*. It is said that this type of game using miniature javelins is one of the world's fastest growing traditional sports [2]. Although there are many references to the use of miniature javelins in various field sports, the game of darts as it is now known in Great Britain started to become popular around the beginning of the 1900s. Games of chance were at that time illegal in the British pubs and in 1908 a Magistrate in Yorkshire took the owner of a public house, called Foot Anakin, to court and prosecuted him for allowing darts to be played on his premises. Keenan [2] tells us that

> *To prove that darts was a game of skill, not chance, Anakin brought a dart board into the court room. After putting three darts into the part of the disc labelled 20 he challenged the court to do the same. The court official took up the challenge but only managed to hit the board with one dart. Verdict, the case dismissed.*

Ardent dart players usually carry their own set of darts with them; a typical set of darts is shown in Figure 5.1(a). A modern dartboard is shown in Figure 5.1(b).

It is interesting to note that during World War I, in the early days of aerial warfare, darts were provided for use in aircraft. The intention was that the pilot could throw them at Zeppelins. The dictionary defines a *Zeppelin* as

> *a cigar shaped airship of the type designed by Count Zeppelin around 1900.*

5.1 Javelins and Snow Snakes

a)
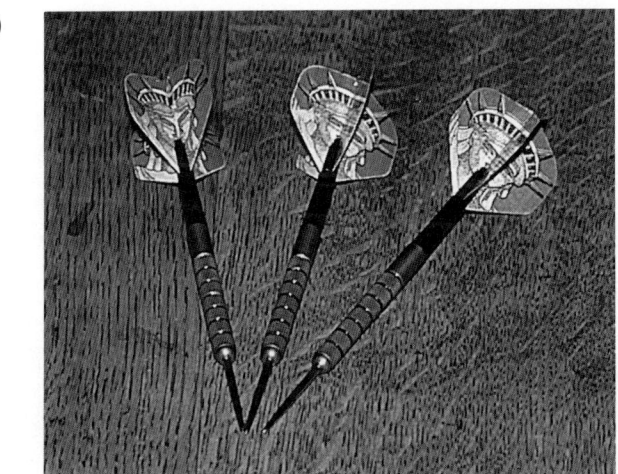

b)

Figure 5.1 Darts is a missile game often played competitively in pubs, as shown on the front piece of this chapter. a) The missiles used in the game of darts have a heavy front portion, to store kinetic energy, with a pointed tip to hold them in place once they hit the board, and flights to stabilize their trajectory. b) The dart board is often made of cork and is divided into sections which score different numbers of points when hit.

Zeppelins were used by the Germans to bomb English cities. I remember my father-in-law describing how the Zeppelins had bombed Nottingham when he was working there as a young man during World War I. The public used to stand in the streets to watch the Zeppelin's slow progress across the sky as the crew threw bombs over the side of the airship. The dart weapon was known as the *Rankin Incendiary dart* after its inventor, Commander Frederick Rankin. Keenan tells us that they were formed from a hollow tin cylinder about the size of a standard candle with a steel needle point at one end and a barbed flight at the other. They were filled with incendiary material. Keenan also tells us that the incendiary darts never proved themselves in combat mainly because it was rather difficult to get close enough to the airship for a direct hit [2].

The simple type of paper airplanes made by bored students during a lecture are really darts although more complicated paper airplanes exploiting aeronautical principles can be made [5].

The javelin is still used in sports and is a recognized Olympic event. A simple javelin is launched with its shaft aligned with the desired trajectory. This is described by javelin throwers as launching the javelin with a zero angle of attack (the angle between the body of the javelin and the trajectory followed by the center of mass). More modern

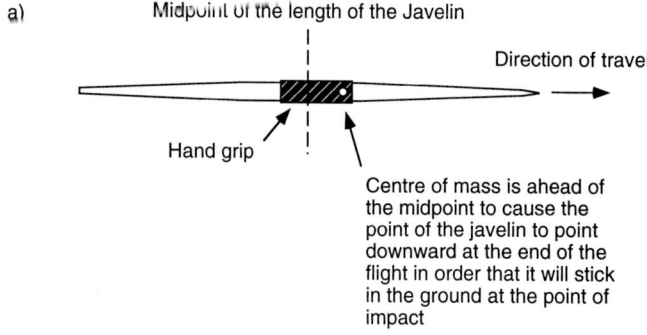

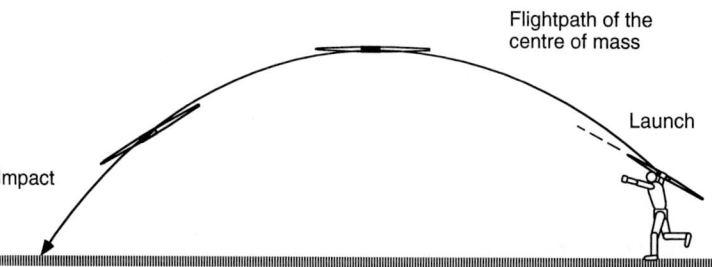

Figure 5.2 While the javelin does not appear to be designed to gain any lift during flight, in fact its shape and balance are crucial to its performance. a) A javelin is designed with its center of mass ahead of its physical center so that it will land point down at the end of its trajectory. b) Typical flight path of a javelin.

aerodynamically designed javelins are often launched at a small angle of attack to the flight path of the center of gravity. A typical javelin is illustrated in Figure 5.2. Brancazio tells us that modern javelins are aerodynamically designed so that there is a slightly higher lift to the rear of the javelin than at the front [6]. This effect makes the javelin tip forward about the center of gravity which helps to improve its trajectory. In effect the modern javelin is designed to behave as an airfoil. In fact the Olympic javelin is one of the few pieces of modern sports equipment which has been made less effective by new construction rules because efficient aerodynamically designed javelins were starting to exceed the distance available in sports stadiums. There were fears that one day a javelin breaking the world record could impale a spectator [7].

To understand what is meant by an airfoil and by aerodynamic design we need to gain an understanding of the phenomenon known as Bernoulli's principle and of the physics of the atmosphere. These are discussed in the next section.

Section 5.2   Bernoulli's Principle, Venturi Throats, and Pitot Tubes

An important phenomenon in physics is known as *"Bernoulli's principle"*[8, 9]. The name honors the Swiss mathematician Daniel *Bernoulli* who lived from 1700 to 1782 [11]. I know from personal experience as a student that I found Bernoulli's principle very difficult to understand. It is another part of physics which is counter intuitive. You must remember that time after time research scientists have been very surprised by their discoveries. Thus Ernest *Rutherford*, the great scientist who explored the structure of the atom, tells us that when he carried out his first experiments to explore the way neutrons bounced off the atom, the results were totally unexpected. He said [10]

> *I was as surprised by the results as if I had seen cannonballs bounce off tissue paper*

We will explain Bernoulli's principle by considering a device known as a *Venturi throat* [8, 9]. Before discussing the functioning of a Venturi throat, it is necessary to learn some atmospheric physics. When the Greeks started to develop their knowledge of the world in which they lived, they started to talk about the spheres of existence. (The Greeks knew that the world was a sphere, a word derived from the Greek word "sphaira" for a ball.) The solid matter of the earth is called the *lithosphere* from the Greek word for stone, "lithos"; the air in which we move and breath is called the *atmosphere* from "atmos", the Greek word for "vapor." The atmosphere extends outwards from the Earth, but its density decreases quickly as we move away from the surface of the Earth. About 75% of the atmosphere lies within seven miles of the Earth's surface. This part of the atmosphere is called the *troposphere* from the Greek word for change, "tropos", the idea being that all the changeable weather takes place in the first seven miles of the atmosphere. Above this troposphere is a layer of still thinner air known as the *stratosphere*.

To discuss the physics behind Bernoulli's principle, we need to have a clear understanding of what we mean by *atmospheric pressure*. To help develop the necessary vocabulary, consider the systems shown in Figure 5.3. In the 1600s, scientists started developing pumps to remove air from vessels to create what we now call *vacuums*. The word vacuum comes from a Latin word meaning "empty." Simple equipment used by scientists studying the atmosphere with their first vacuum pump is illustrated in Figure

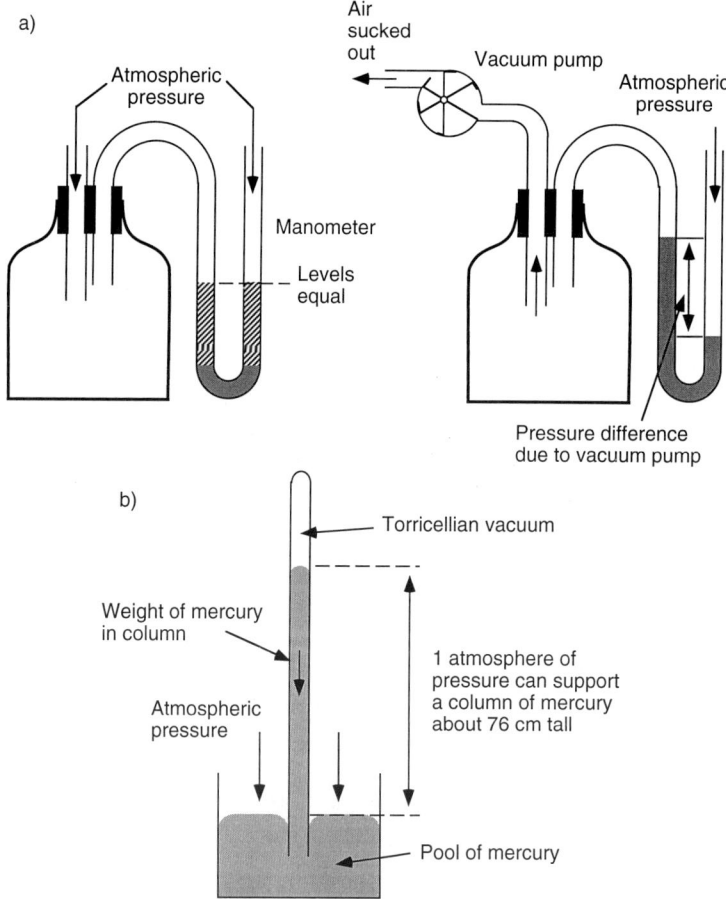

Figure 5.3 The properties of the atmosphere can be explored using vacuum pumps and manometers. a) The level of fluid in both sides of a manometer are equal when connected to a bottle open to the atmosphere. By connecting the bottle to a vacuum pump the pressure inside the bottle can be lowered and the atmospheric pressure on the one side of the manometer forces the fluid in that side down. b) If a closed tube is filled with mercury then inverted in a pool of mercury, the atmospheric pressure on the surface of the pool will be able to support about a 760 mm column of mercury in the tube.

5.3(a). In order to follow the progress of their attempts to empty a bottle of air, scientists developed an instrument known as a *manometer*. The word manometer comes from two Greek root words and means "thinness measurer." Thus, from a knowledge of Greek root words and using logic, one might expect to find a manometer in a diet clinic, whereas we all know that a manometer is actually part of a doctor's standard equipment and is used to measure blood pressure. The manometer was invented by the French scientist *Varignon* (1654–1722). The simple manometer of Figure 5.3(a) is known as a *U-tube manometer*. One branch of the U-tube is inserted into the bottle so that the pressure of the air above that arm of the manometer is that of the air in the bottle. The other arm is open to the atmosphere. The atmospheric air presses down on the surface of the mercury in the manometer. At the start of an experiment, when the air in the bottle is at the same pressure as the atmosphere, the level of the mercury in the U-tube manometer is the same in both arms. If the experimenter connects a pump and attempts to remove the air from the bottle, then as the amount of air in the bottle decreases, its ability to press on the arm of the manometer decreases and the atmospheric pressure outside the bottle pushes the mercury in the tube downwards. The difference in the height of the mercury between the two arms is a measure of the pressure difference between the atmosphere on the outside of the bottle and the remaining air inside the bottle. This experiment, although it appears to be simple, is one that can be dangerous. As the air is pumped out of the bottle, the atmospheric pressure all around the bottle can cause it to *implode*, that is, to collapse inwards (see discussion of implosion and explosion in Chapter 6). When carrying out experiments in which a glass bottle is going to be evacuated, the experimenter should put cloth tape all around the bottle to stop fragments of glass from injuring the observer should the bottle implode during the experiment.

If all of the air could be pumped out of a jar, the height of mercury supported by the atmospheric pressure in the other arm of the manometer would be approximately 76 cm. We say approximately because the atmospheric pressure varies from day to day as the weather patterns move across the surface of the Earth. Also, the atmospheric pressure on the top of a mountain is not the same as that at sea level. When scientists started to develop equipment for measuring the atmospheric pressure, the first unit that they used was named after a co-worker of the famous scientist Galileo, Evangelista *Torricelli* (1608–1647). To test the theory that the atmosphere had weight, as suggested by Galileo and other scientists, Torricelli carried out the following experiment. First, he took a long thin tube sealed at one end and filled it with mercury. He then inverted this tube of mercury under the surface of a pool of mercury as shown in the sketch of Figure 5.3(b). Torricelli found that no matter how long the tube, the mercury always dropped to make a column above the surface of the mercury in the bottom vessel approximately 76 cm high. We now know that the experiment creates a vacuum above the surface of the mercury in the tube with the only gas remaining in the space above the mercury surface being a small amount of mercury vapor. This type of vacuum was originally called a *Torricellian vacuum*, although it really does not differ from any other vacuum except for the fact that it contains a small amount of mercury vapor. When I was a student, Torricelli's experiment was often carried out by students in the physics laboratory. We now know that mercury vapor is extremely poisonous and also that mercury can migrate through the skin if the end of the tube filled with the mercury is held in the fingers. It is not an experiment recommended for today's students. (Note: Scientists think that both Newton and another famous scientist, Faraday, suffered from mercury poisoning late in life because of the experiments they carried out in their private laboratories with open pools of mercury [10].) Of course the same experiment can be carried out with water in the manometer tube, but then we would find that the atmosphere can support approximately 33 feet of water which is a rather long tube for class demonstrations!

Any device used to measure atmospheric pressure is known as a *barometer*. This term comes from the Greek word "baros" meaning "heavy." A barometer measures the heaviness of the air. When pressures were measured with mercury manometers, scientists called a pressure of one millimeter of mercury one *Torr* (short for a Torrecelli; if you want to be famous in science it is best to have a short name!). An absolute vacuum is not possible on the Earth; there is always a very small residual pressure in what scientists call a vacuum. Highly efficient vacuum pumps available today reduce the internal pressure in a chamber to millionths of a Torr and even billionths, e.g., when scientists are trying to simulate outer space. Such space-simulating vacuums are described by scientists as *hard vacuums*. Ordinary vacuums obtained in a typical physics laboratory is sometimes referred to as a *soft vacuum*.

In a new, revised notation, atmospheric pressure is now measured in a unit called the *Pascal*. Blaise Pascal was a French scientist who lived from 1623 to 1662. He was very interested in the atmosphere. In 1646 he asked his strong young brother-in-law to carry a mercury barometer like the one in Figure 5.3(b) up the side of the Pui-De-Dome, the mountain near to which Pascal was born. (Pascal himself was chronically sick, suffering continually from indigestion and headaches.) The brother-in-law climbed about a mile upwards and found the mercury columns dropped three inches because of the drop in the air pressure above him. We are told that Pascal's brother-in-law repeated this experiment five times. We are also told that Pascal repeated Torricelli's original experiment using red wine. Because red wine is even lighter than water, Pascal had to use a 46 foot long tube to balance the weight to the atmosphere.

The Pascal is defined as one Newton per square meter. One Pascal is a very small pressure compared to the pressure of the atmosphere. Therefore, the atmospheric pressure reading given in weather forecasts is measured in *kiloPascals* (1000 Pascals). A typical atmospheric pressure reading would be 100 kiloPascals; a quantity abbreviated to 100 kPa.

Now that we have a working knowledge of the terms used to measure pressure, we can discuss a demonstration of the Venturi throat shown in Figure 5.4(a). This instrument was first investigated by G.B. *Venturi* (1746-1822). In a *Venturi throat*, the fluid moving through a tube is made to flow through a constriction and hence the name throat. Venturi investigated what happened when manometers are placed in the throat and on each side of the throat as shown in the diagram. When students are asked to guess what is going to happen before the experiment is carried out, most of them will suggest that the pressure at the manometer in the middle of the throat will be higher pressure than that at the two outside manometers. In fact, the opposite happens as shown in the sketch. The pressure in the throat drops compared to that on either side. Remember that the pressure being measured is the pressure inside the moving gas and in the sketch of Figure 5.4(a) the flowing fluid is at a positive pressure with respect to the outside atmospheric pressure. One way of understanding what is going on in a Venturi throat is to realize that when the flow rate of the fluid increases, the molecules of

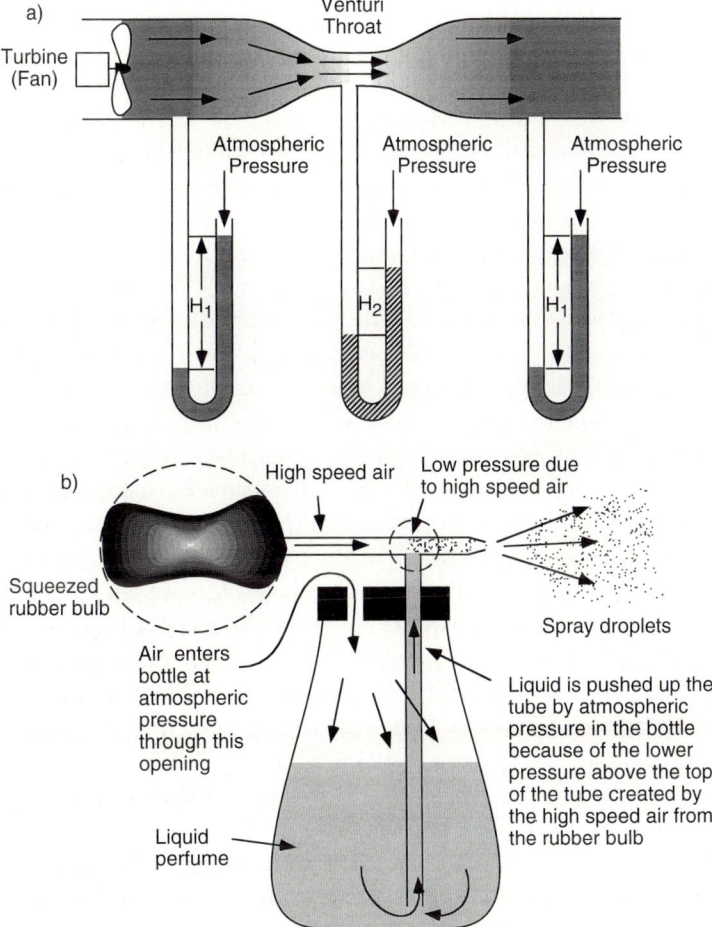

Figure 5.4 Bernoulli's principle can be demonstrated by connecting manometers at various points on a Venturi throat and is manifest in a scent spray bottle. a) The pressure in a Venturi throat is lower than the pressure on either side of the throat because the velocity of the fluid in the throat is higher than in the rest of the system. b) The physics of a scent spray bottle is based on the pressure difference created by a gas moving over the top of the dip tube.

the fluid have a preferred direction of motion and are therefore less likely to collide with the walls of the tube to create the pressure measured by the manometer. The difference between the pressure measured in the throat and the pressure entering and leaving the throat can be used to measure the velocity of flow through the tube.

The behavior of a fluid in a Venturi throat is a physical demonstration of Bernoulli's principle which states,

> *The dynamic pressure inside a moving fluid is less than the pressure in the same fluid in the same condition at rest.*

The pressure in the fluid at rest is known as the *static pressure* and the pressure in the moving fluid is known as the *dynamic pressure*. As the velocity of the flow increases, the dynamic pressure falls. A familiar device which uses Bernoulli's principle is the *scent spray bottle* shown in Figure 5.4(b). In the scent spray bottle, a rubber bulb attached to a tube is squeezed to drive air across a tube that dips into the reservoir of perfume. The dynamic pressure across the top of the tube dipping into the scent, is then lower than atmospheric pressure as predicted by Bernoulli's principle. The top of the surface of the perfume reservoir is open to the atmosphere and is thus at atmospheric pressure. This difference in pressure results in liquid being forced up the narrow tube. At the top of the tube the interaction of the flowing air stream with the rising liquid results in the liquid being dispersed into a spray as illustrated in the diagram. The same principle is used by the gardener who attaches his pesticide bottle to a hose. The rushing of water from the water supply across the top of a tube dipping into a reservoir of pesticide draws the pesticide into the moving water, dilutes it the appropriate amount, and then sprays the mixed liquid out of the nozzle.

A demonstration of Bernoulli's principle which surprises most people is the effect of blowing down a drinking straw onto a small piece of paper. Most people expect the air stream down the straw to make the piece of paper stick to the surface on which it has been placed. In fact because the air in the straw is moving faster than the air under the piece of paper the dynamic pressure above the piece of paper is less than the static pressure underneath the piece of paper. As a consequence the piece of paper is lifted up from the surface toward the end of the drinking straw. This fact is exploited in modern computer factories to provide fingers for robots to pick up delicate computer circuits. Thus the finger of the robot contains a tube down which clean air is blowing. When the robot finger is positioned over the computer chip it is able to pick up the chip because of Bernoulli's differential between the dynamic air pressure above the chip and the static air pressure under the chip. The robot can then swing its fingers through space and, by subsequently reducing the air flow through the finger, drop the computer chip into its required position.

A more mundane application of Bernouilli's principle is observable in the home. You may have noticed that when a shower is turned on with the curtain placed across the shower cubicle the shower curtain is sucked inwards towards the shower of water. This is because the spray of water moving down from the shower head constitutes a moving fluid so that the air in the shower stream is at a lower pressure than the ordinary pressure of the air of the bathroom. The same effect is observed when one places a tarpaulin over a load on a truck. Thus Cutnel and Johnson [9] state that

A tarpaulin is a piece of canvas that is often used to cover a cargo on a truck. When the truck is stationary the tarpaulin lies flat but it bulges outward when the truck is speeding down the highway. This behavior is a manifestation of Bernouilli's effect. When the truck is stationary the air outside the cargo is stationary so that the air pressure is the same in both places. When the truck is moving, the outside air rushes over the top surface of the canvas and this means that the moving air has a lower pressure than the stationary air within the cargo area. The greater pressure inside generates a greater force on the inner surface of the canvas and the tarpaulin bulges outward.

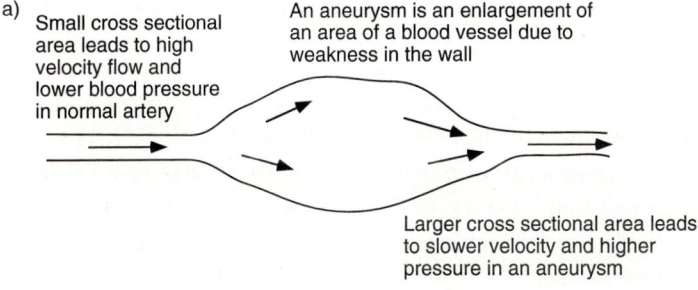

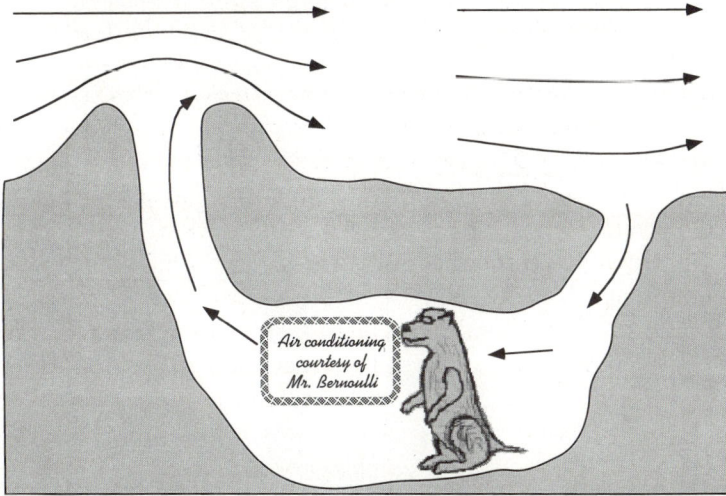

Figure 5.5 Many occurrences and applications of Bernoulli's principle are to be found in nature. a) Weakened blood vessels can result in the occurrence of an aneurysm. b) Prairie dogs make use of Benoulli's principle to ventilate their burrows.

Cutnel and Johnson point out that pressure changes in a moving fluid predicted by Bernouilli's principle increase the danger to a patient suffering from an aneurysm. An *aneurysm* is an abnormal enlargement of a blood vessel. The structure of the aneurysm is the reverse of the Venturi throat. As the blood rushes from the ordinary artery into the enlarged portion of the blood vessel the pressure increases and this causes further hazard to the patient. The situation in the damaged blood vessel is illustrated in the sketch of Figure 5.5(a). Another more pleasant application of Bernouilli's principle is to be found in the way in which prairie dogs ventilate their burrows as shown in Figure 5.5(b). They contour the two ends of their burrow differently to create the dynamic air pressure differential which pulls air through the burrow. Similar devices have been used to ventilate human dwellings [9].

The *Pitot tube* exploits the Venturi throat effect to measure the speed of aircraft moving through the air or of a torpedo moving through the water. The basic design of the Pitot tube is shown in Figure 5.6. It is named after its inventor Henri *Pitot* (1695–1771). Pitot published a scientific paper in 1735 describing his device. It is still a standard instrument for measuring flow rates of fluids as different as blood in the body and air moving past an aircraft. The front of the tube faces into the flow. The fluid moving towards the device is stopped by the tube and is static in the static pressure tube. The sides of the tubes have openings (dynamic pressure ports) that lead to the other arm of the manometer. The dynamic pressure of the air moving past these openings is registered by the other arm of the manometer. Thus the pressure difference is a measure of the speed of the moving fluid. Calibration of the instrument allows the instrument engi-

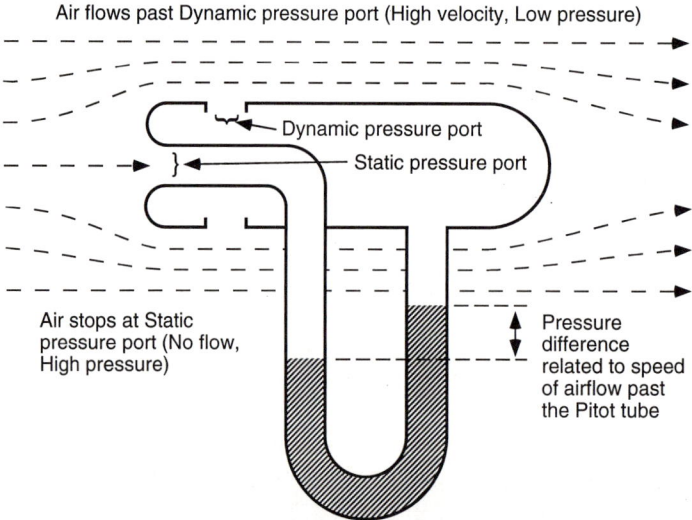

Figure 5.6 A pitot tube exploits Bernoulli's principle to measure the air speed of an aircraft or water speed of a torpedo.

5.2 Bernoulli's Principle, Venturi Throats, and Pitot Tubes

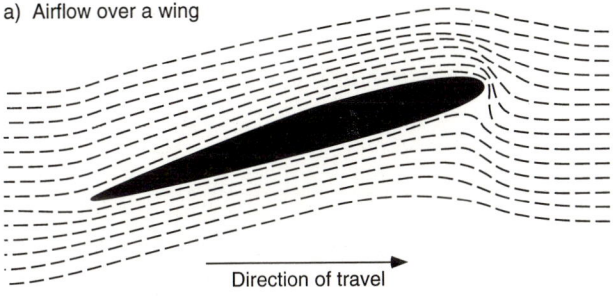

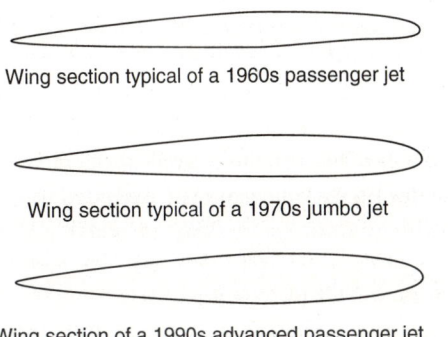

Figure 5.7 An airfoil, such as that used for aircraft wings, concentrates the airflow above the wing resulting in higher flow velocity. The high flow velocity results in lower pressure above the wing than below it and consequently gives lift to the wing. a) Typical normal airflow over a wing. b) Some typical cross sections of wings used for various aircraft over the past decades.

neers to mark off the scale of the manometer directly in velocity of the liquid or gas [8, 9].

The typical cross section of an airplane wing, shown in Figure 5.7, is described as an *airfoil*. As this type of object moves through the air, or liquid in the case of a *hydrofoil* boat, the fluid is deflected further in flowing over the top of the surface as shown in Figure 5.7(a). Thus the velocity of the fluid above the airfoil is higher than underneath, and there is a lift generated by the associated pressure difference. This is the main source of lift in aircraft. In Figure 5.7(b), the airfoil shapes of the wings of three important commercial aircraft are shown. In subsequent sections of this chapter we will discover that many missiles used in sport and warfare obtain lift from Bernoulli's principle because of the way in which they deflect air as they move through it. The performance of many airfoils depends upon the local atmospheric pressure and the performance in sports in which airfoil lift is important, such as the discus throwing, varies according to the local height above sea level due to the drop in air density as elevation increases. Thus, sports results would be expected to vary depending on whether the meet was held at Denver, the mile high city, or in Holland, where many parts of the land are below sea level [6].

Section 5.3   Stone Discs and Flying Dish Pans

One of the oldest sports recorded in history is discus throwing. We know the ancient Greeks used to participate in this sport in the early Olympics. The original *discus* used by the Greeks was a flat stone plate roughly a foot in diameter. Until equipment was standardized, a disc could weigh anywhere from 1.36 to 5.45 kilograms (3-12 pounds). In modern discus throwing, the disc is specified as a wood and metal disc 22 centimeters (8.75 inches) in diameter, about 4.5 cm (1.75 in) thick at the center, weighing 4.4 pounds, and tapered towards the edge to give it an aerodynamic profile. The discus thrower can launch his or her discus from a circle about 2.5 m (8 feet) in diameter. He or she pivots about one foot with one arm fully extended to give the discus maximum distance from the point of rotation at the time of release. A good discus thrower can launch the disc at speeds of up to 83 kph (55 mph). The men's world record for discus throwing currently stands at 71.2 m (233.5 ft) [4].

The way in which the discus is thrown is illustrated in Figure 5.8. The *angle of attack* is the angle between the plane of the discus and the line of its trajectory. The *attitude angle* is the angle between the plane of the discus and the horizontal. An experienced discus thrower takes into account the direction and speed of the wind when launching the discus. The flight of the discus is stabilized by spinning it to create gyroscopic forces. A typical throw will begin with the disc spinning at 400 revolutions per minute. Because of this gyroscopic stabilization of flight, the attitude angle of the discus remains virtually constant throughout its flight. In the absence of wind, sports experts recommend an attitude angle of about 25°. Brancazio points out that during the latter half of the flight, the air will be hitting the discus on its underside producing the greatest amount of vertical lift [6]. This lift is not from Bernoulli's principle, but is the same type of lift that a kite experiences in being pulled into the wind. If the discus is too steeply inclined on its descent, that is, if the angle of attack exceeds 30° it will stall and fall rapidly. Brancazio tells us the interesting fact that a discus will travel a greater distance when thrown into the wind than with the wind. Thus a discus will travel about 7.5 m (25 ft) further when thrown against a 40 kph (25 mph) wind than it would if thrown with this wind. The reason is that the air speed relative to the discus is higher when the discus is thrown into the wind giving more lift from Bernoulli's principle.

In general, the aerodynamic lift and drag on a discus depends on its size, mass, and the density of the air. A discus will travel slightly farther in denser air so that record

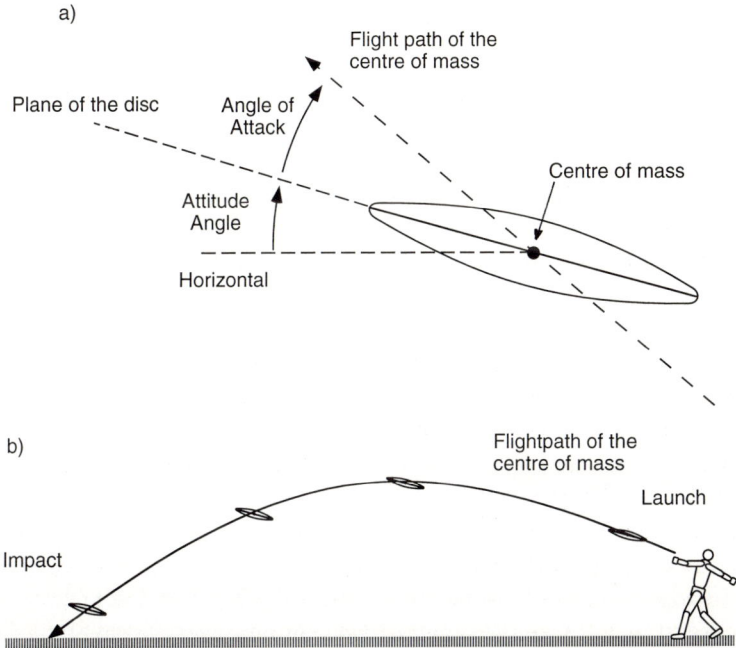

Figure 5.8 In discus throwing the shape of the missile is such that it gains lift from Bernoulli's principle. The discus is gyroscopically stabilized in flight by its spinning action. a) A discus and its principles of flight. b) Typical flightpath of a discus.

discus throws are more likely in Holland than in Mexico or Denver where the atmospheric pressure, and hence the density of the air, is less. Brancazio also points out that the discus used by women throwers is somewhat smaller in diameter and half the weight of a men's discus with the result that the world record in women's discus throwing, 71.6 m (235 ft), exceeds that in the men's event. High technology is moving in the on discus. Bjerlie tells us that beginning in the 1960s manufacturers began to experiment with different materials and weight distributions in the discus. By 1993 designers were using a plastic shell the edges of which were lined with lead weights to increase the moment of inertia of the disc. As a result of these changes in the disc construction, the world records for discus throwing increased by 25% over the period 1961–1986 [7].

Brancazio indicates that discus throwing is not a top participation or spectator sport, but that a closely related activity, the throwing of *Frisbees*, is one of the most popular recreational sports in North America. There is some controversy over the origin of the Frisbee. It appears to be an old sport dating back to the late 1880s when Yale University students invented a game using pie plates, after they had consumed the pies, made by one William Russell *Frisbee*. The plastic Frisbee developed later from the original used by Yale University students.

Actually, any shallow, lightweight disc with an airfoil shape will fly like a Frisbee so it is a sport which can also be played with flying plastic saucers. Spinning saucers is one of the favorite tricks performed by entertainers. This trick looks very clever to the audience but is a straightforward application of the gyroscopic stabilization of a missile. In this trick, the performer balances a series of spinning plates on top of thin sticks. By spinning the plates as they are balanced, the performer creates gyroscopes. A good performer can keep up to twenty plates spinning simultaneously.

William Russell Frisbee opened his pie factory close to Yale University in 1871. Frederick Morisson, an enthusiastic pie plate thrower before the war, saw the potential of making plastic flying missiles similar to the original pie plate and made the first plastic flying saucer in 1948 from a plastic known as butylstearate. The plastic flying saucer, patented by the Wham-O Company, was first marketed in 1957 [11–13]. In a review of the history of the Frisbee, Schuurmans describes how over the years different types of flying plastic discs evolved with various features to enhance their flying properties. However, all of the different versions concentrated as much of the mass as feasible at the rim of the disc to maximize the gyroscopic stabilization obtained by spinning the disc. After the disc was taken over by the toy company, some ridges were added to the top surface by Ed *Headrick*. In adding the ridges it was his intention to copy the vortex spoilers added to aircraft wings to stop the turbulent flow over the top of the disc. However, some Frisbee players maintain that the *lines of Headrick* do not improve the dynamics of the Frisbee [11]. In Figure 5.9 there are sketches showing the various physical aspects of the flight of the Frisbee. Anyone who has played with a Frisbee knows that it has a tendency to tip and swerve to the side when the player spins the disc into the wind. This is because the relative velocity of the air over the side of the disc spinning into the wind is higher than that at the other side of the disc whose motion is away from the wind. This means that the lift on the two sides of the disc is unequal. This unequal lift tries to tip the disc over but because it is already spinning it behaves like a gyroscope pushed to one side and starts to precess.

By the early 1980s heavier Frisbees became available for long-distance throwing; the heaviest disc is usually about 200 grams and the distance record of 190 meters was achieved with a Frisbee weighing 175 grams. Other Frisbee players compete in discipline known as *self-caught flight* in which the disc should take as long as it can to return to the thrower. In July 1990, the maximum time aloft for this type of throwing was 16.72 seconds. In designing a disc it is important to remember that giving the disc a high spin may make it stable, but that rapidly spinning discs can be difficult to catch. In 1972 the first match between two teams of players was played in New Jersey between the universities of Princeton and Rutgers. In this game, players of the one team throw the disc to one another, trying to keep possession as they move towards the opposition's end of the field. A player scores points for a team by catching the disc in the opposition's end zone. Tackling other players is not allowed in the game.

In the late 1970s Allan Adler, a lecturer in engineering at Stanford University in California, and a disc sport enthusiast, began to explore the possibilities of throwing rings

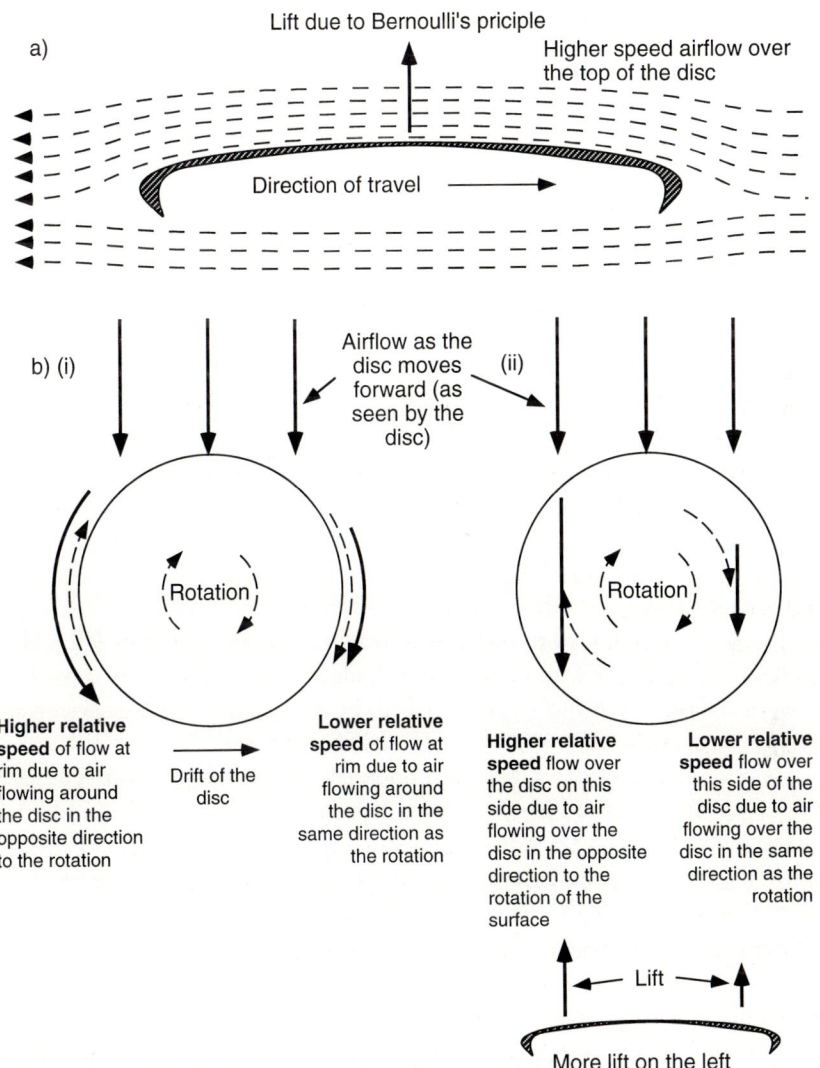

Figure 5.9 The flight of a Frisbee involves the interaction of the gyroscopic action of the spinning disc and the airflow around the disc. a) As with other airfoils, higher velocity airflow over the top of the disc provides lift. b) (i) The differential airflow around the edge of the disc causes it to drift preferentially in one direction. (ii) Differential airflow above the disc causes it to tip to the side.

without a center. After ten years of experimentation, he had designed a throwing ring which he called the *Aerobie Ring* [11]. With the use of a computer to simulate the flight dynamics, Adler designed a flying ring that remains perfectly balanced at all speeds. The outer edge of the disc has a short inclined rim that extends above and below the ring's main body. The main part of the ring is made from polycarbonate to give it a tough flexible backbone whereas the edges are made from soft rubber to make the ring comfortable to catch. Scott Zimmerman, a former world champion in disc sports, threw the Aerobie Ring a distance of 382 m (1256 feet) in 1987. This is twice as far as has ever been achieved with a Frisbee [14].

Section 5.4 Killing Sticks and Boomerangs

In 1957 I visited the Atomic Weapons Testing Range in Maralinga, Australia. (The name is very appropriate since *Maralinga* in Australian Aborigine means "the voice of thunder.") One of the objects that I acquired while in that part of Australia, known as the Nularbor Plain, was a wooden missile which could be a killing stick or a boomerang. I know it was made by the Aborigines of Southern Australia. In fact, I visited the Aborigine location where such sticks were made and I watched an Aborigine shaping the mulga wood piece by hand using a sharpened flint stone. I am not sure if it is a *killing stick* or a *boomerang* because I have never thrown it, being anxious to preserve its elegant appearance. The difference between a killing stick and a boomerang is that the killing stick was intended to be thrown in a horizontal manner to kill a bird or a small animal. The boomerang, on the other hand, was intended to be thrown over a distant area to disturb and flush out animals to be hunted with killing sticks and is intended to return to the thrower. Although the Western world associates boomerangs with Australia, archaeologists have found boomerangs and killing stick type weapons as far apart as Arizona, the New Hebrides, Northern Europe, Egypt, and India. It is difficult to decide whether the different sticks discovered are killing sticks or boomerangs since many of them have been distorted with age and often they are very fragile and cannot be tested in flight. Killing sticks are described as *Kylies* by the Australian Aborigines and scholars believe that only a few tribes of Australia actually had returning killing sticks as part of their weaponry [15].

To date, the oldest killing stick discovered by anthropologists was discovered in Poland in 1985 [16]. Amongst a collection of prehistoric human artifacts an ivory mammoth tusk carved into the shape of a boomerang was discovered. Carbon dating established that the object was 20 thousand years old, thus originating from the last ice age. The boomerang-like object was 61 cm (2 ft) long and weighed nearly two pounds. Recent tests with a plastic replica show that the object was in fact a killing stick. The experiments showed that the object traveled an average of 27.4 m (90 ft) when thrown with the wind but, when thrown into the wind, it traveled 37.4 m (123 ft) due to Bernoulli's lift. Pawel Vald'Nownuak of the Institute of Archeology and Ethnology in Krakow, Poland (*ethnology* is the study of the races of man from the Greek word "ethnos" meaning "a nation") thinks that the killing stick was probably used to kill reindeer.

Boomerangs were even found in the tomb of Tutankhamen (Pharaoh of Egypt 1361–1352 BC). These had gold caps on each end to increase the stabilization effect of spinning the boomerang when it was thrown, i.e., to increase the moment of inertia of the missile [15, 17].

Killing sticks have a roughly *lenticular* cross section, which means they look like a split, dried pea, curved evenly on the top and flattened on the bottom. A typical killing stick is about a meter long. It is thrown horizontally. As it swerves through the air, the Bernoulli's lift of the lenticular cross section of the stick is sufficient to counteract gravity and it appears to fly horizontally without dropping. It is thus easier to aim than a spear and, in the article "Why boomerangs boomerang" [15], we are told that a killing stick can be accurate up to about two hundred meters which is about twice as far as a man can throw a stone or spear. It is suggested that the boomerang may have been discovered accidentally by someone who was heating and twisting a killing stick over a fire and in the process twisted it more than was originally intended.

The boomerang is basically an L shaped flying wing which is thrown with a twist. Modern boomerang blades have a cross section similar to that of an aircraft wing, with a leading edge and a trailing edge, a convex upper face and a flat or slightly concave lower face [17-19]. There are left- and right-handed boomerangs depending upon the twist given to the boomerang when it is thrown. A left-handed boomerang spins and travels clockwise. People new to boomerang throwing try to throw their boomerangs horizontally whereas the boomerang must be launched with a tilt angled to the vertical of between 0 and 30 degrees (see Figure 5.10(b)). Jacque Thomas [15] describes the following technique for launching a boomerang.

> *The thrower's arm works like a catapult. At first it is bent backwards over the shoulder then it is stretched rapidly forward until it makes an angle of ten to twenty degrees with the horizontal plane.*

This catapulting gives the boomerang the energy necessary to achieve its flight. Film studies have shown that a boomerang weighing 100 grams or more can be launched with a speed of 100 kph. At the moment of release, the boomerang snaps over the forefinger to accelerate its spin up to ten revolutions per second. It is essential that the upper convex face of the blade faces the inside of the curved path. Thus the upper face is turned to the left of the thrower for a right-handed boomerang. Even the lightest breeze effects the boomerang's flight and air movements must always be taken into account. In general, right-handed boomerangs must always be launched to the right of the direction the wind is blowing from, at an angle of about 45°; this is known as the *wind angle*. In a book on aerodynamics [20] John Allen has a very good description of the aerodynamics of a boomerang:

> *A boomerang leaves the thrower's hand at a speed of about 27 meters per second (60 mph) rotating at about 10 times a second. It should initially be held nearly vertical with its flat surface away from the right-handed thrower (and vise versa for a left-handed thrower). The curved surface should point towards the left of a right-handed thrower. Thus the spin is initially about a horizontal axis. The advancing blade meets*

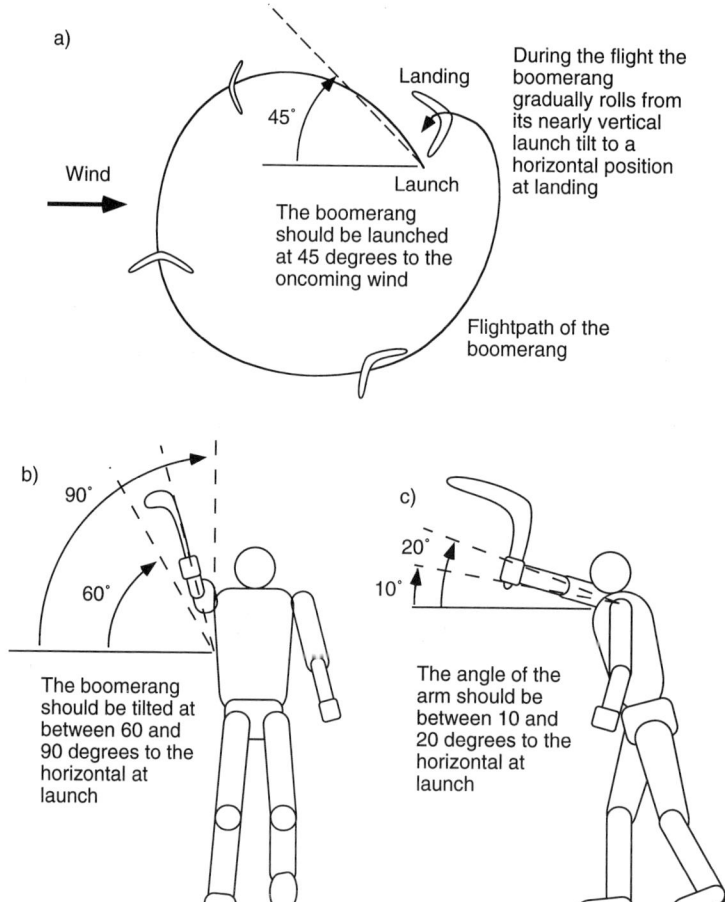

Figure 5.10 The trajectory of a boomerang can surprise the uninitiated observer. a) The boomerang changes its orientation in flight and returns to the thrower, landing horizontally. b) Launching a boomerang involves releasing it in a nearly vertical orientation and at a small angle to the horizontal.

the air with its rounded leading edge and as in an aircraft wing it generates lift. As it retreats backwards from the general speed in the line of flight its lift reduces and may even disappear. The other blade advances in a similar way. Hence the lift is not distributed evenly over the circular area swept by the rotating blade. This unbalanced lift creates a torque which acting on the rotation of the boomerang causes it to precess as a gyroscope does. Thus after a period of direct flight away from the thrower it tilts over towards the horizontal and begins a circling arch. Eventually it lies horizontal spinning about a vertical axis. This continues until the forward velocity is lost and it descends nearly vertically at the feet of the thrower.

However, Allen goes on to point out that

Most amateur flights are miserable failures.

The record boomerang flight stands at 99 meters (108 yards) out and returned.

John Bell tells us that, in Australia, the design of a boomerang is often the secret of the individual player. The boisterous nature of Australian boomerang throwing is shown by the rules and curses apparently prevalent in Australia. Thus in the "Mudgeeraba Creek Emu Racing and Boomerang Throwing Association," one rule states "the decisions of the judges are final unless shouted down by a really overwhelming majority of the crowd present. Abusive and obscene language may not be used by contestants when addressing members of the judging panel, or conversely, by members of the judging panel when addressing contestants (unless struck by a boomerang)."

One Aborigine curse from the Nularbor of South Australia is "may the fleas from 100 wombats live forever in the armpits of any dingo (a name for an Australian wild dog or a rotten human being) who copies the boomerang of another bloke." (Wombats are small bear-like marsupials.) For an entertaining and detailed discussion of the dynamics of the boomerang see Jacques Thomas [15]. The actual dynamics of a boomerang are a combination of the lift from Bernoulli's principle and the differential force created by the stabilizing gyroscopic spin given to the boomerang when it is launched.

Section 5.5 Flying Toys of Tomorrow

In an article in the New York times June 20, 1995 [21], Warren E. Leary reviews the public's fascination with aerodynamic toys such as boomerangs and Frisbees. He reports on the efforts being made to devise new and better toys which fly through the air. Leary states [21]

Whether soaring or diving, dashing or fluttering, zipping away or coming back millions of throwing toys in diverse shape, size and colors go aloft each day as testimony to human fascination of flight and competition.

Leary states that some of the newer devices push the limits of toy aeronautics and puzzle even the experts as to how or why they fly.

Increasingly flight toys depend upon precision design and manufacturing tolerances within a few thousands of an inch to seemingly defy physics and fly better than their predecessors.

In his review Leary describes how Adler the inventor of the Aerobie continues to develop his toys (Mr. Adler is the owner of Super Flight Incorporated, a maker of flying toys in Palo Alto, California). In 1994 Adler introduced the Aerobie Superdisc. Another toy, the *Whoosh Flying Ring*, is produced by Aoddzon Products Incorporated of California. The rings have an outer ring of polycarbonate plastic covered by a circle of spandex fabric held taut by an inner rubber band. The design allows the ring enough lift to fly but the porous fabric also lets enough air through to reduce the unwanted side lift caused by gyroscopic precession. Silverglate the maker of the Woosh Flying Ring claims that the leaking of air through the holes in the fabric also keeps it from rolling so much. Leary [21] describes the *Toobee Flying Can* (sold through mail order by Toobee International, Milwaukee) as

> *an aluminium device which looks like a cut-off beer can with an open rounded leading edge which can be tossed 86 m (283 ft). The Toobee apparently exploits the Magnus effect to gain its lift when spinning along its trajectory.*

Another new toy called the *X-Zylo* is a hollow plastic cylinder weighing less than an ounce and measuring 9.5 cm (3.75 in) in diameter and two and an eighth inches long has been thrown 200 m (655 ft). The device has a weighty metal plastic ring at its front end that makes its leading edge thicker. It appears to fly due to a combination of aerodynamic and gyroscopic forces. Eric Darnell of Norwick, Vermont has developed a three bladed *tri fly boomerang*. In this device the paddles are narrower towards the hub and wider and flatter at the ends. A hole is placed in the tip of each blade to create turbulence at the tips to slow the new type of boomerang down at the end of the flight.

Over the next few years we can expect many developments in aerodynamic toys for recreation to increase the range of missiles available on the sports field [21].

References

[1] *Chambers Etymological English Dictionary*, W. & R. Chambers Ltd., 11 Thistle Street, Edinburgh, Scotland, 1967 Edition.
[2] G.P. Keenan, "The History of Darts", *British Heritage*, June/July 1983, pp. 10–15.
[3] P.H. Blythe, "The Technology of Ancient Warfare", *Science Spectrum* No. 1–7, 1975, Issue 2.
[4] "Ancient Indian Sports Touted to Join Winter Olympic Lists," *Sudbury Star*, August 27th, 1985 (Canadian Press Release).
[5] K. Blackburn, J. Lammers, *The World Record Paper Airplane Book*, Workman, New York, 1994.
[6] P.J. Brancazio, *Sport-Science*, Simon & Schuster, New York, 1984.
[7] D. Bjerklie, "High Technology Olympians," *Technology Review*, January 1993, pp. 23–30.
[8] D.H. Fender, *General Physics and Sound To Advanced and Scholarship Level*, English Universities Press, 1957.
[9] J.D. Cutnel, K.W. Johnson, *Physics*, 3rd Edition, John Wiley & Sons Inc., New York, 1995.
[10] I. Asimov, *Biographical Encyclopedia of Science and Technology*, Doubleday, Garden City, New York, 1948.
[11] M. Schuurmans, "Flight of the Frisbee", *New Scientist*, July 28, 1990, pp. 37–40.
[12] M. Gold, "The Fairytail Physics of Frisbees." *Science*, June 1982, pp. 76–78.
[13] Newstory (Reuter) This year marks 30th birthday of the invention of the Frisbee. *Sudbury Star*, Wednesday, August 5, 1987.

[14] R. Mestel, "Flying Saucer That Anyone Can Navigate." (A Brief Review of the Improved Aerodynamics of the Aerobie Superdisc, *New Scientist*, July 1, 1995, p. 8.
[15] J. Thomas, "Why Boomerangs Boomerang and Killing Sticks Don't." *New Scientist*, September 22, 1983, pp. 838-843.
[16] "The Killing Stick", *Discover*, June 1995, p. 28.
[17] F. Hess. "The Aerodynamics of Boomerangs," *Scientific American*, November 1968, pp. 124-136.
[18] J. Walker, "More on Boomerangs Including their Connection with the Dimpled Golf Ball." The Amateur Scientist section of *Scientific American*, April 1979, pp. 180-189.
[19] D. Robson, "Many Happy Returns: A Discussion of the Dynamics of Boomerangs" *Science*, March, 1983, pp. 100,101.
[20] J.E. Allen, *Aerodynamics; the Science of Air in Motion*, Second Edition, Granada Publishing Limited – Technical Books Division, Saint Albans, Herts, England, 1982.
[21] W.E. Leary, "Lift, Drag, Spin and Talk: Sending Toys Aloft," Science Times, *The New York Times*, Tuesday June 20, 1995, pp. B5, B10.

Chapter 6

Pea Shooters, Rockets, and Rifles

Chapter 6

Pea Shooters, Rockets, and Rifles

Section 6.1 Peashooters and Blowpipes

Before the days of sophisticated electronic toys many a small boy reveled in his ability to annoy unsuspecting passers-by by blowing small missiles through a tube. Since a favorite ammunition for such a device was small dried peas, the instrument was known as a *peashooter*. To be effective, a peashooter must have a relatively long tube of diameter only just greater than the diameter of the missile. The missile (a dried pea, or small, round, ink-saturated pellet of paper) is accelerated along the tube by air pressure created by blowing into the mouthpiece. A plastic drinking straw can function as a peashooter and as a suction device! The air blown into the tube is known as the *propellant*. If the gap between the pea and the tube is too large, the air rushes past the pea instead of accelerating it along the tube.

When Europeans went to South America, they encountered warriors who used rather long peashooters, described technically as *blowpipes*, to fire darts at their opponents. The darts were tipped with a potent poison which paralyzed their victims. This poison is known as curare. A dictionary definition [1] of this substance is

> *a blackish resinous substance named from a Tupi word "urari" which means "he to whom it comes, falls."*

The Tupi are a South American Indian tribe. Curare is sometimes used in modern medicine to aid muscle relaxation.

Missile launching took on a whole new significance when gunpowder was invented. The use of gunpowder as a propellant to fire missiles at opposing army and defensive positions was developed by the Chinese some time before it became available to western military experts. Edward DeBono the noted lecturer on innovation and creativity [2], makes the following statement:

> *Gunpowder ended the permanent mediaeval world in a way that no other invention — except perhaps printing — could do. Gunpowder killed both knight and serf and in doing so proved them equal.*

We will now study the important role of explosives in the development of modern missile technology [3, 4].

When presented with a chemistry set, many aspiring chemists start their experimental careers with a bang by trying to make gunpowder. A typical recipe for gunpowder is as follows:

Chemical Substance	**Chemical Formula**
Potassium Nitrate 74%	KNO_3
Charcoal 15.6%	C
Sulfur 10.4%	S

This type of gunpowder is known as *black powder*. The old name for potassium nitrate was *saltpeter*, a name made from the Latin rootwords "sal" meaning salt and "petra", rock. A salary was originally an allowance given to Roman soldiers to enable them to buy salt. (If this seems a silly name for a monetary allowance, remember that pin money, originally for buying pins, became the term in Victorian times for the wife's money to spend on her personal projects.) When gunpowder was introduced into the western world from China, the lack of a good source of pure saltpeter was a major problem and at first prevented the widespread use of gunpowder. Eventually, chemists found that they could make the substance by fermenting urine in large pits. After the fermentation process, the saltpeter could be scraped off the walls of the pit [5].

In looking at the recipe for gunpowder, it could be asked how anyone thought up such a strange mixture of ingredients. We know that long before the discovery of gunpowder, the Chinese had learned how to fumigate ships with sulfur fumes to rid them of rats and other vermin. They also fumigated libraries with sulfur fumes to destroy insects that attacked the books. *Charcoal* (which is mainly carbon) and sulfur were both known to have medicinal and/or cleansing properties. Sulfur was used to treat skin diseases (and it still is). Saltpeter was used to treat fevers and stomach ailments as well as to break up internal accumulations of blood. As recently as the 17th century, gunpowder was still classified as a medicine because of its use in treating ringworm sores, insect bites, and eczema [1, 4]. Given that the Chinese knew the medicinal/cleansing qualities of these three ingredients, is it surprising that eventually they found that a mixture of these could be ignited?

The Chinese name for gunpowder indicates the Chinese knew about the double nature of this mixture. Chinese writing differs fundamentally from that of western cultures. It is not a written representation of sound but, rather, it is a collection of patterns of lines which are recognized as agreed-upon symbols representing objects and ideas. Because Chinese writing is based upon pictures, it is easier for Chinese scholars to read ancient documents than it is for Westerners to read the documents of their ancestors. For example, English people have considerable difficulty with Chaucer's English, which is only from the 14th century, whereas Chinese scholars can read much older documents. It is interesting to note that Westerners who suffer a stroke in the speech area of the brain lose the ability to speak and read whereas Chinese who suffer the same type

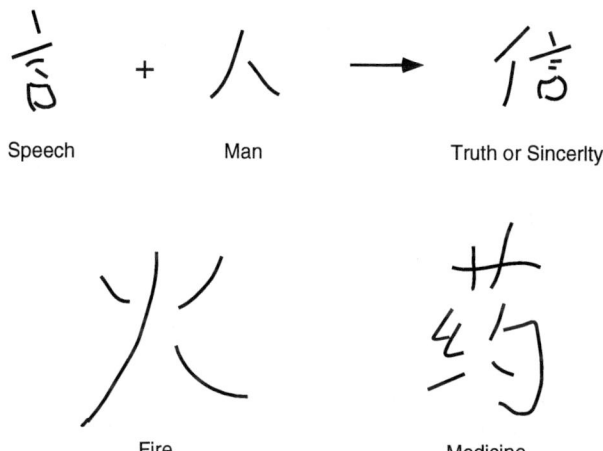

Figure 6.1 The Chinese use pictographic symbols to suggest words and ideas.

of speech damage can still read and communicate by writing even though they lose the power of speech [3]. Chinese who do not understand the spoken words of the different dialects can still write messages to each other.

To understand the way in which Chinese writing has developed, consider the sketches of Figure 6.1. The two lines representing a man are obviously a greatly simplified sketch of a human figure; a kind of "stickman". This type of simple pattern is a *pictograph*, literally a picture drawing. To depict something like speech, the pictograph for mouth is combined with some lines to show the idea that speech is what comes out of the mouth. (Note: This is not unlike the balloons drawn from the mouths of characters in western cartoons.) The Chinese scribes joined together the pictographs for man and speech to form the *ideogram* (drawing of an idea) shown in Figure 6.1 which depicts the idea that true sincerity and honesty can be represented by a sketch of a person willing to stand by what comes out of his mouth. In classical Chinese writing, the symbol for fire is essentially a pictograph. It is a sketch of a fire made out of sticks with flames leaping into the air. The Chinese combined the pictograph for fire with an ideogram for medicine to describe a mixture of ingredients that would become gunpowder. This graphic representation, which translates into "fire medicine" in English, indicates that the Chinese knew the mixture could burn as well as heal [6].

There is reason to believe that the Chinese first used early forms of gunpowder as poison-gas smoke bombs. Thus pictures dating back over 1000 years show smoke bombs or grenades giving off poisonous fumes. A major difficulty with any type of chemical or bacteriological warfare is that the people launching the grenades were just as likely to receive the noxious fumes as the people they were attacking. In Chinese manuscripts the soldiers launching the poison gas grenades were told to suck black feathers and licorice to combat any poisonous fumes blowing back onto them. We know that this type of chemical warfare was developed in large part to neutralize the effectiveness of defensive armor worn by attacking soldiers. The difficulty of making gunpowder and controlling its explosiveness probably accounts for the fact that it was not initially used as a propellant to deliver missiles but rather as an ingredient of a bomb or a grenade. A grenade is defined as a small hand-held bomb which can be thrown at the enemy [7].

As the technology of explosives and the use of gunpowder in weaponry developed, a whole new vocabulary became necessary. The histories of several of the words used in early warfare are shown in Figure 6.2. The path leading to the development of a technical term can be long and tortuous as illustrated by the history of the words fusilier and fuse. To track down the meaning of fuse we must start with the Latin word "lens" for "a small seed". (See discussion of the lenticular shape of killing sticks in Chapter 5.) The first lenses used to create optically magnified images of an object were probably small beads of glass made accidentally by a glass blower. The word "lens" (genitive "lentis") lingers in modern English in the word *lentil*. Early investigators of the properties of "small glass beads that look like seeds" found they could modify the image made by the beads by altering the curvature of the surface of the glass. DeBono suggests that *spectacles* were in regular use by the end of the 13th century [2]. By this time, the

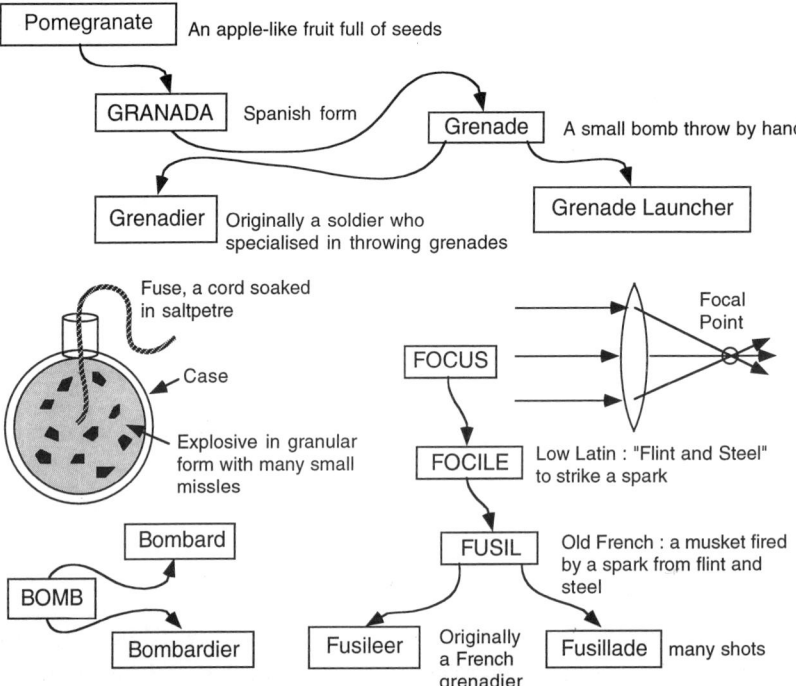

Figure 6.2 As gunpowder technology advanced, the military had to develop a whole new vocabulary to describe the processes they used.

glassmakers had found out that the older people became, the more curved their lenses needed to be in order for them to see clearly. People experimenting with lenses found that, if they pointed the glass bead type of lens shown in Figure 6.2 at the sun, there was a point behind the lens at which all of the energy of the sun became concentrated. Because the sun's energy could start a fire at such a point this became known as the *focal point.*, from the Latin word for the fireplace, "focus". The focal point was also the place where the sharpest image of a distant object was formed by the lens.

When the central authority of the Roman Empire started to crumble, the dialects and slang expressions of various regions became the normal patterns of speech of ordinary people. Some scholars refer to the intermediate forms of Latin spoken in such regions, before they developed into languages such as French and Spanish, as *low Latin*. Other scholars call it *vulgar Latin* from the fact that vulgar originally meant "of the people." In low Latin, the word "focus" became "focile" and this word was used to describe equipment used to make a fire by striking a piece of steel with a piece of flint to create a spark. This then became the name of an early form of gun which was activated by a spark from a flint and steel. The name of the gun then became the name of the soldier who fired the gun — *fusileer.*

The first bombs using gunpowder were spherical containers filled with gunpowder and small missiles, such as fragments of glass or small lead balls, and equipped with a firing device which came to be known as a *fuse*. This was usually a piece of cord soaked in saltpeter to keep it burning slowly. The term fuse came from the fact that it was a slow fire lit with a flint and steel. One of the important design features of such a bomb was a fuse of such a length that the bomb could be thrown a distance. The fuse length was designed so that the grenade would explode immediately on landing. If the fuse was too long, the enemy could always pick up the bomb and throw it back; this technique of inadvertently supplying the enemy with ammunition was to be avoided!

Scholars who study the origins of words describe a word such as *bomb* as being an *echoic* or *onomatopeic* word. *Onomatopeia* is defined as the creation of the name of a process or an object from the sound created by the activity or object. The word comes from the Greek "onoma", "a name" and "poieein", "to make." The Greek word "echoic" means "sound" and hence was the alternative word for onomatopeia [7, 8]. In this book we will use the word echoic for the process of word making that generates words such as bomb. The term bomb is obviously a verbal imitation of the sound made by the bomb. Other echoic words are *clap, splash,* and *crack*. Sometimes an original echoic word loses its obvious derivation in the same way that an optical lens of today bears very little resemblance to the pea from which it took its name. The word *zoom lens* comes from the early days of aerial warfare. A fighter or bomber pilot who wanted to take a closer look at the target he was hoping to hit would go into a nose dive to swoop near the ground. To achieve such a maneuver, a pilot would rev up the engine to make sure that after he had come close to his target he would be able to recover and fly upwards. As a result of this practice, people started to talk about "zooming in" on a target from the increasing noise of the revved up engine of the diving aircraft. When optical specialists developed a special lens to enable the photographer to take a close look at an object without moving the camera, this special type of lens became known as a zoom lens even though it did not make a noise as it "zoomed in".

As shown by the word webs of Figure 6.2, the soldier who became a specialist in throwing small bombs became known in some countries as a *bombardier;* he would bombard the enemy with his bombs. To trace the origin of the word grenade; it is necessary to take a look at the history of the use of gunpowder. When soldiers first started to use gunpowder, they soon found out that if they made a mixture of the powder and transported it any distance they would often have a very inefficient explosive because the ingredients of the powder would segregate during the vibration of the travel. They also found that if they attempted to overcome this problem by creating an intimate mix of the ingredients, the mixture would often explode in the manufacturing process. Thus, the first written account in ancient Chinese with regard to the manufacture of gunpowder (written around the 9th century) warns workers:

> *Those who mix the ingredients of gunpowder — especially with the addition of arsenic — have had the mixture catch fire, singe their beards and burn down the buildings in which they were working [4].*

Eventually the manufacturers of gunpowder minimized both the segregation of the ingredients and the explosion hazard by mixing the ingredients as a paste which was then made into granules by pushing the paste through a sieve. These granules looked like grains of wheat and, when dry, could be handled relatively simply and safely. The first use of such grains of explosive in warfare was to make what the Chinese called a "fiery pomegranate, shot from a bow." The basic aim of this weapon was to set the enemies' defenses on fire. The word *pomegranate* comes from two Latin rootwords which together mean "an apple full of seeds." This word moved into the Spanish language as the word *granada* which gave us the word for *grenade.* Thus, a grenadier was originally a soldier who threw hand grenades full of seeds of gunpowder (Figure 6.2).

The Chinese actually developed rockets propelled by gunpowder before they developed guns using missiles driven by exploding gunpowder. No one is quite sure how rockets came to be developed, but some scholars have suggested that the evolution of the rocket came through the use of gunpowder in mining and blasting. We do know that the Chinese used to use cylinders of paper full of gunpowder grains to blast rocks in mines and in the construction of roads and canals [5]. Perhaps one day, one of these cylinders of gunpowder became ignited accidentally and it was observed that the tube of gunpowder moved away rapidly as it was propelled by the gases of the burning gunpowder escaping from the end of the cylinder. The Chinese pictograph for an early type of rocket can be translated literally as "flying crow with magic fire." Some scholars think that this type of weapon developed out of the use of birds trained to set fire to the enemy defenses. It is interesting to note that these early rockets were built like birds to take advantage of the aerodynamic features of wings. Winged missiles were used in China long before comparable weapons were developed in the western world.

Gunpowder can be used to propel missiles through a tube because it burns quickly at the closed end of the tube creating high gas pressure which pushes the missile out of the tube. The development of guns based upon the use of gunpowder was slow because of the difficulty encountered in the making of consistently good gunpowder. In modern technology, gunpowder is described as *low explosive* because it does not explode in the same way as nitroglycerine, a modern *high explosive*. The burning rate of gunpowder deteriorates during storage because the grains of powder fuse to each other, reducing the surface area available for burning. If the gunpowder was old, it was never a certainty that a gun was going to fire. High explosives *detonate* when triggered by a pressure wave which is often created by a *percussion cap*. Whether an explosive burns or detonates, the essential process is one of combustion in which carbon, hydrogen, and other chemical atoms in the explosive combine with oxygen to create a large quantity of high pressure gas which either propels a missile or destroys a system to be blasted. Figure 6.3 shows some of the basic molecular structures encountered in the manufacture of explosives and gases which can spontaneously explode. The gas which causes explosions in coal mines is *methane,* made up of one carbon and four hydrogens. As shown in the sketch of Figure 6.3, carbon likes to form four bonds. When all of the bonds of a carbon atom are connected to other carbon atoms to form a network, the resulting carbon compound is known as *diamond.* Carbon forms diamond when it crystalizes under intense pressure. The way to make synthetic diamonds is to dissolve carbon in molten rock and let it solidify as the rock cools. Making diamonds this way mimics the way in which diamonds are made in nature. The so-called *diamond pipes*, which are cylindrical channels in rock of earth containing diamonds found in the diamond fields are actually the solidified throats of old volcanoes.

Diamond is just a particular form of carbon and like coal will burn (oxidize) to form carbon dioxide. *Carbon dioxide* is the gas which gives champagne and soda pop their fizz. When methane builds up in the working areas of a coal mine, it can combine with oxygen to create carbon dioxide and water molecules. The energy released and the pressure built up by this combustion causes an explosion. The word used by miners for methane is "*fire damp.*" The explosions in a coal mine are often made worse by the fact that the primary explosion of the methane gas stirs up coal dust which then burns explosively to perpetuate the explosion. *Acetylene* is a gas with the molecular structure shown in Figure 6.3(b). In this molecule, the carbon atoms share three bonds. This type of bonding is very unstable (chemists say the bonds are strained) and so acetylene will burn explosively with oxygen, releasing a lot of energy. A welder exploits the large amount of energy released by burning acetylene to produce a very hot flame that can melt the surfaces of two pieces of metal to be fused together [9].

Another very different compound of carbon and hydrogen is what is known as a ring compound or an *aromatic compound.* The name comes from the fact that many ring compounds of carbon have a pleasant or fruity odor. The simplest ring compound that can be built is the benzene ring shown in Figure 6.3(c)(ii). The structure of the benzene molecule was a mystery for a long time. The problem was solved by the Belgian chemist

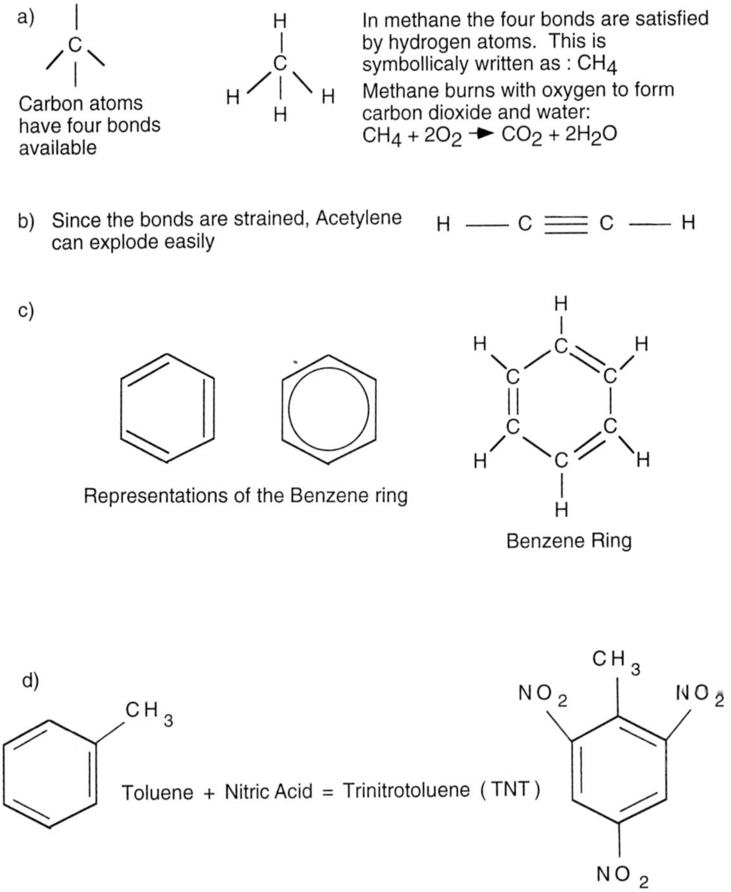

Figure 6.3 Representations of the chemical structure of several compounds of hydrogen and carbon which are used in explosives or directly as explosives. a) Methane is a major ingredient of natural gas and the main cause of coal mine explosions. b) The structure of acetylene. c) Many explosives contain aromatic compounds, i.e., compounds with one or more benzene rings. d) The chemical structure of trinitrotoluene (TNT).

Kekulé (1829–1896) [10]. The benzene molecule contains six hydrogen atoms and six carbon atoms. If these formed a chain compound of the type shown in Figure 6.3(c)(i), benzene should be highly reactive and explosive. The compound had been discovered in 1825 by Faraday. The name benzene was given to the molecule by the German chemist Mitscherlich. Asimov reports the story told by Kekulé about how he was inspired to solve the problem of the benzene molecule [11]. Kekulé told an audience of scientists that one day he was riding on a bus and, while half asleep, he seemed to see atoms whirling in a dance. Suddenly, the tail end of a chain attached itself to the head end and formed a spinning ring. Asimov comments, "If Kekulé had been Archimedes, he would have sprung off the bus and run the street crying Eureka!" (See the discussion of Archimedes' principle later in this chapter.)

Another story told about the inspiration that solved the structure of the benzene ring is that one day Kekulé was watching a line of children playing in the street. As he watched, they suddenly joined hands to form a ring. Whatever the true story, it seems to have been a flash of creative imagination that made Kekulé realize the compound would be a lot more stable if it formed a ring of the type shown in the diagram. Traditionally, the benzene ring is shown with three single and three double bonds (Figure 6.3(c)). A more modern way of showing the structure is as in the neighboring drawing. It should be noted that when the carbons in the ring of Figure 6.3(c) are attached not to hydrogen but to other carbons in neighboring hexagons, the compound that results is known as *graphite*. The layers of carbon hexagons in graphite are very stable and the bonding between the layers relatively weak so that they glide over each other very easily. That is why graphite is used as a lubricant to reduce the friction between two moving surfaces. Another result of the weak attachment between layers is that they flake off easily. For this reason, a stick of graphite can be used as a pencil, hence the name of the compound, from "graphein", meaning "to write." Modern *"lead" pencils* do not have any lead in them; they are made from a mixture of graphite and clay.

If one of the hydrogens of the basic benzene ring is replaced by what is known as a *methyl group* (formula CH_3) a compound results known as *toluene*. The way chemists write the formula for toluene is shown in Figure 6.3(d). The word toluene comes from the Spanish word "tolu". Tolu occurs in nature as a fragrant balsam. This balsam is a gum that can be collected by making small cuts in the bark of the tolu tree. This tree takes its name from the town of "Santiago de Tolu" in Columbia where the tree grows in sufficient numbers for the gum to be harvested commercially. *Balsam* is a general word for perfumes based on naturally occurring gums taken from plants. If a chemist treats toluene with nitric acid, an experiment that *should not* be carried out by the amateur, the compound *trinitrotoluene* is created. This compound is better known as *TNT*. Again, note that all the atoms in the molecule of TNT are such that when combined with oxygen they can produce many molecules of carbon dioxide, steam (H_2O), and oxides of nitrogen which can build up tremendous pressure when the explosive compound is detonated, i.e., reacted with oxygen.

Nitroglycerine and *guncotton,* the two chemical compounds which form the basis of modern explosives technology, were discovered at about the same time in Germany and in Italy. The Italian chemist Ascanio *Sobrero* (1812–1888) discovered the compound known as nitroglycerine in 1847 [11]. In his experiments, he slowly poured glycerine into a mixture of sulfuric and nitric acids and produced a compound which exploded in a very different way from traditional gunpowder. Sobrero found that small quantities of the oily liquid exploded much more powerfully than black powder, even when the liquid was unconfined. Realizing the potential destructive power of such an explosive in warfare, Sobrero tried to keep his discovery a secret. Gradually, however, the news of the invention leaked out and other people developed methods of producing it.

In particular, Alfred Bernhard *Nobel* (1833–1896) started to manufacture nitroglycerine. Part of Nobel's education had taken place in the United States and, as Asimov puts it [11],

> Nobel's stay in the United States had given him the vision of a continent about to be tamed. Nobel could see how roads could be blasted out of mountains, canals dug, and foundations laid by using the directed violence of a shattering explosive such as nitroglycerine instead of the weary muscles of countless human beings.

Manufacturing the nitroglycerine, however, proved to be very difficult. There were numerous accidents culminating in the blowing up of his own factory in 1864. His brother was killed in this accident. The Swedish government refused to allow the factory to be rebuilt. Nobel was regarded by his contemporaries as a lunatic bent on destroying the world. Nobel, however, was determined to try to discover a safe way of handling nitroglycerine. For safety reasons, he carried out his subsequent experiments on a barge in the middle of a lake.

In 1866 he noticed that one of his casks of nitroglycerine had been leaking. Nobel saw that the liquid had spilled onto the packing material placed around the container. This material was a powder known in English as *diatomaceous earth* or in German, *Kieselguhr*. (We will discuss the nature of this powder later in this section.) Nobel noticed that the mixture of nitroglycerine and Kieselguhr could be molded into sticks and balls and could not be set off without a detonating cap. Once detonated however the power of the adsorbed nitroglycerine was the same as that of the raw liquid. Nobel called the mixture of nitroglycerine and kieselguhr *dynamite* from the Greek word for power. Kieselguhr is a very fine, inert, powder. It appears that when the nitroglycerine is adsorbed onto the powder it is not as sensitive to low shock waves, but that once the detonation is initiated, the chemical activity of the material is unhindered by the presence of the powder. (The word **ad**sorb used in the previous sentence means to cause to cling or stick to the surface of a solid; to cling onto, rather than **ab**sorb which describes the movement of a substance into the interior of another substance). Anyone who has watched Hollywood movies about the use of dynamite in the Old West or in modern oil fields knows that *old dynamite* sweats. That is in old dynamite sticks the nitroglycerine for some reason starts to accumulate on the surface of the old sticks of dynamite. The sweating dynamite stick becomes as sensitive to shock as the original nitroglycerine.

Some people would say that Nobel discovered the recipe for dynamite by accident. However, Nobel only discovered the value of the mixture of powder and nitroglycerine because he was looking for safe methods of handling the chemical. Many other people could have had leaking nitroglycerine adsorbed by other powders and would never have drawn the important conclusion that this was the basis of a method for creating a safe explosive compound. The scientific community recognizes that important discoveries aided by chance occur to the prepared mind on the look out for the importance of fortuitous events. The word *serendipity* is used for such discoveries. A dictionary will define serendipity as, "the gift of finding interesting things by chance." This word was created by the English author Horace Walpole (1717-1797) in a letter to a friend written in January 1754. In the words of Klein, "Walpole created the word from the title of his tale 'The Three Princes of Serendip,' all of whom were endowed with the same faculty." Klein notes that "Serendipor Serendib" is the former name of Ceylon, now known as Sri Lanka. (Note that Shipley defines serendipity as, "the happy faculty of finding what one did not seek" [8].)

Nobel also invented *blasting gelatine* and became wealthy by manufacturing and selling explosives. Asimov comments that

> *Dynamite did indeed open up the American West, and explosives had, and generally still have, myriad peace time uses. However, Nobel was seen in the eyes of the world as the inventor of horrible tools of war. Nobel himself thought that dynamite would actually outlaw war by making it too horrible.*

When Nobel died in 1896, he left a fortune of over nine million dollars with which he endowed the *Nobel prizes* for excellence in the areas of physics, chemistry, physiology or medicine, literature, economic sciences, and the promotion of peace. The first scientist to be honored with the Nobel prize for physics was *Roentgen* whose discovery of X-rays opened up new ways of treating diseased and injured bodies (1901). Roentgen's discovery of X-rays involved a great deal of serendipity as readers can discover for themselves by reading about his life [11].

Later in this chapter, we will discuss security measures to prevent the smuggling of explosives onto aircraft and we will discover that the problem of detecting dynamite is compounded by the fact that nitroglycerine is used medically to dilate blood vessels. Blood can flow more easily through dilated than undilated blood vessels. Nitroglycerine is used to prevent and to treat *angina*, a heart disease. If scientists developed a bomb sniffer sensitive to the presence of nitroglycerine molecules in the air, any elderly patient on nitroglycerine tablets would carry a sufficient supply of the chemical to trigger the alarm. Therefore, scientists must look for other compounds that indicate the presence of explosives.

Workers who manufacture dynamite must be protected from small amounts of dynamite dust which may be generated during the manufacturing process. Workers who inhale small amounts of dynamite dust during the week have dilated blood vessels, a condition to which their bodies become adjusted. On the weekend, in the absence of the dynamite dust, their blood vessels *constrict* and they may suffer from severe headaches. Workers who begin to suffer from the problems of inhaling dynamite dust have to be given nitroglycerine medication when they are not at the factory.

The diatomaceous earth (Kieselguhr) that Nobel used in his development of dynamite is made up of the skeletons of tiny plants that flourish in the lakes and oceans of the world. Their skeletons are almost pure silica (SiO_2). When they are alive, diatoms are described as *phytoplankton* from "phyton" meaning "plant" and "plankton" from a Greek word that means "wanderer." Thus phytoplankton are able to wander freely in the water where they live. (The root word plankton has given us the word *planet* for the objects in the sky that seem to be free to wander whereas the other stars appear to stay in fixed positions.) It requires several hundred billion diatoms to feed a humpback whale and a half a tonne of diatoms to make half a kilogram of seal meat [5]. When the phytoplankton die, their tiny skeletons sink to the bottom of the lake or ocean where they collect in large quantities. In some parts of the world, such deposits have been lifted above the ocean by the movement of the surface of the earth and engineers are able to mine the diatomaceous earth for use in many industrial applications. The largest of the diatoms are a millimeter across and the smallest need to be viewed with a microscope. In Figure 6.4, enlarged photographs of typical diatom skeletons are shown. There may be as many as 10 000 living species of diatom. When viewed at high magnification as in the figure, it can be seen that the diatom skeletons have very high surface area. It has been estimated that a half pound of diatomaceous earth has a surface area equal to that of a football field. Literature associated with the pictures shown in Figure 6.4 states that some of the tiny holes in the skeletons are as small as 200 Ångstroms. An *Ångstrom* is named after a Swedish physicist Anders Jonas Ångstrom (1814–1874). Ångstrom was a pioneer in the study of the radiation leaving the sun and he started to make measurements of the wavelength of light in units which were a ten billionth of a meter. In recent years, Ångstrom has tended to be dropped from science in favor of the *nanometer* which is one billionth of a meter (written nm = 10^{-9} meter). Thus

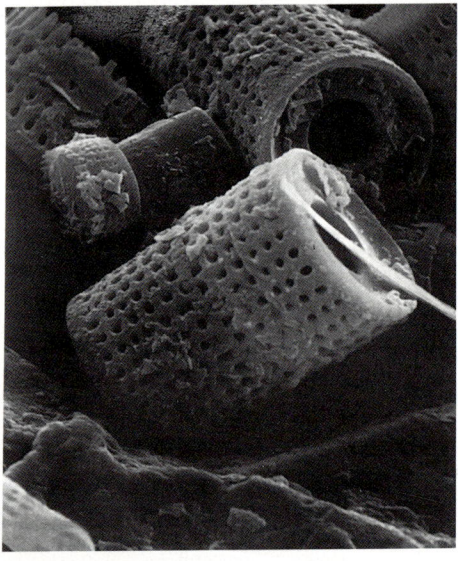

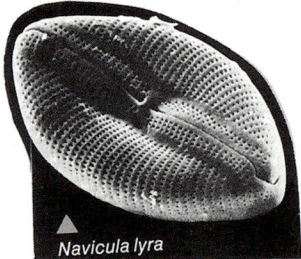

Navicula lyra

Figure 6.4 Diatomaceous earth is a name given to a form of silica formed from the accumulated sediment of the skeletons of tiny animals that "wander" in seas or lakes. The two commercially available forms of diatomaceous earth shown are available from the Eagle-Picher Corp. and Johns-Manville Corp.
© Eagle-Picher Minerals, Inc., 6110 Plumas Street, Reno, Nevada, 89509.

the wavelength of the yellow light given out by sodium street lights can be quoted as 5,700 Ångstroms (Å) or 570 nanometers (nm), or 0.57 *micrometers* (microns, μm).

For some reason that is not fully understood, adsorption onto the great surface area of the diatomaceous earth appears to stabilize nitroglycerine until it is detonated. The name *diatom* comes from the fact that many common varieties of this tiny plant are divided into two nearly separate parts. The Greek words "dia" meaning "through" and "temnein", "to cut" give us the word diatom meaning "cut into two". Note that "insect" comes from Latin root words with the same meaning. The word *atom* used to describe the smallest part that has all the characteristics of a paticular element, comes from the same Greek root word as diatom, with the addition of the prefix "a", meaning "not". Thus atom means "indivisible" or "something that cannot be cut". The name was given by scientists to such things as the hydrogen atom because they thought they had found the ultimate constituents of the universe when the atom was discovered. Only later did they find that atoms can be split into electrons, protons, and neutrons.

Serendipity played a big part in the discovery of another major explosive which is known today as *guncotton*. The essential ingredient in guncotton was discovered by the German-Swiss chemist *Schönbein* (1799-1868). In the story of the discovery of guncotton by Schönbein in 1845, it is said that he was experimenting at home with a mixture of nitric and sulfuric acids in the kitchen — an activity strictly forbidden by his wife, but she was absent at the time! During the experiment, he spilled some of the acid and we are told that, in a panic, he seized the first thing at hand — his wife's cotton apron — and mopped up the mixture. He then hung the apron over the stove to dry before his wife came home. We are told that as he watched, the apron dried out and then exploded. Fascinated by this fact, Schönbein carried out further experiments and determined that he had inadvertently added the chemical group nitro (NO_2) to the cellulose in the apron forming a compound which became known as *nitrocellulose*. Schönbein recognized that this compound could be useful in warfare and described it by a German name compounded from gunpowder and cotton which we now translate into English as guncotton [11].

To understand the chemical structure of guncotton, it is necessary to understand the chemical structure of cotton. To arrive at the structural formula for *cotton*, we must begin with the structure of *glucose*. Sugars such as glucose are made by plants from carbon dioxide and water using the energy of the sun; a process known as photosynthesis. The word *photosynthesis* literally means "put together by light". In Figure 6.5, the structure of glucose, a simple sugar is shown. Again the reader should note the proliferation of oxygen, hydrogen, and carbon in the structure — all of which can make gases when the compound is burnt. In fact, ordinary household sugar, which is made up of two glucose molecules joined together can be a very potent explosive. Factories that make very finely pulverized sugar known as *confectioners' sugar* in North America and *icing sugar* in Great Britain and Canada have suffered from serious explosions with disastrous consequences. Many food compounds and pharmaceutical materials are rich in carbon and hydrogen and are potentially explosive. Thus flour milling facilities and cocoa man-

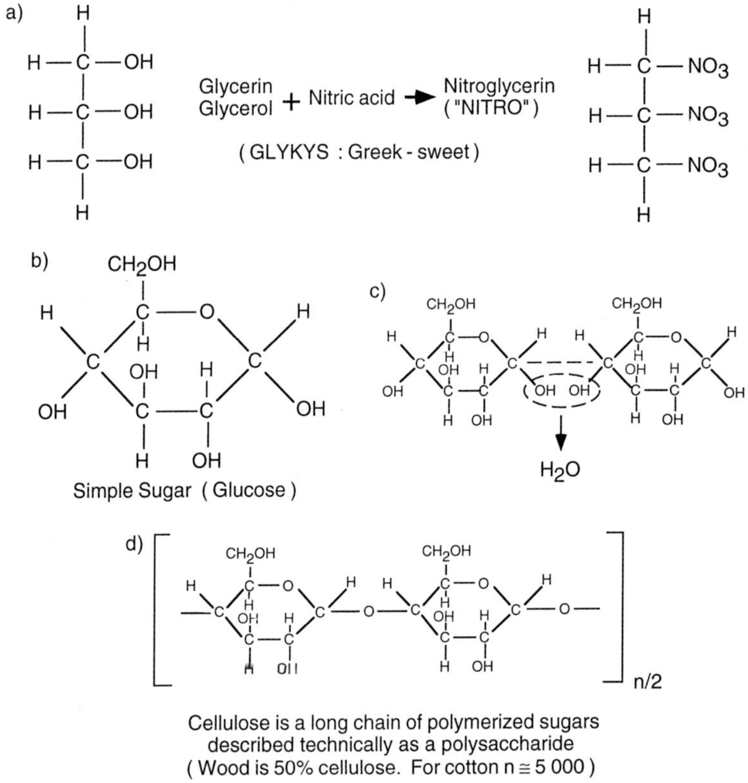

Figure 6.5 Many explosive mixtures contain chemicals derived from glycerine and/or wood.

ufacturing installations all have histories of serious explosions when safety precautions have been ignored or forgotten.

If two glucose molecules are placed end to end as illustrated in Figure 6.5(c), they can be made to combine with each other by eliminating a molecule of water. This reaction is illustrated in the sketch. This linking up can take place between many molecules of glucose and the process of linking simple molecules to produce the long chains of compounds is known as *polymerization.* This word comes from the two Greek rootwords: "poly" meaning "many" and "meros" meaning "part." The resultant compound is often called a *polymer.* The simple molecule which is linked up to form the polymer is known as a *monomer* from the Greek word "mono" meaning "one." A Latin word for sugar was "saccharum". When an artificially sweet compound was created by Fahlberg and List, they called this artificial sweetener *saccharine.*

Cellulose, which is major component of wood and woody type plants, is a long chain of polymerized sugars which is described technically as a *polysaccharide.* The cellulose obtained from cotton is a particularly useful form, available in long fibers. These fibers are made up of a polysaccharide formed from approximately 5000 glucose molecules.

Schönbein's experiment had succeeded in adding nitro groups onto the sugar chains of the cellulose. One of the things that impressed Schönbein as he watched his wife's apron explode was that there was no smoke. When carbon-containing chemicals are burnt; that is, combined with oxygen, the presence of smoke represents unburnt carbon chemicals and therefore lost energy. Schönbein also recognized that smokeless gunpowder would be not only more efficient than traditional gunpowder, but also tremendously advantageous on the battlefield. The use of ordinary gunpowder created so much smoke that it was very difficult to know what was happening on the battlefield. Camouflage uniforms were not necessary until the invention of smokeless gunpowder based on guncotton.

Schönbein pushed the development of his guncotton but the manufacture of the material was very difficult and there were many explosions in the manufacturing installations. For example, in 1847, a factory run by a consortium, of which Schönbein was a member, exploded, killing 21 people. In fact, by the 1860s, the use of guncotton had stopped because it seemed to be too difficult to make. However, the overall advantages of the material were such that other scientists continued working to develop methods to manufacture guncotton safely. The problems were eventually solved by joint investigation and development carried out by the English chemist Fredrick Augustus Abel (1827–1902) and the Scottish chemist Sir James *Dewar*. (Dewar is the same scientist who, in 1892, was the first to construct a double-walled flask with a vacuum between the walls for storing hot or cold liquids. See the discussion of the problems of storing cryogenic rocket fuels later in this chapter.) Dewar and Abel found out that they could mix nitroglycerine and cellulose with some petroleum jelly to make a mixture that was comparatively safe to handle when purified ingredients were used. This jelly-like substance could be extruded through holes to make spaghetti-like strands which were called *cordite*. When the cordite strands were dry, they could be handled safely. Apparently, during the development of cordite, Dewar and Abel had long discussions with Nobel who claimed they had stolen some of his ideas. Nobel sued Dewar and Abel when they claimed a patent for cordite. Nobel lost his case although there was probably some substance to his claims [11].

The fact that wood is very rich in cellulose explains the fact that one of the modern types of explosives used in the mining industry is a mixture of a fertilizer (ammonium nitrate) with sawdust. (The formula for *ammonia* is shown in Figure 6.6(a).) Asimov tells us that ammonia received its name from the fact that it was first discovered on the walls and ceilings of a temple to the Egyptian god *Ammon* built in an oasis of the North African desert. Apparently the priests of the temple used to burn camel dung as a fuel to warm the temple at night and white, salt-like crystals from the burning of the dung collected on the wall. These crystals were called sal ammoniac meaning "salt of Ammon." We now know that *sal ammonia* (the substance in *smelling salts*) is the chemical compound ammonium chloride. Ammonium nitrate is a widely used fertilizer and its formula is also shown in Figure 6.6(a). It is a compound rich in atoms which form gases when burned. Therefore, the combination of ammonium nitrate and the cellulose in sawdust makes a powerful explosive widely used in the mining industry.

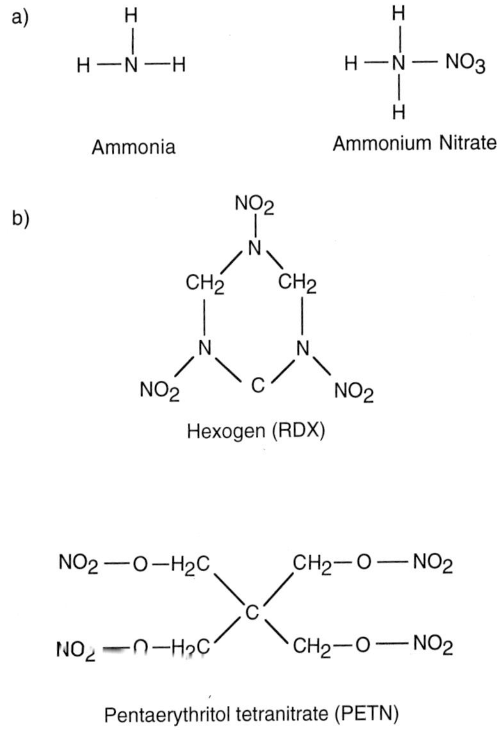

Figure 6.6 The chemical structure of some materials used in the manufacture of explosives.
a) Ammonium nitrate is used in modern blasting procedures.
b) SEMTEX is a plastic explosive made of 44.5% RDX and 44.5% PETN with 11% plasticizer.

In the 1980s, the public became very aware of what is known as *plastic explosive*. Plastic explosive was used by terrorists to blow up several aircraft and to carry out other acts of sabotage. The main type of plastic explosive is known as *Semtex*, a name taken from the name of a small village in Czechoslovakia close to the factory where large quantities of the explosive were made. The two main ingredients of Semtex are the explosives known by the short names *rdx* and *petn*. The chemical formulas for these two compounds are shown in Figure 6.6(b). Semtex was made by mixing RDX and PETN in the amounts shown in the legend of Figure 6.6. The compound described as *plasticizer* is a chemical added to the two explosives to make the mixture into a pliable compound which could be moulded into various shapes. Because of the very large molecules involved in their chemical structure, the two compounds do not give off easily detectable levels of gas molecules. In addition, this material cannot be detected by X-rays. For these reasons security forces find Semtex very difficult to detect, and hence its appeal for terrorists [12, 13].

Section 6.2 From Muskets to Machine Guns

From a technical viewpoint, and as illustrated in Figure 6.7(a), a gun consists of three parts. The first part is the support system to enable the gunner to hold the missile launching device. The second is a combustion chamber. The third is a barrel to enable the explosion in the combustion chamber to drive the missile with high energy in a

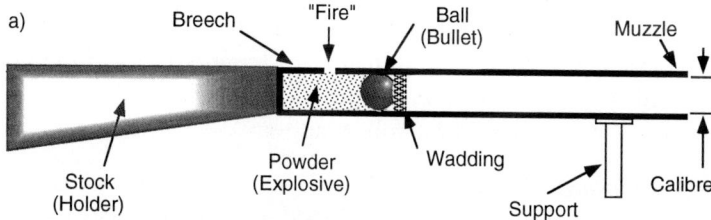

The three basic parts of any gun are:
(i) The holder (stock), to brace the gun against the shoulder in order to steady it.
(ii) The combustion chamber (breech) packed with explosives.
(iii) The barrel, in which the energy of the explosion of the powder is transferred to the bullet.

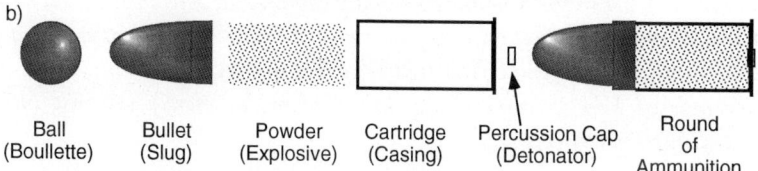

Figure 6.7 A gun is a device for dispatching a missile with high energy imparted from an explosion. a) Construction of a musket. b) Modern guns use self-contained rounds of ammunition ratrher than the separate powder and ball used in the musket.

specific direction. The first guns were actually supported on a Y-shaped stick to free the gunner to use both hands in the aiming and firing of the gun. The origin of the term *gun* is obscure. Brewer suggests that the word is perhaps a shortened form of the old Scandinavian female name "Gunnilder"; "gunner" is Icelandic for war and "hildr" is Icelandic for battle. Brewer [14] goes on to comment that

> The bestowing of female names on arms is not uncommon. Thus Big Bertha of World War I, the long range gun that bombarded Paris, was named after Bertha Krupp, wife of the head of the great armament factory at Essen.

Perhaps we should call the combustion chamber of the gun the propellent chamber since the very earliest record of a gun, from the writing of the ancient Greeks, describes a missile fired from a barrel by steam. It is said that *Archimedes* used a steam-operated gun to defend Syracuse, Sicily from the Romans. At that time Syracuse was a Greek colony ruled by Hieronymus, a relative of Archimedes. Some people believe that Archimedes was the greatest scientist and mathematician of ancient times and, in the words of Asimov [11],

> *His equal did not arise until Newton, 2000 years later.*

Archimedes was born in 287 BC and lived most of his life in Syracuse, after studying in Egypt. He was killed in 212 BC by a Roman soldier. The people of Syracuse had decided to become allies of the Carthaginians, the bitter enemies of the Romans. The Romans attacked Syracuse with a fleet commanded by General Marcellus. Asimov [11] describes the resulting conflict in the following way:

> *Thus began a strange, three-year war of the Roman fleet against one man — Archimedes. According to tradition, the Romans would have taken the city quite quickly had it not been for the ingenious devices that were brought against their fleet by the great scientist. Archimedes is supposed to have constructed large lenses to set the fleet on fire, mechanical cranes to lift the ships out of the water and turn them upside down, and so on. In the end, the story goes, the Romans dared not approach the walls too closely and would flee if as much as a rope showed above them for they were convinced that the dreaded Archimedes had designed some new monstrous device. During the destruction of the city, when the Romans eventually captured it, Archimedes, with a magnificent and scholarly disregard for reality, engaged himself in a mathematical problem and was bent over the geometrical figures he had marked in the sand, when a Roman soldier ordered him to come along, but Archimedes merely gestured imperiously, "Don't disturb my circles"*

For this abstract response he was killed.

Recently a Greek engineer set out to rebuild some of the reported devices of Archimedes; one was a one-fifth scale model of a steam driven cannon. The model the engineer built consisted of a wooden barrel into which a large stone was placed, attached to a cylindrical-shaped, metal steam-chamber of the same diameter. The chamber was heated to a temperature of about 400 °C then quickly filled with cold water. The resulting steam built up enough pressure to launch the stone cannonball which was held against the chamber entrance by a wooden pin. The cannon could accurately fire a three hundred kilogram stone a distance of 80 meters [15].

For a long time, guns and cannons were loaded through the *muzzle*, the technical term for the mouth of the gun barrel. The word muzzle comes from a Latin word, "musus" meaning "beak." To load the early guns, the powder was poured into the combustion chamber down through the muzzle. The ball to be fired would be added, and then the ball-powder system was compacted using a wad of paper rammed down the barrel of the gun with a rod. The gunpowder in the combustion chamber was then ignited with a fuse through a hole in the breech of the gun. There seems to be some confusion over the spelling of the word breech as applied to a gun. The word *breech* is related to an old Germanic word meaning the rear end or the back part of anything. Thus *breeches* (now called *britches* in American English) were the clothes used to cover a person's rear end. To *breach* means to break. Thus when the walls of a castle were broken, the breach in the wall was the entrance to the castle. Modern guns are breech-loaded and the missile and propellant form a *round of ammunition*. The storage place on a gun for rounds of ammunition is called a *magazine*. This latter word comes from an Arabic word "makhzan" meaning "storehouse." The news magazine was originally intended to be a storehouse for knowledge. (Note: In modern French, the term "magazin" is used to describe a shop or store.)

The size of a bullet is referred to as the *caliber*. The caliber is also the inside width of the gun barrel. An important task in the manufacture of guns was to ensure that the caliber of a gun was exactly that specified in the manufacturing instructions. In modern science the term *calibrate* has been extended to mean the act of standardizing the measuring capacity of an instrument. Today, we calibrate thermometers and the everyday ruler that we use is calibrated against a standard meter stored in Paris.

Throughout history a variety of words have been used to describe guns. The term *cannon*, reserved for larger guns, is derived from a Greek word for a hollow reed or rod. In Figure 6.8, the way in which *canon*, a closely related word, has come to mean a church official who administers church rules is illustrated. Humor often depends upon exploiting the double meanings of a word. Thus the story is told of the black-haired bishop whose black-haired wife gave birth to a red-headed child. It is said that the Archbishop went to the church and fired a red-headed canon.

The word *musket* has an interesting history. It comes from the Italian word "moschetto" which means "a little fly." The soldiers used to refer to the iron bolt (missile) delivered by a crossbow as a "moschetto". The name was transferred to the small lead balls used in the first handguns which then in turn became known as muskets. As the armies of the world started to use more and more guns, elite regiments of soldiers equipped with muskets were known as *musketeers*. Through the writings of Dumas, a French novelist, who wrote the famous story about the adventures of *The Three Musketeers*, the word has become widely known to modern readers. I have seen many movies about the three musketeers and strangely enough I have never seen one of them fire a musket. They have always been noted for their exploits of great swordsmanship and their penchant for getting into trouble rather than their ability to fire muskets.

As guns became more readily available to troops, it became necessary to provide better means of loading the gun than pouring the powder from a powder horn. It became common practice to provide a cardboard carton complete with the powder charge. A small ball to be fired from the gun was placed on top of the charge inside the cylinder of cardboard. The cylinder was described as a *cartridge*. The small ball was known in French as "a boulette," which has given English the modern term *bullet*. The energy that the bullet was able to derive from the explosion depended on the closeness of the

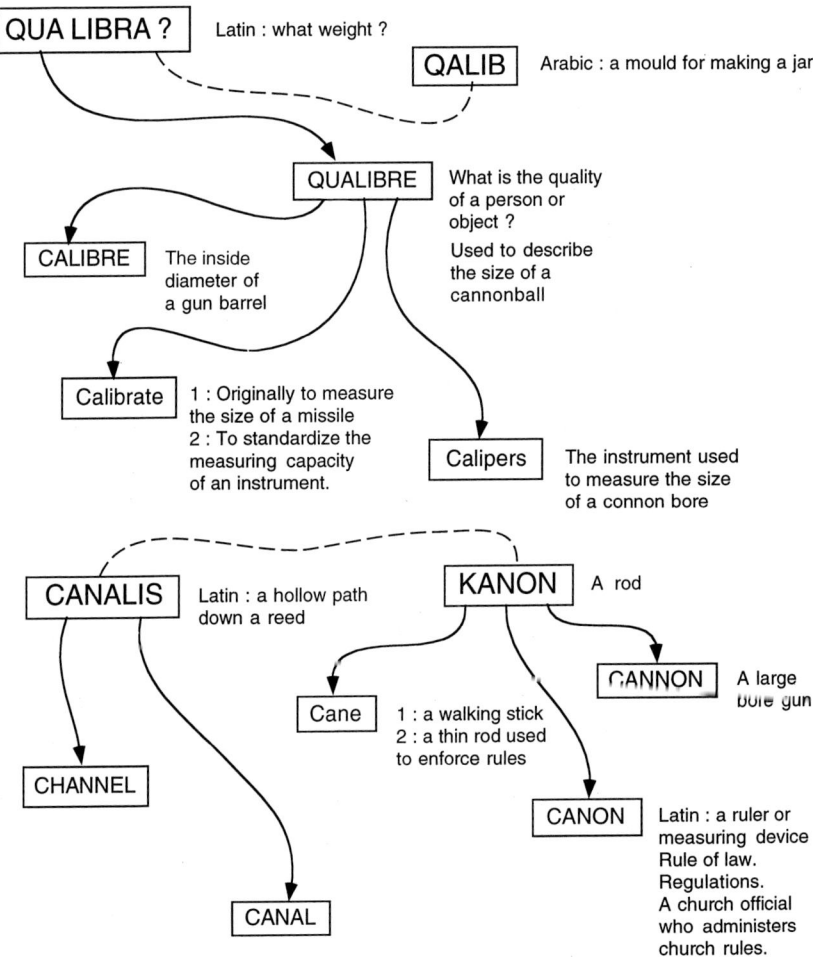

Figure 6.8 The calibre of a gun is the inside diameter of the barrel. The bullet used must fit snugly inside the barrel for safe operation.

fit between the bullet and the wall of the barrel, and the length of the barrel. If the barrel was longer, the explosion was able to push on the bullet for a longer time. The small ball was often greased to make it slide along the barrel and to decrease the effect of any gap between ball and barrel as the explosive pushed the ball the length of the barrel.

Greasing the shot proved important in history; it was the immediate cause of a rebellion against British authority in India known as the *Indian Mutiny*. This rebellion lasted from May 1857 until July 1859. Resistance to British rule by what was known as the East India Company had been growing for several reasons. The failure of the British civil and military authorities to respect Hindu and Moslem religious customs was one of the major sources of friction between the soldiers of the army and the authorities.

The soldiers in the Bengal army of the East India Company were equipped with muzzle-loading *Enfield rifles* (named after the town in England where they were made). The troops were issued with cartridges containing the powder charge and a bullet. In order to load the rifle, the troops had to remove the greased ball from the cartridge with their teeth. It was rumored that the balls had been greased with a mixture of beef fat and lard. The army had both Hindu and Moslem troops. The Hindus objected to beef fat, and the Moslems to lard, which was pig fat. The rebellion began when a company of soldiers refused to load their rifles because of their religious beliefs.

It was during this same rebellion that the British soldiers realized that their red coats made them very conspicuous targets for the Enfield rifles. Upon this realization, they rolled their red coats in river mud to make them blend with the background as they fought the rebellious troops. The word for mud in the the local language of that part of India was *Khaki* and hence the use of that word to describe the color of the uniforms of British and others soldiers. Modern soldiers disguise themselves and make themselves blend in with the background more effectively by using uniforms with patterns which are described as *camouflage*. The word camouflage comes from a French word "camoufler" meaning "to disguise."

One of the technical problems to be overcome before a breech loading gun could be developed was to stop gases from the explosion escaping through the crack of the opening where the ammunition round was inserted. Such escaping gases represented wasted energy which was not used to push the bullet out of the gun. This problem was by introducing a thin brass case for the explosive charge. A percussion cap was placed in the base of this cartridge case so that the explosive would be ignited when this cap was hit by the trigger pin. The pressure of the exploding gases in the brass cartridge caused it to expand against the breech of the gun, sealing off the escape routes for the high pressure gas. Once the bullet had been fired, this pressure was released and the case of the cartridge returned to its original size, leaving enough room around the outside of the shell of the cartridge for it to be ejected from the breech of the gun.

A *rifle* differs from a smooth-bore gun, such as a musket, in that a spiral groove is cut into the barrel of the gun. The bullet is just slightly larger than the barrel and as it moves through the barrel, it is pressed into the grooves causing it to spin, thus giving it a stable trajectory after it leaves the gun. "To rifle" originally meant to scratch or scrape and the term rifle described the act of cutting the grooves in the barrel. The term was then transferred to the new type of gun. The development of rifling greatly increased the accuracy with which a missile could be delivered by a gun. Since the pattern of grooves in a rifle and the scratches from the wearing of the barrel create unique patterns, the marks left on the bullet as it moves though the barrel of the gun can be used to identify the gun from which it was fired [16].

The weekly magazine New Scientist has an interesting feature called "The Last Word". Readers are invited to send in scientific questions to which other people then reply. When questions were asked about why bullets spin there was some interesting information provided by different letter writers. Thus Reginald Titt [17] wrote

> Even a spinning bullet will not travel in a straight line. Besides the parabolic trajectory caused by the fall of the bullet due to gravity there is a second effect due to surface drag accentuated by the grooving of the bullet caused by the rifling of the barrel. The drag will cause the bullet to drift gradually into an increasing spiral. The effects of inadequate spin can be quite spectacular. Toward the end of the Second World War I took part in trials to assess the rifling life of a machine gun by firing some thousands of live rounds continuously. We could see the barrel was worn out when the bullet started going sideways through a target placed only about 200 meters away.

A letter from Finland [18] in the same issue of the New Scientist tells us that

> In addition to accuracy, international law requires bullets to spin. Early American M16 assault rifles from the Vietnam era had inadequately rifled barrels which didn't give the bullets enough spin. This made the bullets tumble into a human target provoking accusations that the Americans were using dumdum bullets to cause greater injury than conventional ones.

(See the discussion of dumdum bullets later in this chapter.)

Anyone who has watched movies about gun fighting must have heard at some time or other the statement that, "You never hear the one that gets you!" This refers to the sound of the explosive in the gun which has fired the bullet. This saying tells us that, at least since the turn of the century, bullets fired from modern guns travel faster than the speed of sound and so the bullet reaches its target before the sound of the discharge does.

As we have discussed earlier, a sound wave is a compression wave in the air. The velocity with which the compression wave travels in the air is 330 meters per second at 0 °C. The speed increases by 0.6 meters per second per °C. On the other hand, the speed of light is 3×10^8 meters per second. Thus when we see a flash of lightning, the light reaches the observer virtually instantaneously whereas the sound has to travel at 330 meters per second. This is why it is possible to estimate how far away the lightning flashed to the ground by counting the seconds between observing the flash of lightning and hearing the sound of the thunder. The *thunder clap* is created by the fact that the tremendous energy in the lightning flash very rapidly heats a cylinder of air along the pathway of the lightning. For a very short time this hot air expands and then as soon as the lightning ceases, the hot air cylinder collapses. As it collapses, the walls of the encircling cool air rush in to fill the space and collide with each other to make a kind of super handclap which is the roll of thunder heard at a distance.

When an object travels through a substance at a speed greater than the speed of sound in that substance, the energy dissipated by the moving object is manifest as a shock wave. Thus if a low-flying aircraft is traveling below the speed of sound, the sound wave can reach the observer on the ground before the aircraft passes overhead. If an aircraft attempts to fly at exactly the speed of sound, some of the noise that it creates stays with it, accumulating to produce a very dangerous situation where vibration and noise can destroy the aircraft. Therefore the aircraft must try to move quickly through the speed of sound, and then as it moves faster than the speed of sound, the noise energy spreads out from the speeding aircraft to form a shock wave which reaches an observer on the ground after the aircraft has passed overhead.

Sometimes shock waves created in the flesh by impact of the human body with various objects during an automobile accident create far more serious damage than any visible external wounds. The shock wave can actually cause the rupture of internal organs such as the liver and spleen with subsequent internal bleeding.

A high speed stroboscopic picture of the shock wave produced by a bullet moving faster than the speed of sound is shown in Figure 6.9. The straight lines leaving the tip of the bullet create a V-shaped pattern indicating the path of the sonic energy leaving the tip of the bullet. The discontinuity in shape at the end of the bullet generates another shock wave produced by the sound created by the turbulence in the wake of the bullet. Notice that the second set of lines do not intersect at the tail of the bullet but are about one quarter of an inch into the wake at the end of the bullet. If we had a

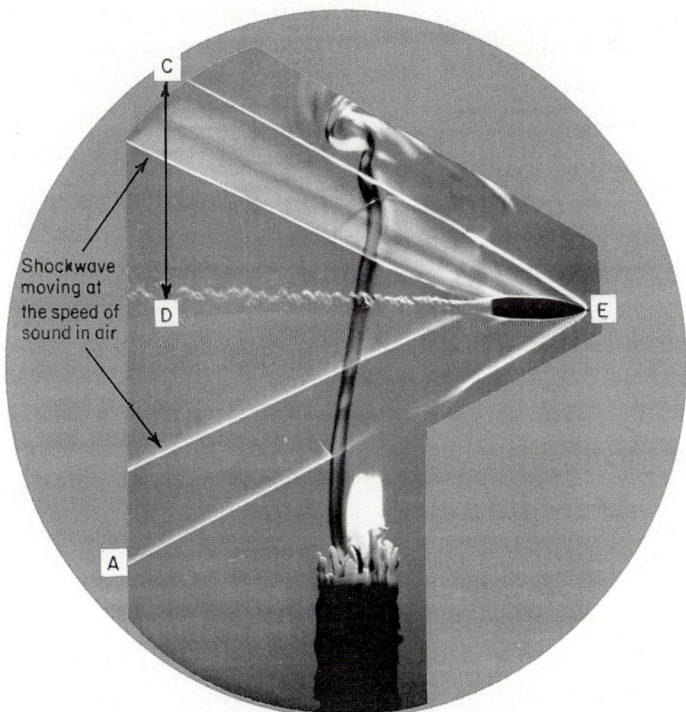

Figure 6.9 Modern photography makes it possible to study the dynamics of bullets in detail. High speed photography can capture the spinning of the bullet created by the grooves (rifling) in the gun barrel. Presented here is a stroboscopic picture of a bullet moving through the heat of a candle flame. (Exposure time 1/3 of a millionth of a second = 0.33 microsecond = 0.33×10^{-6} second.) Courtesy of J. Kim Vandiver and Harold E. Edgerton.

sound detecting device at the level of the edge of the candle, the sound of the bullet going through the flame would only be detected at the point marked A, when the bullet had already passed over the candle flame. We can estimate, very approximately, the speed of this bullet from the structure of the sonic shock wave using the fact that the bright line indicates the noise of the tip of the bullet, which has moved out from the horizontal line at the speed of sound. Thus if we consider the point C, then the sound energy has moved from D to C in the time that the tip of the bullet has moved from D to E. Therefore the speed of the bullet is the distance DE divided by the distance DC times the speed of sound. Using a ruler to make the necessary measurement, we discover that this ratio is 1.9. Therefore, we can say that the bullet is moving at approximately twice the speed of sound. If the gunfighters of the old west had known a little more about physics, they could have been philosophical about the fact that "you don't hear the one that gets you" by explaining that the bullet heading for you is traveling above the speed of sound so that the bullet will have done its damage before the sound of the discharge arrives.

Scientists have developed a way of comparing the speed of an object with the speed of sound in the substance through which the object is moving; the measurement is called the *Mach number*. The Mach number is named after the Austrian physicist Ernst Mach (1838–1916)[1]. In 1887 Mach did some experiments in which he noticed the sudden change in the nature of the air flow over a moving object as it reached the speed of sound. The Mach number of a moving object is the speed of the object divided by the speed of sound. The ratio of the size of the triangle of the shock wave in Figure 6.9 cannot be used to calculate an accurate actual speed of the bullet because the temperature of the air near the candle is unknown. (The temperature at the tip of a candle flame is of the order of 1200 °C!) The fact that there is hot air rising near the tip of the flame of the candle is illustrated by the disturbances visible in the shock wave caused by the different optical properties of the hot air from the immediate surroundings. However, our measurements allow us to estimate that the bullet of Figure 6.9 is moving at about Mach 2. Modern aircraft can travel at speeds as high as Mach 5.

Today scientists would measure the speed of a bullet by measuring the time required for the bullet to travel between two laser beams located at points such as D and E. However, long before the invention of the laser, forensic scientists had worket out a technique for measuring the speed of bullets known as a *ballistic pendulum*.

Forensic science is the use of science in the pursuit of justice in the law courts [16]. The word comes from the Latin word for market place, "forum", because the administration of justice in Roman times was carried out publicly in the market place. See, for example, the trial of Jesus by Pontius Pilate in the New Testament.

---

[1] Mach was not only an experimental scientist but also a philosopher. He insisted that the laws of nature were simply man-made generalizations — conveniences invented to collate innumerable observations. He maintained scientific laws had no independent existence apart from the observations they described; only the innumerable observations themselves had reality. In recent years the development of a new scientific subject called *deterministic chaos* has turned many scientists back to some of Mach's views on reality in science [19].

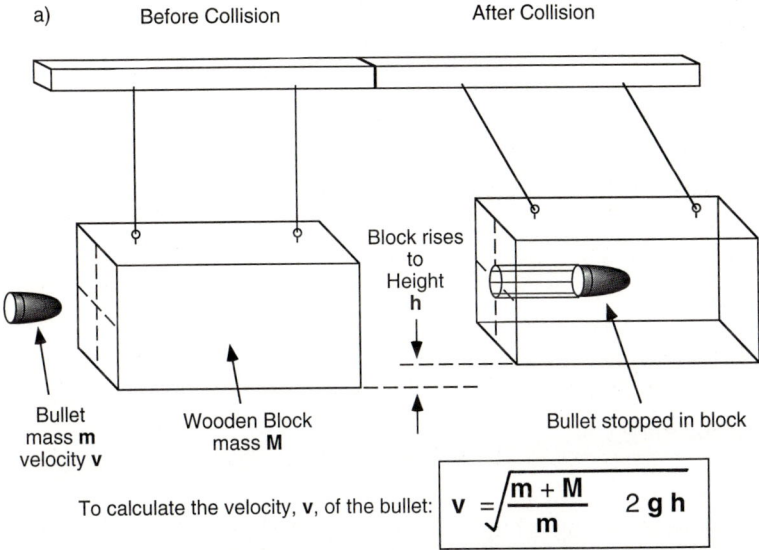

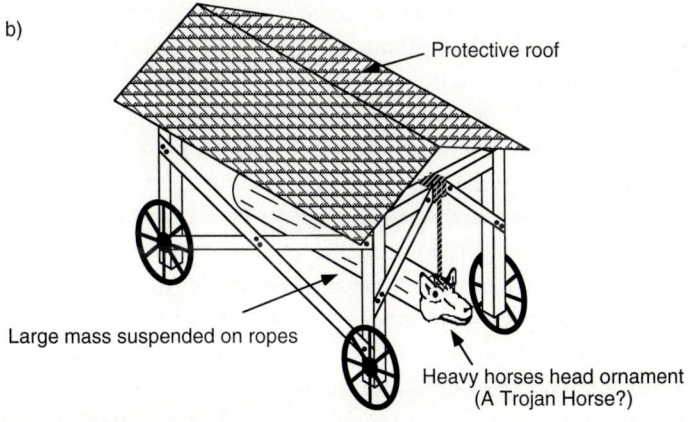

Figure 6.10 The energy stored in a ballistic pendulum can be used in several ways. a) The energy of a bullet captured in a block of wood, suspended on strings, causes the block to rise. This allows us to calculate the speed of the bullet before it entered the wood. b) The principle of the ballistic pendulum was utilized in some types of battering rams.

The basic system of the ballistic pendulum is illustrated in Figure 6.10(a). A block of wood of mass M is suspended from two strings. The bullet, of mass m, is fired into the block of wood and energy transferred from the bullet to the block of wood. This causes the block and bullet system to rise on the strings to a maximum height denoted by h. When in this position the block of wood comes to rest with all the energy in it stored as potential energy. *Potential energy* is the energy of position and can be shown to be:

$$E_p = (m + M) g h$$

where E_p is the potential energy and g is the acceleration due to gravity.

By equating this to the kinetic energy of the bullet before it hit the block, we have the relationship:

$$E_k = \tfrac{1}{2} m v^2 = (m + M) g h$$

where v is the velocity of the bullet and E_k is the kinetic energy of the bullet.

Equating the two energies and rearranging the equation we can show that

$$v = \sqrt{\frac{2(n + M)gh/n}{n}}$$

The principle of storing energy in a swinging block of wood such as that of Figure 6.10(a) was used by ancient warriors to create an effective *battering ram*. The log constituting the ram, often equipped with a head of toughened iron, was hung from several ropes underneath a protective roof. The soldiers operating the ram could approach the wall or gate to be breached with the battering ram under the cover of the roof made of metal or wet animal hides to prevent defenders from setting it on fire. When in position, the soldiers could swing the log backwards to give it energy and then let it swing forwards to batter down on the door. Using ropes in this way to make the battering ram, a ballistic pendulum, is much more efficient than having the men put their arms around the log and run at the gate to be battered down. If the warriors carried the ram, the operators would have to absorb the impact of the collision between the ram and the door, and the jarring of the impact would tend to scatter the people operating the system. I have never been a full believer in the story of the Trojan horse which supposedly concealed Greek warriors hidden in its belly when it was pulled into the city of Troy. The traditional story tells us that once inside the city, the hidden soldiers emerged from the horse to kill the guards and open the city gates.

Historic epics were often written centuries after the original events and, when phrases are used which were unknown to those writing the story down, they sometimes imagined fantastic explanations for strange events. For example, the hill people of Judea who wrote about the events of the Exodus had never seen tidal waters and would find it hard to understand how pharaoh's troops were swamped by an incoming tide as they attempted to cross mud flats. Therefore, what was probably a wind-driven high tide became a miraculous parting of the waters created by Moses — with a little help from God. In the same way, if the storytellers repeating the traditional story that Troy had fallen because of the Trojan horse had no knowledge of a battering ram with a head the shape of a horse, they would not understand that it was the momentum of a swinging log that broke down the gates of Troy. They therefore invented a group of warriors carried in the hollow of a model horse; a much more exciting explanation than a swinging log. We will probably never know the real explanation of the events of the battle of Troy, but regarding the Trojan horse as a rather large ballistic pendulum provides an interesting alternative to stupidity on the part of the Trojans. Recent research suggests that the whole battle of Troy was not really over the kidnapping of Helen (the face that launched a thousand ships), but was an expedition of Greek warriors to break the monopoly Troy held over trade in Black Sea tin needed to make good bronze weapons [20, 21].

After the development of breech loading guns and rifles, many engineers tried to develop rapid firing guns. The *Colt* with its revolving magazine of ammunition, a design feature which gave the gun the name *revolver*, enabled the gun fighter to fire several shots in quick succession before the gun had to be reloaded. The *Springfield repeating rifle* (named after the town in Massachusetts where it was made) added a new dimension to infantry fighting. But it was the advent of the machine gun which totally changed the balance of warfare.

The *Gatling gun*, the first commercially successful machine gun, was invented by the American Richard Jordan *Gatling* (1818–1903). He started out as an engineer, but then became a medical man graduating from Ohio Medical College in 1850. When the civil war between the states started, he began to work on the development of a rapidly firing gun in the belief that such a gun would make it unprofitable to fight. By 1862 he had made a gun that could fire nearly six bullets per second. However, it was not used until the very end of the Civil War. Asimov notes that, "The name of Gatling lives today in the slang word '*gat*' for a gun."

The rapid fire machine gun used during the First World War with devastating effect was the *Maxim gun* named after its American–English inventor, Sir Hiram Stevens *Maxim*, who was born in Maine in 1840 and who died in England in 1916. The basic design of the Maxim gun was perfected by 1883. A novel feature of the gun was that it made use of the energy of the recoil of a fired bullet to eject the empty cartridge and bring in the next round of ammunition. The British army adopted the Maxim gun in 1889. To a later generation, the absolute stupidity of the generals of World War I in sending massed troops into the automated fire of machine guns is absolutely incredible. In the battle of the Somme over one million men died, and at one point, 20,000 men a day were being killed by machine guns. The French were still using red-coated infantry who were slaughtered as they ran into the fire of the machine guns. The *Bren gun* was a small hand-held machine gun so named because it was first manufactured in Bruno, Czechoslovakia and then in Enfield, England. The machine gun dominated warfare until the invention of the tank (Section 6.6).

The development of different small-arms machine guns continues to be a preoccupation with the arms manufacturing countries of the world. In 1986, the West German gun manufacturers Heckler & Koch designed an assault rifle called the G-11 which fires cartridges that have no metal case. The bullet is pressed into a solid block of explosive which also contains the pellet of the ignition system at the base of the explosive. Such *caseless ammunition* weighs much less than the conventional ammunition and is much cheaper. The traditional metal cartridge case has to be machined and the weight of the ammunition increases transport costs. With caseless ammunition, the assault rifle is able to achieve a high rate of fire because there is no need to extract the cartridge case after each shot. Elimination of the movement of the gun to eject shells improves the ability of the soldier to aim and fire the gun [22].

Section 6.3   Shrapnel, Dumdums, and Devastators

Shrapnel, dumdums, and devastators are all types of shells and bullets. The first bullets used in guns and rifles, were simple lead slugs. This is still true of most ammunition used today. Lead has the advantage of a relatively high density which enables it to store larger amounts of kinetic energy as it moves up the barrel, and its deformability under pressure helps a bullet to fit snugly into a rifled barrel when fired. However, the firing of lead bullets results in the vaporization of some of the lead and the subsequent inhaling of the lead fumes by those who fire the weapons. There have been many cases of lead poisoning among policemen who used to attend enclosed firing ranges for gun practice and were unaware of the danger of lead fumes. In modern firing ranges the air is continuously filtered and those practising with their guns are protected from the lead fumes. Sportsmen cause a health hazard for birds because of the lead shots that fail to hit their targets. The unsuccessful lead shots accumulate in the mud and debris of river bottoms and cause lead poisoning to water fowl. The Canadian government has tried to persuade the hunting community to switch to iron or bismuth shot. Iron has a lower density than lead, and for a given velocity, v, at the muzzle of the gun, the kinetic energy of the bullet is

$$E_k = 1/2 \ m \ v^2$$

where m is the mass of the bullet. The density of a material is the weight of a unit volume (e.g., 1 cubic centimeter) of that material. Lead has a density of 11.34 grams per cubic centimeter, iron has a density of 7.86 grams per cubic centimeter. Therefore, for a given type of gun the emerging iron pellets would have less energy than lead pellets. Furthermore, iron pellets cause greater wear in the barrel of the gun because they are harder.

In recent years, the US military has used uranium in antitank (armor piercing) bullets. Natural uranium is a mixture of two types of uranium atoms which are called *isotopes*. (Isotopes are discussed in more detail in Chapter 9.) To create nuclear reactors and nuclear weapons, the most highly radioactive isotope of uranium has to be separated from the other uranium atoms. After the isotope has been taken out of natural uranium, the remaining material is described as *spent uranium*. The density of spent uranium is

19.05 grams per cubic centimeter so that much more energy can be invested in a uranium bullet as compared to a lead bullet of the same size and this gives it the ability to pierce armor [23].

Special-purpose bullets have also been developed. Bullets designed to pierce armor plate can also be made of hardened steel. *Incendiary bullets*, which contain an ingredient that burns easily, were developed in order to enable infantry to attack gasoline-powered vehicles. When the machine gun was developed, the direct aiming of the gun became less important than the ability to know where the spray of bullets was traveling. Therefore to track the path of bullets leaving a machine gun, the *tracer bullet* was developed. The path of this bullet became visible throughout its trajectory because of a burning pellet in the base of the bullet. The tracer bullet enabled the gunner to see the path of the bullet and hence the stream of fire as he aimed at the target.

During the Gulf War the public became familiar with one of the more expensive guided missiles known as the *Tomahawk*, named after the small throwing axe used by North American Indians. Also used during the Gulf War was the *Scud missile*. This was more properly known to the military authorities of Iraq as the *Hussein Missile* and was a modified version of a Soviet Missile known a Scud V. The Iraqis originally developed the Hussein missile in the 1980-88 war with Iran. To increase the Hussein's range as compared to the original Soviet Scud V missile the fuel tanks were lengthened and its warhead was made much lighter. These changes made the missile unstable and caused it to flop belly first as it re-entered the atmosphere, often breaking up in the process. The *Patriot missile* was used to combat the Scud attacks but it could not even detect the Scud missile until it was within 90 miles of its target. Although the Patriot missiles were credited with being a very effective weapon, other analysts independent of the company making the Patriot, estimate that it was far less effective than claimed by the authorities. In particular because the Hussein missile started to break up on entry the various bits unintentionally provided effective decoys to distract the Patriot. Originally the Israeli government refused to have Patriot missiles on their territory because they considered it to be an ineffective and dangerous weapon. It is thought by military analysts that the Israelis accepted the Patriot missile in return for not attacking Iraq during the war since, if the Israelis had joined in the conflict, the Arab allies of the Gulf War force may have dropped out of participation in the war. One analyst points out that, although the makers of the Patriot missile boasted that 42 Scuds had been tackled with 41 of them being intercepted, since every Scud requires at least two Patriots, and even more if the first two go awry, some industry sources estimate that 100 to 120 had been launched at a cost of at least 1 million dollars each [24-26].

The search for weapons to counter missiles continues. One of the main problems during the Gulf War was that the Scud missiles were launched from mobile launchers and the Allies had trouble tracking down the launch vehicles. However, it is also estimated that Iraqi troops launched only a fraction of their Scud missiles during the war because some operators of the missile launchers were reluctant to stop and shoot for fear that they would be pinpointed. A scientific device known as *LIDAR* is being developed in a form suitable for locating the firing point of a missile. This device, which takes its name from the phrase LIght Detection And Ranging, is similar to the old radar in that it sends out a signal and analyses the return time of the reflected signal, but in this case the signal is laser light. When the laser is aimed at a *rocket plume* (the term for the exhaust trail left by the rocket) some of the laser light is reflected off the smoke and fine-particles in the plume to a telescope back at the LIDAR base which detects the reflected light. A computer can then work back from the details of the direction and size of the plume of smoke to locate the original launch point [27]. (See discussion of detectors of volcanic dusts for aircraft safety in Chapter 11).

Developments in armaments frequently give generals the edge over an enemy, but historians often fail to note the difference that superior weapons have made in the outcome of a battle. Desk-bound scholars may attribute success to supreme strategy when often, in fact, strategy was secondary to the balance of power between the armies as represented by their technology. For example, historians discussing the famous battle of Waterloo in which Napoleon was finally defeated by his enemies, fail to mention the role played by the shrapnel shell in the battle. In modern English the term *shrapnel* has two meanings:

(1) An anti-personnel projectile or shell containing metal balls or fragments, which was fused to explode in the air above enemy troops.

(2) Shell fragments from a high explosive shell.

During World War II when I was at school children used to collect and swap shrapnel fragments from shells and bombs dropped over England by the Nazis. The shrapnel shell was named after its inventor, General Henry *Shrapnel* (1761–1842). During his military career, General Shrapnel served in Gibraltar and the East Indies. Records show that on several occasions his quick thinking was a great asset to the army [2, 7, 28]. Thus when the British army was retreating from the town of Dunkirk during the Napoleonic Wars, Shrapnel observed that the wheels of the gun carriages sank into the sand as the men tried to pull them. He suggested that the wheels be locked so that the carriages could be skidded over the sand without sinking into it. (Dunkirk is a town on the coast of France, which was to become the scene of the great military episode of W.W. II when the British army was saved from annihilation by a fleet of small boats.) Shrapnel also devised the strategy of decoy fires. The British set fires a distance from their own troops so that the enemy wasted their ammunition shelling the fires while the British troops were elsewhere resting before the next day's battle.

The word *decoy* has an interesting history; it comes from a Dutch "de kooi" which means "the cage" or "trap." A decoy was originally a deceptive device for trapping an enemy or an animal and has since come to mean something which diverts the attention or the aim of aggression to increase the efficiency of a military or hunting strategy.

Shipley gives us some interesting information about the meaning of *strategy*. He tells us an old Chinese story about a general who, when sending his advance guard through a wooded section where the enemy might lay an *ambush*, ordered each man to take a stone and hurl it into the trees. If birds flew away, there were no men hidden there and the army could proceed. The soldiers threw the stones, the birds flew away, and the army marched through the woods and was ambushed. Apparently the opposing general, hiding his men in the trees, had ordered each to catch a bird which was to be released when the approaching force threw their stones. The overall plan by which a general leads his forces to victory is described as his strategy from the Greek word "stratogos" for "the army leader." The word ambush is from an old French word "embusche" meaning, literally, "to hide in the bushes" [8].

Apart from his contributions to army strategy, Shrapnel worked for a long time on a shell which was intended to explode in the air and scatter small projectiles amongst enemy troops. Henry Shrapnel did not fight with Wellington's army during the final defeat of Napoleon's forces at the battle of Waterloo, April 18, 1815; but Sir George Wood, who commanded the British artillery at the battle, wrote to Shrapnel from Waterloo Village three days after the battle and said,

> Had it not been for the Shrapnel shells it was very questionable whether any effort of the British forces could have led to the recovery of the farm house of Lahaye Saint. And hence, on this simple circumstance hinged entirely the turn of the battle [28].

The Shrapnel shell had been introduced ten years earlier and the British army's response was so enthusiastic that the Duke of Norfolk testified at an inquiry that his troops had derived great benefit from its use and he considered it most desirable that the invention not be made public. Shrapnel was never able to convince the British government that he should receive financial reward for the funds he had taken from his own resources to develop this shell.

The "dumdum" bullet of our section title is defined [7] as

> a bullet with a soft nose, expanding on contact to cause a large, rough-edged wound.

The bullet is named after the place Dum Dum, a town near Calcutta in India where the bullet was first manufactured. Because the nose of the bullet deformed on impact, the wound made by such a bullet was much more severe than a simple wound made by the direct impact of a bullet. Wounds caused by modern high speed bullets include a shock wave in the flesh from the force of the impact. The dumdum bullet is banned by international arms agreements, but in the realm of weapon design and use, international agreements do not always prevent the use of a device. Thus the assault rifle, designated by the Russians as the AK74, fires a bullet which is essentially a dumdum missile. By military standards, the caliber of the bullet, 5.45 millimeters, is small and slender (diameter-to-length ratio of 1 to 4.5). The bullet leaves the gun at 900 meters per second (approximately Mach 3) and is spinning at 4500 revolutions per second to achieve stability in flight. The main body of the bullet is steel, but the nose contains a lead plug behind an air gap. When the bullet hits a soft target, such as a human being, the hollow tip deforms and the center of gravity of the deformed bullet shifts so that the bullet starts to tumble. In the words of a technical article [29] on the bullet,

This creates all the unpleasant effects of a dumdum bullet. While the entry wound is small the tumbling bullet follows a spiral path tearing flesh and creating an extremely nasty injury.

Because of its small weight, a soviet soldier using an AK74 can carry almost twice as much ammunition as a British soldier equipped with a 7.62 millimeter self-loading rifle.

During the late 1970s, when airplane hijacking became a major problem, arms experts developed a bullet known as the *devastator*. In this type of bullet, a small explosive charge is inserted into the nose of the bullet. Upon impact, the explosive charge is supposed to explode. When this happens, the bullet is fragmented. The intent of such a design is to quickly incapacitate the highjacker with a relatively small bullet and to prevent the bullet from passing through the highjacker and damaging the frame of the aircraft. A traditional gun, used at close quarters in an attempt to prevent a highjacker from taking over a plane, could cause serious damage to the aircraft. A devastator bullet was used by John Hinckley in his attack on US President Ronald Reagan in April 1981. Fortunately the lead azide charge in the bullet failed to detonate. History may have been very different if President Reagan had sustained the serious wound that could have been inflicted by such a detonated bullet [29].

The bullets shown in Figure 6.11 were given to me on a visit to a company that made chemicals for vaccinating animals. They are *biodegradable bullets*. Biodegradable means that it can be broken down in the environment by living creatures such as bacteria and insects or by processes occurring in situ in the natural metabolism of the animal. I was told that they were used in Texas to vaccinate cows coming off of rail cars. The hollow center was filled with the chemical or vaccine which was to be delivered to the animal.

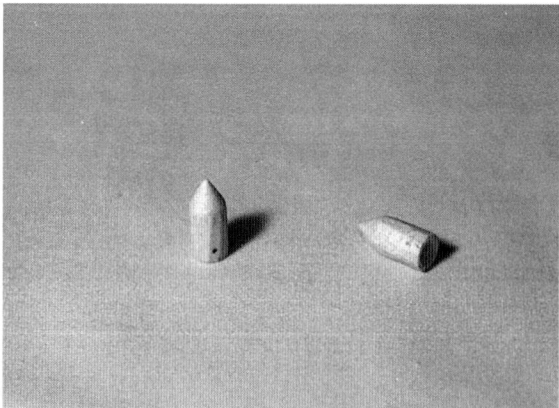

Figure 6.11 Special, small calibre, bullets have been made to vaccinate livestock against disease. These bullets are made of a material which can be broken down in the animal's body, biodegradeable, thus releasing the medication.

The bullets were then fired into the rump of the animals where they lodged and delivered the medication or vaccination material. You will note that the surface of the bullet has angled grooves to make the bullet spin as it travels to its target. I was told that this was a lot less traumatic to the animals than the act of capturing them and holding them still whilst they were injected using a very large needle by a veterinarian.

The same types of bullets have been used to deliver contraceptives to the backside of deer in areas where their population has become a problem. In one trial carried out by research workers at Purdue University in Indiana 250 white tailed deer were treated with contraceptives [30]. Two types of contraceptives were tried. One is *norgestone*, a steroid used as a contraceptive in cattle. The other is a vaccine which stimulates the doe's immune system to inhibit the ovarian cycle. The scientists carrying out this study claimed that biodegradable bullets offer advantages over other ways to administer the contraceptive. Thus Swihirt one of the investigators stated

> *Darts for example tend to fall to the ground where they may injure other animals or people and the biodegradeable bullets pose no such danger as they remain embedded in the deer.*

At the time of writing this book, there was a controversy going on in Ontario about the bullets that should be used by police [31-32]. The police were pushing for the use of *hollow point bullets*. Like the devastator bullets hollow point bullets expand into a flower shape on contact and are designed to remain inside the body of the person who is shot. The police claim that the traditional bullet at close quarters can pass through someone or something without being stopped "After it has been shot it is pretty hard to tell where it is going" said Tomasik a spokesman for the Ontario police. The pressure to bring in hollow point bullets was actually generated by the murder of a police officer in Sudbury, Ontario. A young police constable, Constable Joe MacDonald of the Sudbury Regional Police Service, had stopped two suspects. In a subsequent struggle, he managed to get two shots away into one of the attackers, Clinton Suzak, who then shot Constable MacDonald. A police spokesman states that

> *If we would have had our new service pistol with hollow point ammunition available the circumstances could have been very different.*

Some civil liberties people expressed disquiet that such bullets can kill more easily but here again the police argued that if a gun is drawn on a policeman, the attacker must expect the most drastic consequences as the police officer acts in self defense.

In their attempts to suppress riots, police have used rubber bullets with the idea that such bullets would bruise and hurt but not kill. *Rubber bullets* have been largely replaced by plastic bullets. The term is somewhat misleading since the missiles launched in riot control are much larger than the traditional bullets made from metal. A *plastic bullet* such as that used by the British military in Northern Ireland, and other trouble spots, is made of PVC (polyvinyl chloride) and is a cylinder 38 mm in diameter and 10 cm long. It weighs 135 grams and leaves the muzzle of a riot gun at a speed of

210–270 kph. In the words of Jonathan Rosenhead in an article on the use of plastic bullets,

> This plastic projectile has roughly the same impact as a cricket ball traveling twice as fast as a ball thrown by a good fast bowler.

The police justify the use of plastic bullets on the grounds that they discourage rioters from throwing stones and such things as Molotov cocktails. (The dictionary definition of a *Molotov cocktail* is a glass bottle, or other suitable container, filled with a flammable liquid such as gasoline. It is equipped with a fuse on the outside of the bottle that is lit before the bottle is thrown. As the bottle breaks upon impact, its contents are ignited by the fuse and it constitutes a low cost firebomb. It is named after a Russian statesman who fought in the Russian Revolution and later became the Russian spokesman at many international conferences.) The plastic bullets are not supposed to be able to kill anyone, but during riots in Northern Ireland in the period up to 1985, 43 000 plastic bullets were fired, causing twelve deaths. The fatalities from the use of these riot control devices appear to come from several causes. First of all, the bullets are supposed to be aimed at the lower part of the body, but several children have died of head injuries when accidentally hit during attempts by police and soldiers to control riots. Probably many of the bullets had ricocheted off walls and sidewalks. Rosenhead gives figures demonstrating that rubber bullets killed one person for every 18 000 rounds fired, whereas plastic bullets cause one fatality per 4000 rounds. The British army rules state that plastic bullets should not be fired at a range of less than 20 meters except where the safety of the soldiers, or others, is threatened. However, people are notorious for their inability to judge distances, particularly under stressful conditions. It appears that many fatalities are caused by use of guns at too close range. Measurements have shown that the plastic bullets, when fired under normal conditions, have an energy of 285 Joules. Tests by the US army on animals have identified the danger threshold for the kinetic energy of impact weapons as 122 joules. The plastic bullet at the extreme range of 50 meters still carries 150 joules of energy.

There is an ongoing search for safer weapons for crowd control by the police and security forces. Thus in 1994 the company Armwell in England claimed in a patent that they had developed jelly bullets. As discussed by Fox in the New Scientist [32]

> Normally it would be impossible to fire jelly from a gun because it would gum up the barrel. In the new type of ammunition the jelly is contained in a tube. When the gun fires the explosive charge drives the tube down the barrel. At the muzzle of the gun the tube hits a spring which stops it dead but the jelly inside shoots out of the end of the barrel as a long thin bullet of gunge which splatters all over the target.

Section 6.4 Laser Rifles and Swords of Light.

The invention of the laser in the early 1960s has created a revolution in a subject known as *electro-optics*. This word combines the word electronics with optics. It is a term used to describe the use of optical devices to generate sophisticated electronic equipment. As will be discussed in more detail in the next chapter, light energy and infrared energy are both types of electromagnetic energy, differing only in what is known as the wavelength.

Electromagnetic waves are difficult to explain in simple terms but as the word implies they have both electrical properties and magnetic properties. In a light wave, energy travels in pulses through space. The distance between adjacent pulses of energy is known as the *wavelength* of the electromagnetic radiation. The visible light which the human eye sees has wavelengths in the range of 0.3 to 0.8 micrometers, yellow light has a wavelength of 0.77 micrometers [16]. Scientists seeking to develop powerful sources of light developed a device which has come to be known as a *laser*. This word developed from the acronym Light Amplification by Stimulated Emission of Radiation. In the device an avalanche of electrons triggers an optical wave which itself stimulates the emission of more light, hence the term stimulated in the phrase describing the light. Ordinary light of the type given out by the sun is a mixture of many different wavelengths and one of the special features of laser light is that the energy present is very *monochromatic*, i.e., the radiation has only a single wavelength. The other important aspect of a laser is that the beam of light can be extremely powerful. (We will discuss power levels in some lasers later in this section).

The power of a laser is measured using a unit known as a *Watt*. This unit is named after James *Watt*, a Scottish Engineer and inventor who lived from 1736 to 1819. James Watt invented the steam engine and was one of the first engineers to measure the capacity of engines to do work [11]. He developed the concept of "horse power". In modern scientific units, one horse power is equal to 745.7 Watts. (If you worked a horse at this rate today, you would risk being charged with cruelty to animals.) Incandescent electric light bulbs are rated according to the power that they consume in order to generate the light that they give out. On the domestic scene, the power of lamps is rated by the amount of energy they consume and varies from 3–25 Watts for modern energy-saving lamps for domestic use to as much as 500 Watts for outdoor area lighting. Note that the Watt is a unit describing energy used per unit of time.

Apart from its monochromaticity, another important property of laser light is what is known as its *coherence*. The word coherence in every day language means organized and understandable. It comes from the Latin pair of words which mean "to stick together". In a scientific sense the name coherent is used to describe a system which is organized in a special way. Light energy leaving an incandescent bulb consists of billions of short bursts of light of different wavelengths generated at a multitude of different times. It is said to be *incoherent*. Light from a laser on the other hand consists of billions of wave chains all in step with each other and it is described as *coherent light*. The difference between laser light and ordinary light is the same as the difference between an organized army marching in step and a riotous crowd moving at random. Because of its coherence, laser light can readily be used to generate systems such as holograms.

The third property of laser light which makes it very useful in science is the fact that a laser beam only has a very small divergence as it travels away from the laser. The average distance of the Moon from the earth is 384 000 kilometers (238 000 miles). Before the invention of the laser it was impossible for scientists to reflect light from the surface of the Moon because the best focussed search lights on earth spread out to cover an area of the Moon so large that the intensity of the energy reflected was too low to detect on the Earth. The divergence of a laser beam however is so small that it can be used to bounce a signal off of a mirror left on the moon by the Apollo astronauts. (See discussion of cosmic collisions and the vibration of the moon in Chapter 8).

In military applications of the laser, the monochromatic nature of the light, the power of the high intensity beam, and the low divergence are all exploited in weapon design. Scientists starting to use laser beams as weapons against aircraft soon ran into a major problem. As soon as the laser hit a surface it vaporized some of the material and created a cloud which reflected the laser light and diminished the impact of the energy of the laser beam on the object at which it had been directed.

Just after the movie Star Wars went on general release children could be seen playing in the street with what they called Lightsabers. These were toy versions of a weapon used in the movie by combatants who used beams of powerful light in the same way that the musketeers used to use their swords. Although such applications of lasers are picturesque and imaginative they are probably outside of the realm of reality for the forseeable future. Much more likely is the development of a laser rifle. Thus the Desmans Corporation of France are already marketing what they call a *laser rifle*. In their sales literature they describe it as a weapon designed for the startling of birds. It is said to be capable of dispersing all sorts of birds within a range of 2.5 kilometers. It is powered by a 12 volt battery and uses what is known as a helium–neon laser. Its power is stated to be 5 *milliWatts*. (5 thousandths of a Watt). They claim that the bird is startled by the strong contrast between the ambient (surrounding) light and the red light of the laser beam. The bird is in no way hurt because the power of the laser is so low. The rifle is most effective at lower light levels (when the contrast between the laser light and the surrounding light is higher). They say that it has potential use in hunting, wildlife management, and for airport authorities concerned with the removal of birds which pose a hazard to aircraft (see Chapter 11) [33].

Although the Desmans laser rifle is designed as a super scarecrow it is an ominous symbol of what could be developed in the future. In the future soldiers may use weapons deliberately designed to blind the opposing armies by shining lasers into their eyes. This possibility was discussed in 1992 by Hecht and in 1994 the Red Cross began a campaign to ban the use of battlefield lasers to blind the enemy [34–37].

Hecht points out that the military are actively developing instruments such as the laser rifle essentially on the grounds that they can be used to defend armies against guided weapons. One of these weapons is called the *Stingray*. It weighs 160 kilograms and is built around what is known as a solid-state *niobium laser* (named after an element present in the laser). In one mode, Stingray can scan a wide beam over a broad area searching for reflections from enemy sensors. In another mode it can generate an intense narrow beam to blind or destroy enemy periscopes, night vision equipment, and gun sights. It is claimed by the US authorities that it is only intended as a defence measure against guided weapons, not to deliberately blind enemies. Another weapon under development in the USA, the *Dazer*, is meant to be carried like a sub-machine gun. Few details are available on the system but is intended to jam enemy electro-optical sensors.

Already police patrolling the subway system in New York City have started to carry guns equipped with laser sites so that they can shoot more precisely. The laser produces a red dot at the spot that the gun is aimed. Police are hoping that the laser will deter violent criminals who cruise the underground railway system looking for victims. They hope that the sight of a red dot in the middle of their chest will be a chilling announcement that a bullet could follow if they do not surrender. The officials claim that, if the police officers are forced to shoot, the aim will be more accurate in the dimly lit subway stations, cutting down the risk of innocent bystanders being injured by stray bullets. In discussing the new device Kiernan [38] comments that

> *The gun bears a label warning anyone against looking into the laser to avoid damaging the eye but the bullets however carry no such health warnings!*

Because lasers are very monochromatic it is possible to design goggles to stop a specific wavelength of radiation so that the users of such weapons are protected from reflected rays coming back from the enemy.

The damage that can be done to the eye depends on the wavelength of the laser beam. Infrared beams are particularly dangerous because they are invisible to the naked eye and can nonetheless cause serious damage to the eye. After accidental exposure to such a laser a scientist David Decker described the damage that he sustained. He tells us that when he walked into a laboratory he was inadvertently exposed to a pulse of light of wavelength 1.064 micrometers (red light is around 0.8 micrometers in wavelength and wavelengths longer than or close to this value are known as infrared wavelengths; see discussion in the next chapter). He then describes what happened next [34]

> *When the beam struck my eye I heard a distinct popping sound caused by a laser induced explosion of the back of my eye ball. My vision was obscured almost immediately by strings of blood floating in the eye and by what appeared to be small piece of matter suspended in the fluid of the eye. It was like viewing the world through a round fish bowl full of glycerol into which a quart of blood and a handful of black pepper had been partially mixed.*

He tells us that as a Vietnamese war veteran

I have seen several terrible scenes of human carnage but none effected me more than viewing the world through my blood filled eyeball. I went into shock as is typical in personal injury accidents.

Again he tells us that

The beam struck my retina (the network of nerves at the back of the eye which enables us to see things) between the fovea and the optic nerve, missing the optic nerve by about 3 millimeters.

After six months there was no physical improvement in the functioning of the eye. Laboratories in which lasers of any significant power are used have a sign outside of the laboratory warning people to be careful and wear eye protection when entering the laboratory.

Hecht, in discussing the deliberate use of lasers to blind people, said that it would probably be useless at distances beyond a kilometer, nor would it stop soldiers in close combat. But

The fact that the enemy were using such weapons could serve as a weapon of terror undermining morale and keeping troops pinned down.

Lasers were first used in the battlefield in the 1970s in Vietnam. *Laser range finders* measured distances by firing pulses of light at targets and timing the return of the reflected light. Lasers are used by guided missiles to designate a target by the series of coded pulses bounced off the target which the missile recognizes and takes aim at. During the Gulf war in 1993, public television showed many pictures of the successful hitting of a target using such guidance systems. The low divergence of the beam plus the well-defined wavelength is essential to the success of this type of weapon. After the Gulf War critical analysts challenged the superiority of laser guided weapons as claimed by the military authorities and their suppliers [39]. In the closing section of this chapter we will discuss this aspect of laser guided missiles in more detail.

Section 6.5 Moon Shots

Jules Verne was a French writer who lived between 1828 and 1905. The entry on Verne in a reference encyclopedia [40] states

He struck a new vein of fiction whereby he earned a world wide reputation. He cleverly exaggerated the possibilities of science and gave verisimilitude to narratives of wild adventure.

Movies have made some of these novels widely known to a later generation. For example the film Twenty Thousand Leagues Under the Sea is based on one of his books. In 1865 he wrote a book entitled "From the Earth to the Moon". In this book he envisaged that people would be shot from a supergun to reach the moon [40]. Will tomorrow's astronauts be shot from a gun instead of taking off from Cape Canaveral in a rocket-assisted spacecraft?

In actual fact, there are serious physical problems associated with shooting people to the moon from a gun, such as the problems of the excessive acceleration that would be required to launch the missile to escape the earth's gravity, and also heating problems from friction as the missile left the muzzle of the gun and traveled toward the moon. However a missile specialist in the United States, John Hunter, thinks that superguns may be the most cost effective way of delivering the components of a space station into orbit. He claims that superguns would deliver orbiting components for assembly into an orbiting space station at a cost of about US$ 500 per kilogram (which is twenty times cheaper than using rockets to send such equipment into space) [41].

Hunter has already built a piece of equipment which is known as *Super High Altitude Research Project – SHARP*, which is a giant gun weighing 90 tons having a steel barrel 47 meters long and 10 cm internal diameter. By 1994 this gun had hurled 90 projectiles weighing up to 6 kilograms into a sandbagged bunker at velocities many times greater than the speed of sound. Hunter works at the Lawrence Livermore National Laboratory in California. He wants to build a bigger gun for launching parts of the proposed international space station. In fact, the gun for such a purpose would be a hybrid between a gas-driven missile and a rocket which is defined as a missile which carries its own fuel (see Chapter 7). He has calculated that one could build a gun launching a small single stage rocket to deliver pay loads of up to 3300 kilograms into orbit. He calls it the *Jules Verne Launcher (JVL)*. The JVL would have a barrel 3.5 kilometers long with an internal diameter of 1.2 meters. The rockets would emerge from its muzzle at a speed of 7 kilometers per second (Mach 21). This launch speed would send the missile up to 60 kilometers when a rocket engine would ignite and accelerate the payload to 7.5 kilometers per second to position the object in an orbit 600 kilometers above the Earth's surface. Hunter says that if he can find a sponsor, he can build the equipment in two years at a cost of US$2 billion.

In the prototype, the missile is driven up the barrel by a series of injections of superheated hydrogen gas. As Hunter points out, like any gun, SHARP recoils when it is fired. Two one-hundred ton lumps of steel mounted on tracks absorb the momentum of the driving tube while another ten-ton mass absorbs the recoil of the actual barrel of the gun. Hunter is pursuing work on such superguns, continuing the pioneering work of William Corzier, a scientist at the New Mexico Schools of Mines who carried out experiments in 1944. In a review article on superguns, Crabb tells us that in the 1960s the Canadian born rocket scientist Gerald Bull built three enormous guns for the US army and the Canadian government [41]. One of these had a barrel 36 meters long which, according to Bull, could fire an 84 kilogram projectile to an altitude of 180 kilometers. After Canadian and US authorities lost interest in superguns Bull worked on the guns for other countries. In 1990 he was working on a supergun in Iraq when he was murdered. It has never been proven but is was suspected that Israeli Commandos took care of Bull because his supergun was aimed at Jerusalem. After the Gulf war, when inspectors visited Iraq, they examined the original supergun which was under development at a site 200 kilometers north of Baghdad. The Barrel was 52.5 meters long with an inside diameter of 350 millimeters [42]. According to Iraqi engineers it could shoot a projectile 250 kilometers. This missile was itself a hybrid rocket since the projectiles were launched up the supergun's barrel with packs of explosive separated by bundles of wood. The wood bundles prevented the simultaneous detonation of the separate packs of explosive charges. They went off in sequence like rocket stages as they moved up the long barrel of the gun. The United Nations inspectors, when they visited Iraq after the Gulf war, discovered that the Iraqi government was building an even bigger gun which had a barrel 1 meter in diameter. Apparently the parts had been smuggled into Iraq, in defiance of embargoes, under the guise of components for a large oil pipeline [42].

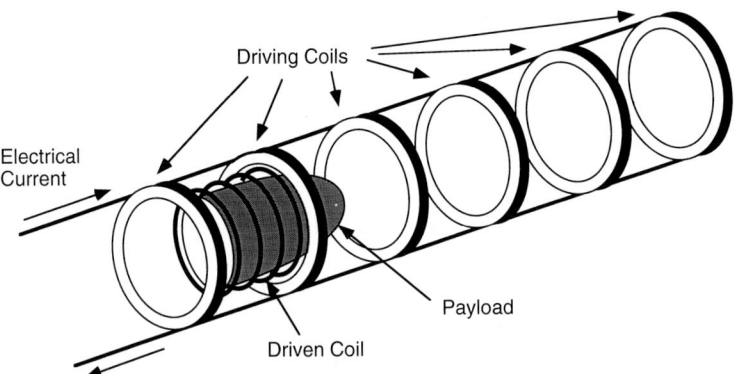

Figure 6.12 A new type of gun, called a rail gun, which uses electric currents to drive a payload to very high speeds along its "barrel" has been proposed as a possible system to launch small objects into Earth orbit.

Another type of supergun is being built by workers in Brooklyn. This type of gun uses electrical forces to drive the missile. Its developer, Zivan Zaber, points out that for a projectile to be shot directly into orbit it would have to leave the muzzle so fast that it would burn up in the atmosphere. The electrical currents needed to propel the missile would be of the order of a million Amperes and would require very specialized high-speed switches. (An *Ampere*, abbreviated "amp", is a unit of electric current named after the French scientist Andre Marie Ampere who lived from 1775 to 1836.) [43].

Other high-speed guns are under development in Scotland. These guns work on what is known as a *rail gun principle* [44, 45]. The basic system is shown in the sketch of Figure 6.12. The projectile sits on top of a special coil which travels along the barrel of a gun. There is one rail taking an electrical current to the coil and another on the opposite side for the return of the current. When the current flows in the coil a magnetic field drives the missile and the coil up the barrel. Special power sources are being developed for such guns. One already built produces 6.7 megaJoules of energy using a direct current of 750 kiloamps. The two copper rails feeding the driving mechanisms have to be replaced after several firings because of the heat produced during firing. A small scale rail gun has successfully fired a three millimeter ball-bearing at a speed of 4.2 kilometers per second. One gun under development at the Culum Laboratory in Great Britain is being used to study the effect of high-speed space debris on the integrity of a satellite when such debris is encountered in outer space (see Chapter 8).

In a letter to New Scientist on August 21 1993, a British physicist J.F.Allen points out that it is unlikely that such rail guns will ever be useful in active warfare because of the power requirements and the difficulty of operating them. He concludes his letter [46] by referring to the Scottish rail gun experiment

> From my experience I would not put any money on the gun's success but it will undoubtedly keep the team of scientists and engineers in hot dinners for some time.

Apparently, during World War II, the German's tried to develop a supergun, which they called a V3, to fire shells across the English channel at London. (In the next section we will look at V1s). The supergun that they developed was buried in French cliffs for protection against bombing. They used a sequential set of charges similar to those used by Bull in his supergun. British intelligence heard of the development of the weapons and the site was bombed using ten-ton earthquake bombs developed by Barnes Wallis. These successfully knocked the barrels out of line [47].

Section 6.6 Tit for Tat in Missile Development

In our review of the way in which the gun has evolved over the centuries we see a constant struggle in the missile industry to produce weapons of attack and defence. We have already noted that the machine gun completely altered warfare even though it took the generals a long time to appreciate the vulnerability of infantry when faced with such weapons of mass destruction. In turn the tank protected infantry from the machine gun and created mobility in warfare. Now in the late 1990s it is recognized that the last big tank battle has already been fought and that the role of tanks in future warfare is going to be limited by the accuracy and power of guided missiles. The battle which confirmed the vulnerability of tanks to new guided missiles was during the Israeli–Egyptian conflict of 1973. Herzog, who later became President of Israel, was involved in that conflict as a Major-General in the Israeli army. The Egyptians destroyed many Israeli tanks with Soviet-made *Sagger missiles*. These missiles were guided by joysticks and lengths of wire. Herzog tells us that [48, 49]

> As we crossed the Suez canal hundreds of guiding wires of anti-tank missiles lay strewn across the road as if a giant spider's web had collapsed.

The Egyptians, although they destroyed many tanks, also suffered heavy casualties because the type of guided missiles they were using required that the guidance person had to keep in the line of site of their target for as long as it took to guide the missile towards the tank. Their visibility made them vulnerable to the tanks' weaponry. The different types of shells developed to penetrate tank armor are beyond the scope of this book but the interested reader can find an interesting discussion in Refs. [48–51].

To reduce fatalities among operators, new guidance systems have been developed in which the missile is given an "eye" in the form of a camera. The missile is connected to base by a link that can send pictures back to the operations area. TV cameras which work in infrared light equipment can see through smoke as they travel to their target [49].

Armor plating is heavy and expensive so that it is not usual to put heavy armor on the top of a tank. This fact has been exploited in a new weapon called *Sadarm* It consists of a missile that releases small canisters filled with explosive charges above enemy tanks. The canisters float down towards the battlefield on three parachutes. Because of the way in which they are mounted, the canisters are spun around by the parachutes. An infrared sensor at the bottom of each canister searches for tanks during the descent. When a device senses a tank within firing range the sensor aims and fires a shell to destroy the tank by shooting an explosive projectile at high velocity through the relatively unprotected turret of the tank [51].

Another device known as a *smart mine* uses similar technology to the Sadarm missile. This smart mine is called *WAM* (for *Wide Area Mine*). It is equipped with noise sensors that can detect the sounds of a passing tank. When WAM detects a tank the mine launches a canister into the air which again descends slowly using parachutes and a sensor just like the canisters of Sadarm. The great military advantage of such wide area mines is that they do not have to wait for the tank to pass directly over them. Each mine covers a wide area making it far more difficult for soldiers to clear the area of mines and they are also easier to lay in the battle zone.

In response to the development of special shells to penetrate armor military vehicles have been equipped with new types of armor. In one type, known as reactive armor, the vulnerable parts of a tank are covered with a layered material containing plastic explosive. When the tank is hit the explosive fires and lessens the effect of the impact of a missile on the tank [50, 51].

Another good example of the latest developments in armor to protect equipment from shells is the ceramic armor which was fitted to aircraft flying relief missions to Sarajevo in the war between Bosnian Moslems and Serbs in the 1990s [52]. The new type of armor which is described as *ballistic armor* is called *Armortek*. It is made from glass ceramic tiles bonded with adhesives to a multi-ply laminate which can be formed into flat sheets or moulded to produce curved surfaces. The glass armor is half the weight of steel, a tremendous advantage with regard to fuel consumption. Before this material was available, the crew of a Hercules aircraft was protected only by titanium sheets fitted under their seats. The new armor covers the floor and sides of the flight deck as well as the crew seats. When a high velocity round hits the ceramic layer, the energy of the bullet is absorbed by the pulverization of the ceramic. After pulverizing the ceramic material the remaining energy is absorbed by the multi-ply laminate which deforms but contains all the fragments. As a consequence the area hit by the bullet has only a small entrance hole on the outside of the armor and a wide but shallow bulge on the inside [52].

In the early days of aerial combat during World War I (1914–1918), aircraft flew at 60 mph and gunners on opposing sides manually fired machine guns at each other. By the outbreak of the World War II (1939–1945), aircraft were flying at several hundred miles per hour and it became a difficult task to fire guns from the ground to hit an approaching aircraft. The best an anti-aircraft gunner could do to establish his target was to estimate the position of the aircraft after the time required for the shell to travel to the target area. From experience, the human operator tried to take into account the observed speed of the aircraft, the distance, the direction of the wind, and the known speed of the projectile aimed at the aircraft. In practice, it became impossible for the operator to make these decisions very efficiently. Therefore, in order to improve the success rate of ground fire against enemy aircraft, the governments of the western world started to develop computers and other techniques for automatically aiming guns at moving aircraft.

One of the experts who was called in to work on the problem was Professor Norbert Wiener (1894–1964) of the Massachusetts Institute of Technology [11]. From his study of the need for machine control and target prediction for anti-aircraft guns, Norbert Wiener developed a subject which today is known as *cybernetics*. Norbert Wiener coined the word cybernetics from the Greek word "kybernetes", the name given to the man who sat at the end of a ship and used a big oar to pilot the ship. The large oar at the end of the ship eventually became the rudder. The word rudder itself comes from the old English word for an oar, "rothor". Thus a cybernetic system is able to steer towards a goal, and, if necessary, correct its own trajectory, or path, if it wanders off course or needs to adjust to meet its target. The cybernetic system in a missile or aircraft replaces the human helmsman or pilot. We will continue with this fascinating topic shortly, but there is an interesting side trip that must be taken from this starting point.

The Latin equivalent of the Greek "kybernaein", "to steer, to control or to govern" is "gubernare" also meaning "to control, or to govern". The Latin in turn gives us the

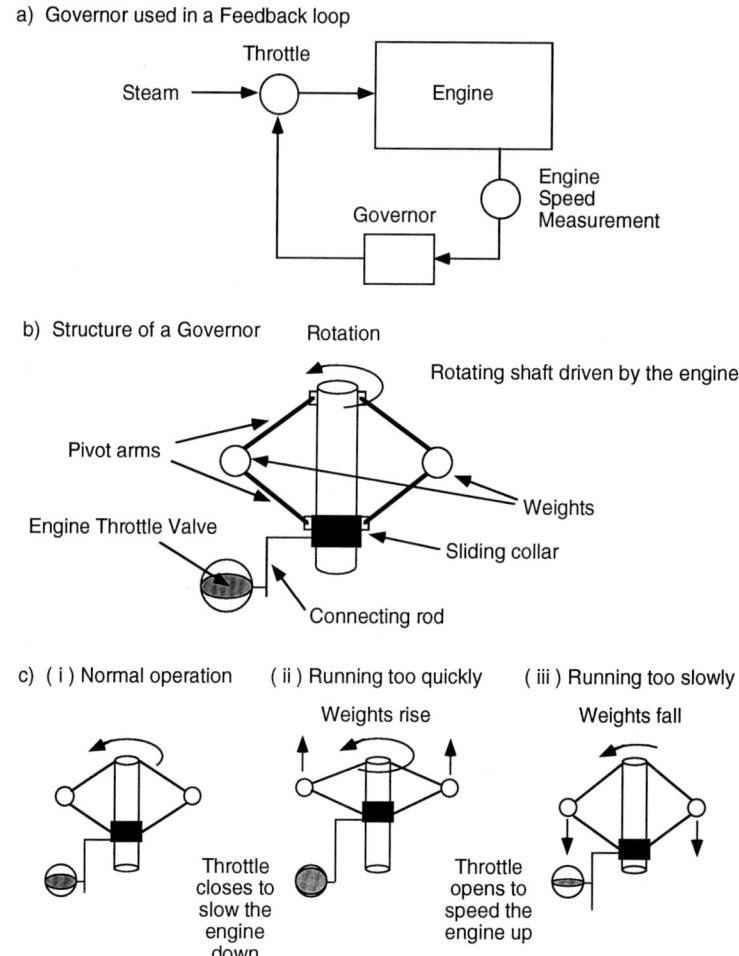

Figure 6.13 The operation of a simple governor used to control the speed of a steam engine. a) Schematic of the feedback loop using a governor to keep an engine at a preset speed. b) Details of the construction of a governor. c) The governor opens or closes the throttle as required to maintain a steady engine speed.

French "gouverner" from which are derived the English words to *govern*, *government* and *governor*. The word governor has two meanings. Science and engineering are concerned with the governor described as a device for controlling the speed of an engine. The dynamics involved in running an engine are sketched in Figure 6.13: Steam (or power) is delivered through a throttle to the engine to produce speed which in turn is controlled by a governor that opens or closes the throttle as necessary. The operation of a spinning ball governor to control the speed of an engine is detailed in Figure 6.13(b),(c). Some description will help the reader envisage its operation. As the engine starts up, the balls attached to a sliding collar around the rotating engine shaft swing outwards because of centrifugal force. The same principles are in play as with the bola balls of the Gaucho's equipment which spin outwards as the Gaucho stores energy in the system before throwing the bola at an errant beast. As the spindle or shaft of the motor speeds up, the balls attached to the collar swing farther out and move the collar up the shaft. This upward sliding of the collar decreases the power to the engine by closing the throttle valve. The closing of the valve causes the engine speed to fall, the balls to drop closer to the shaft, and the collar to slide back down the shaft. All of these then cause the throttle to open up again, increasing the steam feed to the engine once more. To keep an engine running smoothly at a given speed, the governor mechanism must stabilize its operation such that the throttle valve maintains a constant amount of power to the shaft. To further smooth out the variations in the running of such an engine, a flywheel is added to store excess energy so that there is less variation in the output of the engine. Thus we see that two important control systems, governors and cybernetics, are better understood by knowing the origins of their names.

In modern technical English, a cybernetic system is one which can efficiently seek the attainment of a goal. In anti-aircraft technology, a cybernetic missile is one that can home in on the target aircraft, its intelligent systems enabling it to take into account any avoidance flight patterns adopted by the pilot of the aircraft. Experts in cybernetics and automation engineering use two different types of strategies to attempt to achieve their goals. In the first strategy, known as *feed-forward control*, the engineer tries to anticipate the future configuration of a system and sets the control of a system to achieve that goal in the future. In Figure 6.14(a) and (b) the early attempts of engineers to shoot down attacking aircraft using feed-forward control are illustrated. The human engineer attempted to predict the future position of the aircraft from his observations and aimed the shell at the anticipated position. In *feedback control*, the engineer collects a continuous stream of information on the movement of the target and attempts to predict the target position using the more detailed information on the movement of the target. To make use of the continued stream of information from the moving aircraft, a computer is needed, often one that tries to use what is known as *heuristic programming*. A heuristically programmed computer learns from its experience. It discovers new patterns of meaning which are not always obvious to the human programmer. The word heuristic comes from the Greek word "heuriskein", meaning "to discover."

I am reminded at this point of Archimedes' famous, attributed utterance, "Eureka!" meaning I have discovered it! Most science students are told early in their career about Archimedes employing a heuristic technique to discover the solution to one of the first forensic problems discussed in history. The king of Syracuse had had a crown made out of gold and he wished to know if the goldsmith had cheated him by making the inside of the crown out of lead. As we discussed earlier in this section, the density of lead is 11.34 grams per cubic centimeter. Gold has a density of 19.3 grams per cubic centimeter. Therefore, if the crown had a density intermediate between lead and gold, it was not made of solid gold. The real problem facing Archimedes, as he pondered how to determine what was inside the crown without damaging it, was how to measure the volume

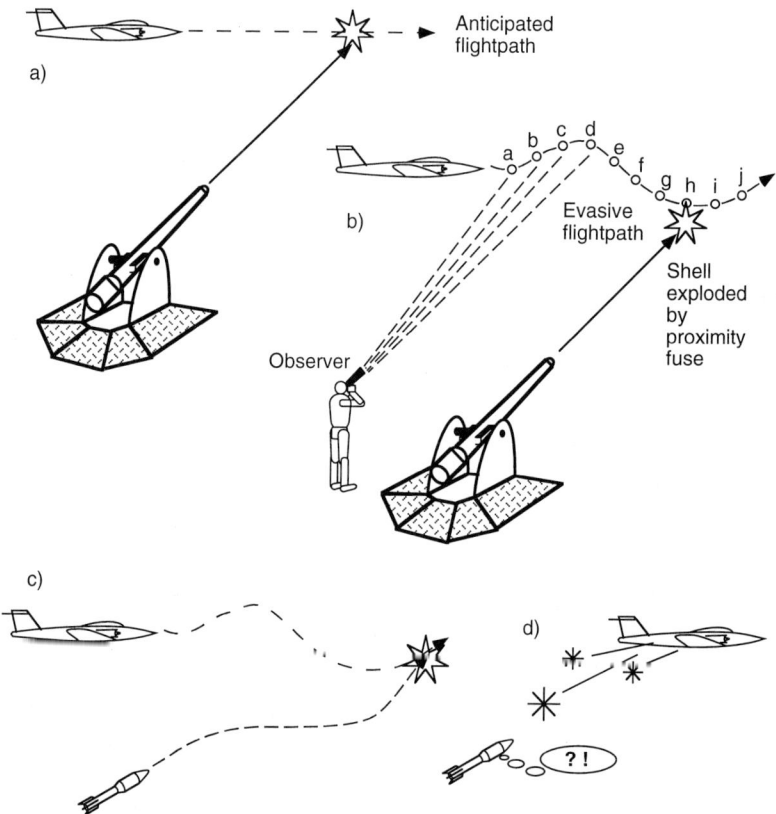

Figure 6.14 Cybernetic systems can use feedback information as they move towards a goal to correct any drift away from their target. a) Human gunners attempt to use "feed-forward" control to hit an enemy aircraft by guessing where the plane will be when the speed of the plane and shell are taken into consideration. b) A cybernetic artillery system uses feedback information sent to it on the flightpath of the plane to gain better accuracy. c) Heat-seeking missiles, equipped with infrared "eyes," are cyborgs able to follow an aircraft without human intervention. d) A heat-seeking missile needs improved electronic eyes in order to avoid a "nervous breakdown" when its target aircraft drops hot decoy rockets or flares.

of such a complicated object. He knew how to find the volume of a regular-shaped object like a ball or a cube. But the crown had a more complex structure. It is said that one day when Archimedes was thinking about this problem, he stepped into a bath of water and as he did so he noticed that his body displaced the water up the side of the bath. He realized that the volume of water displaced by his body was the same as the volume of his body. Thereupon he was able to measure the volume of the crown by carrying out a very simple experiment: He took a cylinder of water and noted the level of the water. He then suspended the crown in the water by a thread and the water rose by the amount equal to the volume of the crown. This amount could then be calculated from a knowledge of the cross section of the area of the cylinder multiplied by the rise in height of the water. The density of the crown was easy enough to calculate; it was given by the weight of the crown divided by the volume. We are told that Archimedes was so excited about discovering a solution to his problem that he ran naked through the streets of Syracuse yelling "Eureka! Eureka!", "I have discovered it!" Today scientists are usually a little more restrained in the excitement that they express upon discovering a new pattern in the data from their experiments. Archimedes' subsequent measurements established that the goldsmith had cheated the king and the goldsmith became the first victim of Archimedes' principle.

Archimedes' principle is the scientific statement of his basic technique for measuring the volume of a complicated or complex object. In formal terms an object suspended in a fluid displaces an amount of fluid equal to the volume of the object. Note, however, many textbooks *wrongly define* Archimedes' principle as the apparent loss of weight of a body totally or partially immersed in a liquid being equal to the weight of the liquid displaced. This is, in fact, the corollary derived from the basic discovery of Archimedes and correctly describes what happens if we weighed the crown in air and in a liquid: the *difference in the weight* would represent the weight of water equal to the volume of the crown. The weight of the crown suspended in the water is described as the *buoyant weight* of the crown. (Corollary, incidentally, means natural consequence or result.)

A ship is also a good example of buoyant weight. When a ship is placed in water, it sinks to a given depth determined by the fact that the weight of the water displaced by the hull of the ship will equal the weight of the ship itself. If a ship could not displace enough water to equal its weight, it would sink!.

Ships are designed to be most stable at a given position in the water. After the cargo is removed from a ship, if it is unable to obtain return cargo, it is often filled with *ballast* to submerge the ship down to its water line so that it has the same stable sailing characteristics as when carrying cargo. (Ballast probably comes from "barlast", an old Swedish word meaning "minimum load"). Human beings with fully inflated lungs can float on the surface of the water if they do not thrash their arms and legs. This fact is also used in developing therapeutic programs for handicapped people. If the handicapped exercise in the water, they do not have to use as much muscular energy to move their limbs as when they are on dry land. Fish have air bladders inside their bodies which are just the right size to enable them to float at any depth they choose in the water. This *neutral buoyant state* helps fish to minimize the effort required to swim. Submarines have air tanks inside them. In order to dive, e.g., when under attack by surface vessels, they use seawater to drive air out of their buoyancy tanks, altering their density so that they sink. To rise up to the surface again, they have to generate air in order to expel the water from their tanks.

Because salt water is denser than freshwater, it is easier for humans to float in seawater. The difference between the buoyancy of a ship in fresh water and seawater has to be taken into account when loading ships which will cross the ocean and then travel on freshwater inland waterways. The waters of the Dead Sea, a large lake situated between Israel and Jordan, has so much dissolved material in it that it is impossible for a human being to sink in the water. The word *buoy* is from the Latin word "boia" meaning "a collar"; the practice of making a floating collar or ring to help floundering human beings is an old technology.

But, back to the matter of missiles, the basic idea exploited in the heuristic programming of an anti-aircraft gun is illustrated in Figure 6.14(b). Information on the movement of the aeroplane from the points *a*, *b*, *c*, through to *g*, could cause the computer to speculate that the trajectory being followed by the aircraft was a sinusoidal path taking it to the point *h* in the time required for a shell to travel from the gun to the target area. (The sine wave is described in Chapter 4.) The use of modern laser tracking with high speed computers makes it quite feasible for a computer, without human intervention, to detect a pattern in the evasive action being taken by the pilot of the aircraft and to adjust the firing pattern of the gun. To improve the effectiveness of modern anti-aircraft weapons, the shell is usually equipped not with a time fuse but with some device which detects when it is in the vicinity of the aircraft and detonates the shell. This type of device is known as a *proximity fuse*. In the military technology of the 1990s, the anti-aircraft missile has become a *cyborg*. The term cyborg was originally an abreviation of *cybernetic organism* and implies that the cybernetic system is a self-contained traveling system able to take certain decisions en route to its target without human intervention. In some of the earliest cyborg missiles, the head of the missile was equipped with infrared eyes that could detect the heat expelled by the engine of the aircraft. Using its infrared eyes, the heat-seeking missile could observe the target as it moved about in the sky to avoid the missile. A computer on board the missile could operate small navigation engines to change its path and keep it moving towards its target as shown in Figure 6.14(c). After the introduction of the heat-seeking missile, aircraft being pursued by such missiles would fire a series of small rockets which when ignited burned fiercely and created decoy hot bodies. These gave the heat-seeking missile a nervous breakdown as it tried to decide which hot body to pursue. This situation is sketched in Figure 6.14(d). The latest generation of anti-aircraft missiles now have more sophisticated vision systems which can actually image the aircraft and use on board computers to track into the aircraft itself. The stealth aircraft developed during the late 1980s had special structures so that the heat of the engine would dissipate throughout the frame of the aircraft. This meant there would be no particular hot spot giving out lots of heat energy to be used by the heat-seeking missile to seek out and destroy the stealth aircraft.

During the battle of Okinawa fought in early 1945, the Japanese introduced what was in effect a sophisticated guided missile. Short on aircraft and suffering heavy losses from US artillery, the Japanese began to use what were called Kamikaze planes. *Kamikaze* is the Japanese word that means "divine wind." During a medieval conflict between China and Japan, the Japanese forces were saved by a typhoon that wrecked much of the Chinese fleet and the idea was that this was a divine wind that saved the Japanese people. Recognizing that they were suffering heavy losses from anti-aircraft fire, the Japanese called for volunteers who would fly their aircraft directly into the American ships on suicide missions. Military strategists calculated that they lost fewer pilots this way than if they sent waves of aircraft against the US ships as in traditional warfare. In fact the Kamikaze attacks on the US ships were very successful. (Thirty vessels were sunk in the Okinawa campaign.) This success was one of the reasons the US military decided to use the atomic bomb on Hiroshima. The Americans felt they could not justify the casualties which would be sustained from Kamikaze attacks during air attacks on the Japanese mainland. The fact that the Japanese were able to use Kamikaze pilots during conflicts indicates how important culture is when generals oppose each other on the battlefield. Even though the Kamikaze strategy was very efficient militarily, it could never have been used by the United States forces since public outcry would have been enormous back in the States.

There is a story told about two generals discussing how to deal with the problem of a mine field. One general, who was very mathematically inclined, pointed out that fewer troops would be lost by marching the army straight through the mine field than by attempting to clear the minefield using tanks and other devices. The other general, an American, had to point out that, although the opposing general was mathematically correct, socially it was a tactic unacceptable to Western culture. (The word *tactic* comes from a Greek word "tassein", meaning "to arrange for warfare" in the manner opposing armies used to face each other over a set battlefield. The way in which a leader arranged troops could often affect the eventual outcome of the battle. Such arrangements made at the beginning of the battle were the tactics.)

In the West, the first use of robot missiles was by the Germans in 1944 when they sent out pilotless aircraft loaded with bombs. I have a very clear memory as a school boy of being instructed by our teachers that upon hearing the distinctive engine sound of these V1 missiles as they were called (Doodlebugs was the popular name for the robot aircraft) we should listen closely. If we heard the engine cut out, that was the signal the bomb was going to land nearby. We were then to immediately get under our desks or go into the windowless corridors. I only remember two or three of these pilotless aircraft coming over my home city, but then we were in the north of England and not a prime target for bombing. The missiles we heard were strays! They did not prove to be very effective because the British Royal Airforce soon developed strategies for dealing with them. These included not only shooting the missiles down but also having British pilots fly very close to the German craft and tip them over with the wings of the British aircraft. This action diverted the doodle bugs from their target back out over the sea [53].

In an earlier chapter we briefly discussed the new science of ergonomics which is the science of putting man and machine together in the most efficient manner. Ergonomics is very important in the design of systems for firing guns from on board a moving platform such as a tank moving over rough territory. First of all, if the gunner has to look through sights at a target he is going to fire on, the movement of the tank over rough terrain can make the gunner seasick. Also, the variations in position caused by the rough terrain make it almost impossible to be efficient at firing a traditional gun as distinct from a guided missile. Therefore, in a modern tank, the gunner is suspended in a gimbal system rather like the gyroscopic compass discussed in Chapter 4. Thus the gunner remains fixed in space as the tank moves up and down and around. The stabilization of such a system is achieved by computer control of devices keeping the tank gun in a stable position. Thus the gunner is isolated from the disturbance of the movements of the tank and he is able to aim the gun without worrying about it. The human eye has an amazing capability to compensate for the movement of the head as one looks around. The reader may not have noticed but when moving the head, the surroundings do not move. If, however, one looks through a pair of long cardboard tubes, while riding in a car for example, so that one is looking in a directed manner, one will soon start to feel sick. This is one of the reasons why some people suffer from travel sickness if they sit in the back of the car where their vision is restricted by the windows in front of them. Sometimes a person suffering from travel sickness is much better in the front seat with an unrestricted view which seems to minimize the effect of the motion.

Section 6.7 Manufacturing with Missiles

The enormous pressures and temperatures which can be generated at low cost by firing a missile onto a target can be exploited to manufacture new materials. Thus even prior to 1969 Kennametal compacted cemented carbide by inserting it, sealed in a rubber bag in the chamber of a 14 inch naval gun barrel that was partially filled with water [54]. Detonation of a propellant charge generated the needed pressure to create a cemented material. In a review article on the use of explosives in materials manufacture Wolkomir describes how explosive technology is being used to make new superconducting materials and large industrial diamonds [55]. To make the diamonds the following technique is used: Diamond powder is packed into 6.5 square centimeter (one square inch) stainless steel capsules, a dozen such capsules are placed in a metal holder like eggs in an open carton. A flyer plate (the missile) and a special explosive charge are set up above the powder. As the explosive charge goes off the *flyer plate* is driven into the capsules at about 2.5 km per second (1.5 miles a second). The powders are subjected to a pressure about 1 million times that of the atmosphere. When the capsules are opened jumbo industrial diamonds half an inch in diameter and 85% as hard as those created by nature are found in the compressed material. The laboratory manager of manufacturer, Ed Roy, says of these large diamonds:

They are dark and are ugly but they are perfect for industrial jobs like slicing metal.

In the same article techniques for making armor for tanks are described. It is pointed out that

The explosives needed for the job cost a mere 20 cents a pound and just a few pounds are needed to make the material.

We can expect to see many innovations in the manufacture of exotic materials using the compressive forces available with missiles.

Section 6.8 Fatal Fiesta Frolicking?

We have already mentioned that New Scientist has a question and answer page at the back of every issue. One of the questions posed was "What is the danger of falling bullets when people, in a festive mood, fire guns into the air", as the questioner put it "with great exuberance and a seeming disregard for the welfare of themselves and the others?"

In answer to the question two writers from the Royal Military College of Science in Swindon, Wiltshire, England said that the firing of hand guns into the air causes injuries with a disproportionate number of fatalities. They gave the figures that for a typical modern 7.62 millimeter caliber bullet fired vertically into the air the bullet would have a velocity of about 840 meters per second as it leaves the gun. It will reach a height of about 2400 meters in 17 seconds. It will then take another 40 seconds or so to return to the ground. It will normally fall with the base of the bullet downwards since this is the stable trajectory for the bullet. When discussing the hitting of a tennis ball it was pointed out that there is a hover point when the tennis ball stops rising and begins to fall [56]. In the case of the bullet fired straight up, the writers to New Scientist said that this is reached after about 8 seconds. At heights between about 2300 and 2400 meters, the vertical velocity of the bullet is less than 40 meters per second and at this speed it will be blown sideways by any wind. It will normally return to the ground at a speed of some 70 meters per second. The writers say that because of the predominance of cranial injuries from falling bullets the number of deaths and serious injuries as a proportion of the number of gun shot wounds is surprisingly high and typically 5 times greater than is observed in normal firing.

Other writers quoted different types of bullets but they all agree that the person actually firing the bullet is in less danger than those around him. One writer quotes from a book by Hick on the Theory of Rifle and Rifle Shooting that recorded an experiment in which a machine gun was fired into the air. Of 500 bullets fired into the air only 4 returned to earth near the gun and that many of them fell as much as 25 meters away.

References

[1] *Webster's Medical Desk Dictionary*, Merriam Webster Inc., Springsfield, MA, 1986.
[2] E. DeBono, *History of Inventions*
[3] J. Needham, *Science and Civilization in China*, Volume 5, Part 7. Cambridge University Press, Cambridge, 1986.
[4] For a review of Ref.[3] see A. Briggs, "The Epic of Gunpowder," *New Scientist*, May 28, 1987.
[5] See I. Asimov, *Words of Science and the History Behind Them*, A Signet Reference Book, New American Library, 1969.
[6] For a discussion of the pictographic and ideogram basis of Chinese writing see R. Chang, M.S. Chang, *Speaking of Chinese*, W.W. Naughton, New York, 1978.
[7] *Chamber's Etymological English Dictionary*, edited by A.M. McDonald, W.&R. Chambers, Edinburgh, 1967.
[8] T.J. Shipley, *Dictionary of Word Origins*, Littlefield Adams, Trotowa, NJ, 1970.
[9] S.C. Bevan, S.J. Gregg, A. Rosseinsky, *Concise Etymological Dictionary of Chemistry*, Applied Science Publishers, Barking, Essex, 1976
[10] L. Hogben, *The Vocabulary of Science*, Heineman, London, 1969.
[11] I. Asimov, *Bibliographic Encyclopaedia of Science and Technology*, Doubleday, NY, 1948.
[12] The formulas for many of the common explosives and techniques for analyzing explosive mixtures are given in the publication by I.S. Crull and M.J. Camp, Analysis of Explosives by High Performance Liquid Chromatography, *American Laboratory*, May 1980, pp. 63-76.
[13] W.C. Davis, "The Detonation of Explosives", *Scientific American*, May 1987.
[14] I.H. Evans in *Brewer's Dictionary of Phrase and Fable*, Centenary Edition, Cassell and Co., London, 1970.
[15] D. Diaconou, "Engineer Copies Archemedes Gun", *Toronto Globe & Mail*, May 28, 1981, p. 1.
[16] B.H. Kaye, *Science and the Detective; Selected Readings in Forensic Science*, VCH, Weinheim, 1994.
[17] Letter from Reginald Titt in *New Scientist*, February 4, 1995, p. 77.
[18] Letter from Pekka Haussalo, Espoo, Finland, to *New Scientist*, February 4, 1995, p. 77.
[19] See discussion of deterministic chaos in B.H. Kaye, *Chaos and Complexity; Exploring the Surprising Patterns of Science and Technology*. VCH, Weinheim, 1992.
[20] Homer, *The Iliad*, Translated by E.V. Rieu, Penguin, London, 1972.
[21] A. Ager, "The Ore that Launched a Thousand Ships". *New Scientist*, June 20, 1985, pp. 28,29
[22] "Taking Aim with the GII" (The gun which fires cartridges that have no metal case), *New Scientist*, May 15, 1986, p. 38.
[23] J. Hecht, "Impenetrable Armor May Be Fire Risk," *New Scientist*, March 24, 1988.
[24] W. Biddle, "The Untold Story of the Patriot," *Discover*, June 1991, pp. 74-80.
[25] D. Charles, "Patriot Missiles Mislead by Accidental Decoys", *New Scientist*, February 15, 1992, p. 20.
[26] V. Kiernan, "Beyond the Patriot," *Discover*, June 1991, pp. 80-82.
[27] V. Kiernan, "How Lasers Can Run Rockets to Ground," *New Scientist*, November 27, 1993, pp. 18,19.
[28] See "Shrapnel" in N.C. Soreland, E Sorel, *Word People*, American Heritage Press, McGraw-Hill, 1970.
[29] P. Johnson, Thompson News Service Release, July 1, 1995 (*Sudbury Star*, July 1, 1995).
[30] V. Kiernan, "Americans Shoot to Sterilize Policy", *New Scientist*, September 25, 1992, p. 8.
[31] News story by D. Brazeau, "Hollow Point Bullets, Police Expect Government to Deliver on Promise," *Sudbury Star*, July 4, 1995, p. 1.
[32] B. Fox, "Jelly Bullets", *New Scientist*, September 1994, p. 24.
[33] Information on the Laser rifle available from Desmans A.R.L., Ste. Marie, Campan, 675 Campan, France.
[34] J. Hecht, "Laser Designed to Blind", *New Scientist*, August 8, 1992, pp. 27-31.
[35] V. Kiernan, "Ban Cruel Laser Weapons Says Red Cross", *New Scientist*, November 19, 1994, p. 11.
[36] B. Anderberg and M. Wolbarsht, *Ways of Weapons. The Dawn of a New Military Age*, Plenum, New York, 1992.

[37] J. Hecht, "Weapons for the 21st Century", *New Scientist*, August 8, 1992, pp. 21,22.
[38] V. Kiernan, "Beware of the Red Spots Warns New York's Finest", *New Scientist*, January 14, 1995, p. 4.
[39] R. Atkinson, *Crusade: The Untold Story of the Persian Gulf War,* Houghton-Mifflin, Boston, 1994.
[40] *Funk & Wagnalls Standard Reference Encyclopedia,* Standard Reference Work Publishing Company, Inc., New York, 1965.
[41] C. Crabb, "Shooting at the Moon," *New Scientist*, August 6, 1994, pp. 27–41.
[42] News story, "Inside the Barrel of Iraq's Next Super Gun," September 7, 1991, p. 20.
[43] T. Folger, "The Guns of Brooklyn," *Discover*, August 1992, p. 14.
[44] D. Clery and D. MacKenzie, "Scotland to Host RailGun Test Bed," *New Scientist*, January 27, 1990, p. 24.
[45] News item, "Railguns on The Right Lines", *New Scientist*, July 17, 1993, p. 22.
[46] J.F. Allen, *New Scientist*, August 21, 1993.
[47] Letter from R. Allen to *New Scientist*, December 12, 1992, p. 53.
[48] News item, "The Evolution of Anti Tank Missiles", *New Scientist*, July 23, 1987, pp. 47, 48.
[49] News item, "Fibers Guide the Smartest Missiles in the West", *New Scientist*, May 15, 1986, p. 39.
[50] M.W. Browne, "America's Mightiest Tank", *Discover*, June 1982, pp. 21–26.
[51] S. Connor, "Anti-Tank Missiles that Fall From the Sky", *New Scientist*, January 7, 1989.
[52] G. Clayton, "Flying Glass to Foil Serbian Snipers". *New Scientist*, August 21, 1993.
[53] For a discussion on how the British developed technology for dealing with Doodlebugs see I. Buderi, "The VI. Menance: Secret Weapons that Saved Britain", *New Scientist*, June 4, 1994, pp. 28–32.
[54] See J. Dieter, "Powder Fabrication" in *Modern Science and Technology*, Ed. by R. Colborne, D. Van Nostrand Company Inc., New York, 1965.
[55] R. Wolkomir, "Boom Town" (A discussion of the use of explosives to make new materials), *Discover*, August 1989, pp. 77–81.
[56] See the contributions of various correspondents in *New Scientist*, July 8, 1995, p. 105.

Chapter 7

Rockets: From Fireworks to Trans-Galactic Messengers

Chapter 7

Rockets: From Fireworks to Trans-Galactic Messengers

Section 7.1 Rockets and Newton's Third Law of Motion

I have a very clear memory of the night I heard a new sound during an air raid on my home city during World War II. We were used to the thunderous crack of the anti-aircraft gun as it launched its missile but the new sound we heard was a quiet swoosh. I later learned that the sound was that of a rocket launcher. My cousin took me downtown the next day to see it. I learned later that one of the big advantages of rocket powered missiles is that many of them can be fired one after the other without requiring massive energy absorption due to the recoil of the gun. The problem of overheating of the gun from rapid firing is also avoided. Rapid firing can cause guns to jam as the metal parts expand from the heat generated by the firing of multiple shells. To begin our studies of the science of rockets we must explore in detail Newton's third law of motion.

Consider the experiment which could be carried out with the system sketched in Figure 7.1. A gun about to fire a bullet is hung from two threads to create a ballistic pendulum such as that used to measure the speed of a bullet. If we observed this gun after the bullet had been fired by a remote control, a simple task these days with our ability to beam signals to the suspended gun, we would observe that as the speeding bullet left the muzzle of the gun, the gun would swing backwards and up to a height h. This would indicate that during the firing, the gun itself had acquired momentum as the bullet left the muzzle. A simple measurement would soon show that the product of the mass of the bullet times the velocity of the bullet equals the mass of the gun times its velocity. Note that the product of the velocity times the mass of a moving object is known as the *momentum* of that object; this is a strictly scientific use of the term momentum which is used more generally in everyday speech to describe real or metaphorical energy of motion. Thus if we say that the movement to ban the ownership of guns in America is gaining momentum, we would be referring to the fact that more and more people were beginning to push the project rather than the projectile! From our experimental measurement (see Figure 7.1) we could develop the following formula:

$$m_g v_g + m_b v_b = 0 \,.$$

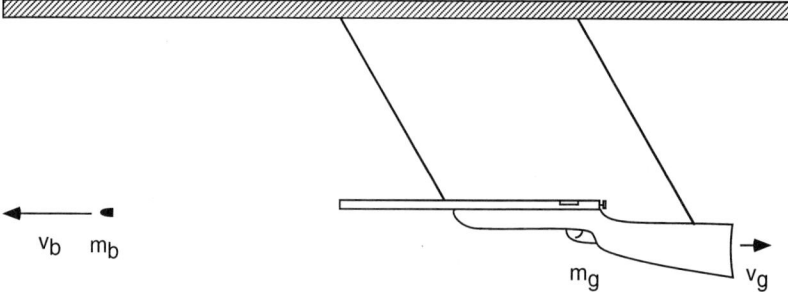

After firing, the gun (mass m_g) recoils (moves back) at velocity v_g.
The bullet (mass m_b) moves forward with a velocity v_b.

momentum equals **velocity** times **mass**

Conservation of momentum states that the total momentum of the system must remain the same.

Since, before firing, the momentum of the system was zero (niether the gun nor the bullet were moving), the momentum after firing must also be zero.

Therefore, the momentum of the gun plus the momentum of the bullet must equal zero. Symbolically:

$$m_g v_g + m_b v_b = 0$$

by rearranging this equation we can find the speed of the bullet:

$$v_b = -\frac{m_g}{m_b} v_g$$

For example, if:
The gun has a mass of 3 kg (m_g = 3 kg)
The bullet has a mass of 10 g = 0.01 kg (m_b = 0.01 kg)
and the gun recoils at 2 m/s (v_g = 2 m/s)

Then the velocity of the bullet is: $v_b = -\frac{3 \text{ kg}}{0.01 \text{ kg}} \cdot 2 \text{ m/s} = -600 \text{ m/s}$

the negative value indicates that the bullet is moving to the left (i.e. right is positive)

Figure 7.1 One can use a ballistic pendulum to demonstrate the physical significance of Newton's third law of motion.

This formula expresses the conservation of momentum. The momentum of an object is equal to its velocity multiplied by its mass. After firing, the gun (mass m_g) recoils at velocity v_g. The bullet (mass m_b) moves forward with a velocity v_b. Since, before firing, the momentum of the system was zero (neither the gun nor the bullet were moving), the momentum after firing must also be zero. The above equation expresses this symbolically. It can be rearranged to give the speed of the bullet as

$$v_b = -m_g v_g / m_b \, .$$

From this formula we discover that the velocity of the gun after firing the bullet was the velocity of the bullet multiplied by the ratio of the weight of the gun to the weight of the bullet as shown. If we consider the dynamics of a typical modern rifle, we can assume that the weight of the gun was approximately three kilograms. Typically it would shoot a ten gram bullet with a muzzle velocity of 600 meters per second (almost Mach 2). If we use these figures, we can show that the gun recoils, moves backwards, with a velocity of 2 meters per second [1–3].

The word *recoil* is compounded from two Latin root words, "re" which can mean "backwards" and "culus" which can be politely translated as "the hind parts of an animal." Culus moved into Latin to give us the French word "cul" and this has given us the English phrase *"cul-de-sac"* which in England is the term used for a dead end street. It literally means a "bottom end of a sack." Strangely enough in bilingual Canada the term is not used; instead Canadians use the term *impasse*, meaning no passage. This is a pity since cul-de-sac seems a much more colorful term, although one of my French speaking colleagues tells me that in Northern Ontario French the word cul-de-sac has to be translated as the "sack's arse" — which may be too colorful for some [4, 5].

When we say that a rifle recoils, we are saying that it backs away from the bullet. The energy of recoil has to be absorbed in the case of a human gunner by impact of the rifle upon the shoulder. People unprepared for the recoil energy when firing a gun for the first time can cause themselves serious injury. The recoil of the gun is also a factor that makes aiming the gun difficult for people who are not experienced in firing techniques. In technical terms, we say that when the gun pushes the bullet, there is a *reactive force* of the bullet on the gun. The fact that the momentum of the bullet will equal the momentum of the rifle is usually described by scientists in the statement known as *Newton's third law* which reads:

> *Action and reaction are equal and opposite.*

(We met Newton's first and second laws in earlier chapters [1–3]).

The fact that the prefix "re" can have two meanings has led to the fact that some words such as reaction can have quite different meanings in different situations. The two meanings of "re" are "intensive," and "back." Thus in one situation when we put "re" in front of another word in order to signify intensive, this means that what happens not only takes place but does so with lots of energy. The word "re" is combined with the Latin word "agere" (past participle "actum") meaning "to set in motion," to create many words in science which look alike but can have different meanings. The origin and meaning of some of these words is illustrated in Figure 7.2. Thus the term *react* can mean that actions take place vigorously. When two chemicals are mixed together and explode, we call it a *reaction*; one part of the chemical combination is often referred to as a *reagent* and the vessel in which the chemical or physical reaction occurs is known as a *reactor*. Thus in chemical factories, the system in which chemicals are mixed to produce a third component is called a reactor; in physics, a *nuclear reactor* is a system in

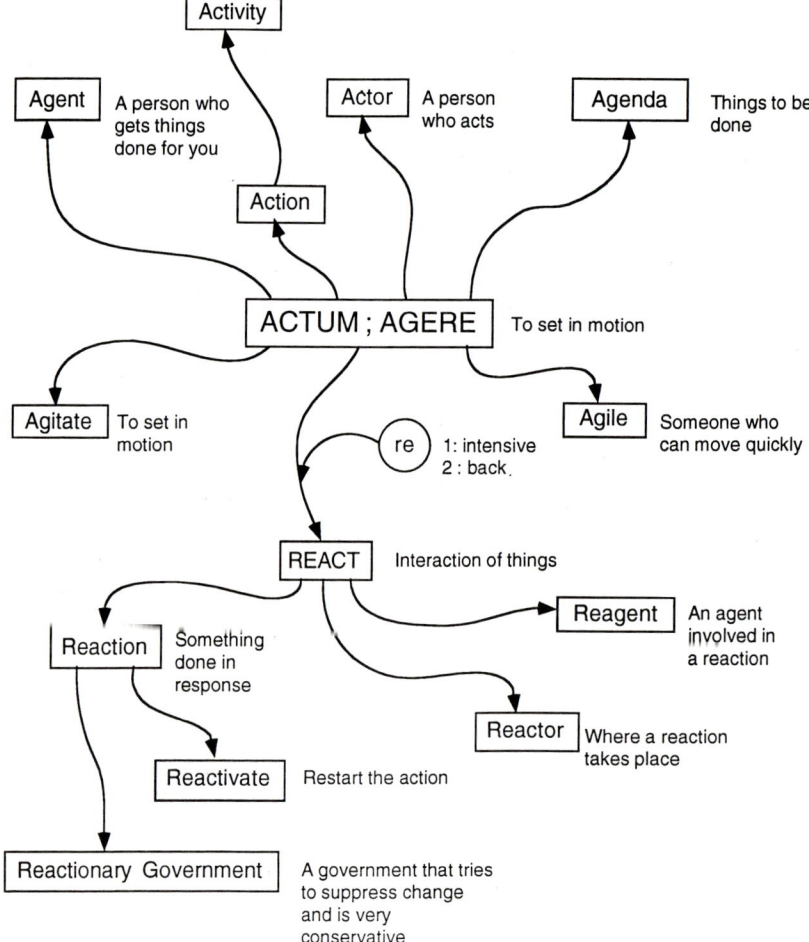

Figure 7.2 The two meanings of the prefix "re" can lead to two words that look alike but have very different meanings.

which *nuclear energy* is released. On the other hand, if we use the word reactivate, we mean that we go back to a system and get it going again. Reaction can mean "the response to something which happens". But a chemical reaction can be quite different from the reactionary government that tries to go backward in time to reinstate an old system of government. Other words that come from the verb "to set things in motion" are *agenda, actor, agent, activity, agitate,* and *agile* as illustrated in the diagram [4, 5].

The absorption of the energy of reaction when a gun is fired is a major problem in the design of artillery. Anyone who has watched a large gun being fired will have noticed that the gun goes backward as the shell is fired forward. On a ship, the recoil of the guns could actually cause the ship to overturn if precautions were not taken to ab-

7.1 Rockets and Newton's Third Law of Motion

sorb the energy of recoil in a series of springs and ropes, or hydraulic systems. One of the big advantages of a rocket is that the energy used to fire this type of missile is not all released at the moment of launch, and the launching system only has to absorb a very small amount of energy relative to that which will eventually be invested in the missile. The rest of the recoil energy is exerted by the continuously burning rocket which pushes on the air as it propels the missile.

Physics textbooks often introduce students to the problems of action and reaction forces as described by Newton's third law by discussing a problem such as that of a man on a sled sitting in the middle of a sheet of ice with no means of propulsion but with a pile of bricks on his sled. The man has to devise a strategy for getting off the ice without touching the ice. It is assumed that the friction between the sled and the ice surface is negligible. From Newton's laws of motion it would follow that if the man threw a brick from the sled in the opposite direction to which he wanted to travel, then the sled plus man and bricks would start to move in the desired direction. If the friction was really zero, the man would only need one brick to overcome his problem. The velocity of the sled system could be calculated from the size of the brick and the velocity with which it was thrown from the sled. Even if this velocity was very small, if there was no friction between the sled and the ice, the man could carry on traveling indefinitely, or at least as far as his patience, or the ice, would sustain him. By throwing more and more bricks, however, he could accelerate his movement, and, in gaining higher speed, move off the ice more quickly. In fact, this is just how a rocket takes off from the surface of the Earth, but instead of throwing solid bricks, it throws high speed gases out of the tail of its engine to give it thrust upwards. (We will be looking at action and reaction forces again when we discuss the way in which a car crashes into a tree in Chapter 10.) In the same way, a man on skates in the middle of the ice could move off the ice by firing a machine gun in the opposite direction to which he wanted to move. As the bullets left the muzzle they would push him toward his desired goal.

Guy Fawkes is celebrated in England on the 5th of November; a festival also known as bonfire night. That night an effigy of Guy Fawkes is burnt in a bonfire and fireworks are set off. *Guy Fawkes* was one of the conspirators who planned to blow up the Houses of Parliament on the 5th of November, 1605. The conspirators placed barrels of powder in a vault under the House of Lords (part of the British Houses of Parliament) and Guy Fawkes was to fire the fuse. However, one of the conspirators warned a relative who was a member of the House of Lords so that he could avoid the catastrophe, the relative told the authorities, and Guy Fawkes was captured. In Figure 7.3 (a) a simple launching system for a fireworks rocket is shown. I remember as a boy of six, launching such rockets from a milk bottle on Guy Fawkes' night. The rockets of bonfire night carry a payload of chemicals used to create colorful star bursts in the sky, but the rockets of warfare carry a more lethal payload, or warhead, on their missions.

The origin of the word rocket suggests that the military missile appeared in the west as a complete, ready-to-operate system since the word is a metaphor created by comparing a typical rocket with what used to be a familiar piece of domestic equipment.

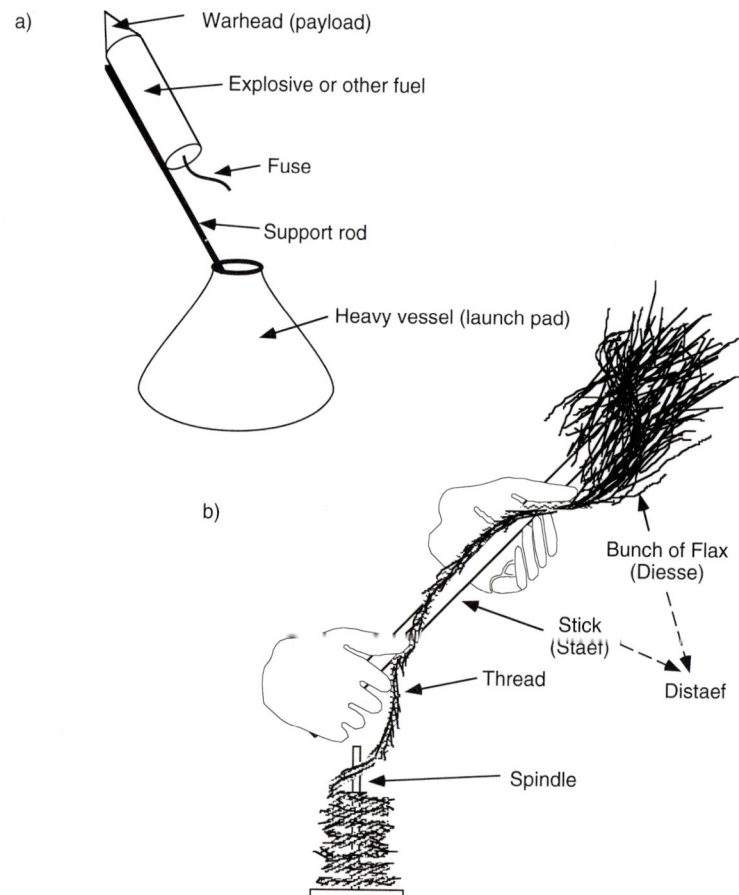

Figure 7.3 The word rocket was originally a metaphor coined because the missile ready to be launched looked like the distaff used in spinning thread from a bunch of fibers. a) A simple arrangement for launching a small rocket. b) Thread used to be spun from flax fibers held on a stick with the thread being wound onto a spindle.

Figure 7.3 (b) shows how linen thread used to be spun from a bunch of flax placed on top of a stick. In the act of spinning, a bundle of fibers is pulled from the bunch and twisted together. The twisting is aided by the fact that at the other end of the growing thread there is an object known as a *spindle* which the operator spins with their fingers. Every now and again they give it a flick so that the thread is not only spun by the spinning spindle but wraps itself around the spindle. The bunch of flax on a stick looks very much like a rocket ready to be launched. The Anglo-Saxon word for the equipment used to spin thread is a *distaff* from the Saxon word for a bunch of flax fibers and the word for a stick which has given us the word stave in modern English. The word distaff is still used in modern English to describe woman's work and affairs, as well as the maternal ancestry of a person, since the spinning of thread was typically a woman's responsibility. "On the distaff side they are descended from Queen Elizabeth" is a phrase that illustrates this usage of the word distaff. (In medieval households an unmarried lady stayed at home and became a major producer of thread and she became known as a *spinster*.) The Italian word for a distaff was "rocca" and the Italian for a small distaff would be a roccetta which has given us the English word rocket. As mentioned, the device at the end of the thread is known as a spindle. This has given us words in English such as *spindly*, meaning tall and thin. The word comes from the old English word for spinning, "spinnan." The Latin word for a spindle, "fusus", has given us the word *fuselage* for the long thin body of an aircraft to which the wings and tail are attached [4].

Early on in the development of the technology of spinning thread, it was realized that if a flywheel was added to the spindle, rather like a potter's wheel, energy stored in the flywheel would enable the spindle to spin faster and more evenly. The flywheel of the spindle gradually evolved into the technology of the spinning wheel.

We know that the Chinese were using rockets in military operations by the year 1232 when they were used in the siege of a city known as Pien-ching. In that siege, rockets were used to set fire to tents and wickerwork fortifications which had resisted penetration by traditional arrows. Within a few years the rocket was also in military use in Europe and North America. However, by the 15th century, European records only show the rocket being used in fireworks displays. (One of the first recorded fireworks displays took place in 1650 in the German city of Nuremberg.) In the late 18th century the ruler of Mysore in India had a permanent army of rocket throwers which, under his son Tipu, was expanded to 5000 men. Their large rockets made of bamboo had a range exceeding 1000 yards. The use of rocket-firing troops was a critical factor in an Indian win over the British at the battle of Seringpatam. When news of the Indians' rockets reached Great Britain, a British artillery officer, Sir William *Congreve*, decided to investigate the use of rockets in warfare. At his own expense he bought a batch of fireworks rockets for research. Within a few years he had improved the ordinary fireworks rocket and converted it from a toy to a weapon with a range of 2000 yards. Congreve's rockets were first used during the Napoleonic wars in a British attack on the French port of Boulogne in 1805. This first use was not particularly effective because of stormy weather. In 1806, Congreve and his troops used a new rocket model with a sheet iron case, carrying a seven pound charge of incendiary material. Its overall weight was 32 pounds and it had a range of 3000 yards. These rockets were used with great success in 1806. In 1807 the city of Copenhagen and a large French fleet in its harbor were almost totally destroyed by a naval attack in which 25 000 rockets were used. Congreve used various types of warheads on his rockets and some of these rockets were used against the United States in the War of 1812 as witnessed by the phrase "In the rockets red glare" in the American national anthem "The Star Spangled Banner." This refers to an attack on Fort MacHenry by the British rocket ship, Erebus.

After the death of Congreve in 1828, rockets were not as enthusiastically pursued by the military establishment; although in 1846, William Hale, an English artillery officer, invented a rocket which was stabilized in flight by means of spirally arranged fins. The

United States government used these Hale rockets during the Mexican War; especially in the taking of Vera Cruz. The invention of the breech-loading, rifled artillery system with its great accuracy caused a decline in the use of rockets. In 1865, a harpoon-carrying rocket was invented for the whaling industry. Whaler rockets were also used in shipwreck rescues; a rocket was used to fire a rope to a ship on the rocks to establish a life line to carry men from ship to shore.

During the second world war, rockets were rapidly developed, since the advantages of low recoil energy had been recognized. For example, one commentator on military warfare states that the barrage rocket gave landing craft the firepower of several destroyers when they were used in amphibious warfare in the landings on islands in the Pacific Ocean. In essence, all that is needed to launch a rocket is a tube which can be held in a fixed position. In the early 1940s, a one man antitank rocket-launching system was developed by the US army. In 1943 Major Zeb Hastings of the US army called the rocket projectile gun a *bazooka* because it resembled the musical instrument. Shipley [4] records that,

> Both the musical instrument and the rocket launchers are made of straight tubing open at both ends and both have a more or less devastating effect.

The original bazooka rocket had a range of two hundred yards. Later modifications gave it a range of six hundred yards. The Germans developed two types of anti-aircraft rockets: one known as the "Nebelwerfer" rocket caused heavy losses among allied aircraft during an air battle over Schweinfurt in the Second World War.

The use of rockets during World War II was mainly the outcome of the work of two men who carried out research on rocket development at a time when this area of work was not fashionable. The first of these pioneers was Robert *Goddard*, an American physicist (1882–1945). Goddard received his Ph.D. in Physics at Clarke University in Worster, Massachusetts. He taught at Princeton University in New Jersey but returned to Clarke in 1914 and remained there for thirty years. Asimov notes that "Goddard had a mind daring enough for a science fiction writer and he was firmly grounded in science"[6]. While still an undergraduate, Goddard described a scheme for a railway line between Boston and New York in which the trains traveled in a vacuum under the pull of an electromagnetic field and completed the trip in ten minutes.

Apparently Goddard was interested in rocketry even as a teenager. By 1914 he had obtained two patents involving rocket equipment. In 1923 he was the first to employ a new type of rocket engine using gasoline and liquid oxygen. In 1926 Goddard's first rocket, about four feet high and six inches in diameter, was launched. Goddard had difficulty raising funds for his research because in official circles he had a reputation as a crackpot. *Lindbergh*, who was the first man to fly across the Atlantic, interested Daniel Guggenheim in Goddard's work. Guggenheim gave Goddard a grant of $50 000. By 1935 Goddard was launching rockets that attained speeds of up to 550 mph and heights of 1.5 miles. He developed systems for steering a rocket in flight using a rudder-like device to deflect the gaseous exhaust of the rocket and using gyroscopes to keep the rocket heading in the desired direction. Over his lifetime he accumulated a total of 214 patents. During World War II, the US government financed Goddard's work on small rockets to help navy planes take off from ships. Asimov [6] records that,

> When German rocket experts were brought to America after the war and were questioned about rocketry they stared in amazement and asked why American officials did not inquire of Goddard from whom they had learned virtually all they knew.

In the 1960s the United States Government retrospectively issued a grant of one million dollars for the use of Goddard's rocket patents. Half of this money went to the Goddard estate and half to the Guggenheim Foundation. The Goddard Space Flight Center in Maryland is named in his honor.

The leading German rocket scientist brought to the United States after the war was Wernher *von Braun* who was born in Germany in 1912 [7, 8]. He obtained his Ph.D. in 1934 at the University of Berlin. Von Braun's interest in rocketry was stimulated by his reading of science fiction stories regarding interplanetary travel. In 1930 he joined a group of German enthusiasts who experimented with rockets. One of the leading members of this group was Willy *Ley* (1906–1969). In fact, it was Ley who introduced von Braun to the group. Originally Ley was studying zoology at the University of Berlin, but discovered a book on rocketry that caught his imagination. He wrote a popularized version of this book that met with wide acclaim. For over 40 years afterwards he remained the most successful popular writer on rocketry in the world. It is also interesting that Ley was consultant for the science fiction movie "Frau im Mond" in which the dramatic countdown for launching a missile or exploding a bomb from 10 to 0 was introduced. When Hitler became the ruler of Germany, Ley left because he was very anti-Nazi and in the words of Asimov, "He was not content to pursue his rocket studies regardless of the political atmosphere, as was von Braun" [6]. Ley immigrated to the United States in 1935 and took American citizenship in 1944. Ley was one of the stimulators of the science fiction movement in North America and helped to develop support for rocket research.

The rocket research work being carried out by the Ley group of enthusiasts was taken over in 1932 by the German army. When Hitler came to power in 1933, he was an enthusiastic supporter of rocket research. In 1936 the rocket research center in Peenemünde on the Baltic shore was started. Von Braun joined the Nazi party in 1940 and by 1944 his group had developed a missile which came into combat use. The weapon was the rocket known as the V-2. (The V in this weapon and in the buzz bombs discussed in a previous chapter stood for the German word Vergeltung meaning "vengeance".) The V-2 rocket fired in North Germany could hit England and other distant targets. It used to go high into the upper atmosphere and then drop down at great speed. Since the sound of the approaching rocket could not reach the area where the rocket would land before the rocket impacted, it struck without warning. The V-2 rocket was designed as a terror weapon against civilians; it was too inaccurate to be directed at military targets. (In the same way, the *Scud missiles* used by Iraq in the Persian Gulf War of 1991 were not military weapons in that they could not hit a target accurately but instead were intended to scare and demoralize civilian populations.) As the armies of the allies and Russia came into Germany from opposite sides, the rocket experts decided to surrender to the American forces. Von Braun warned his workers that they might face war crime charges because they had worked on a weapon intended for use against civilians; but as they soon discovered, they were too useful to the Americans and would not face charges. In military science, justice sometimes takes second place to pragmatic need.

Intercontinental rockets were described as *ballistic missiles* because of the way in which they were thrown, without guidance systems, whereas most of the closer range rockets used today are guided missiles. The origin of these words and the distinctions among the various words used in missile technology are illustrated in Figure 7.4.

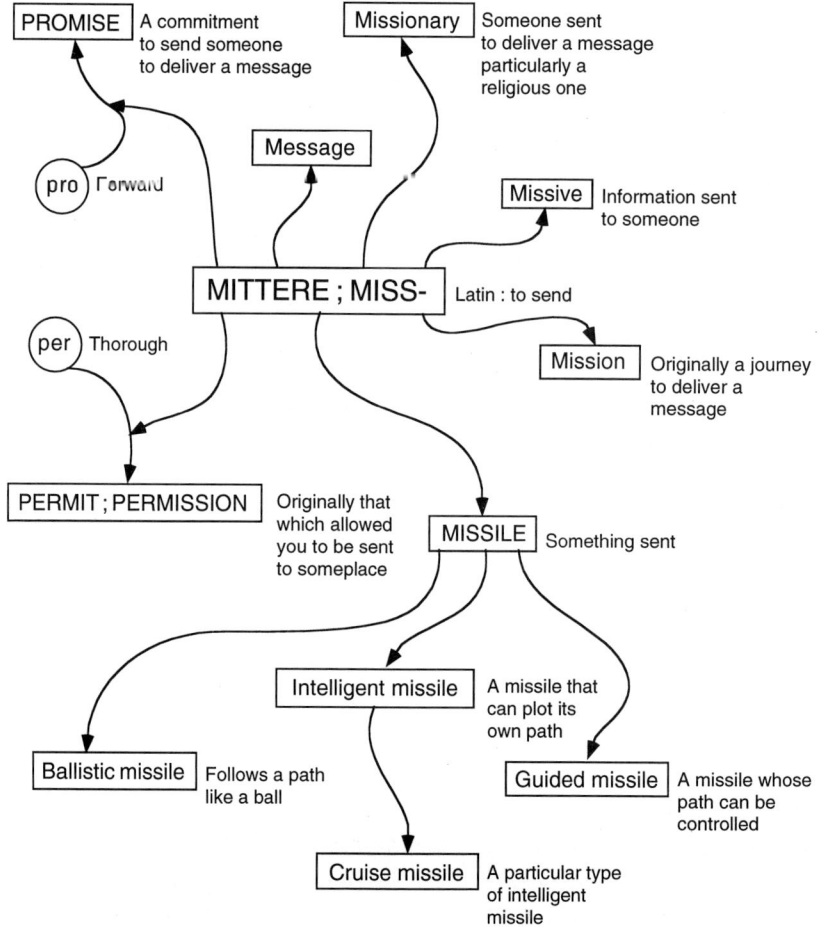

Figure 7.4 Using a missile to deliver a message requires the use of words based on the Latin word "mittere", to send.

Section 7.2 Getting the Rockets off the Ground

As already mentioned, the first rockets used gunpowder as a fuel. Goddard was the first rocket builder to use gasoline and liquid oxygen. In today's technology, gunpowder would be described as a *solid propellant* and Goddard's later rockets would be described as driven by *liquid propellants* [9]. Modern rocketry uses both solid and liquid fuels. Basically, liquid fuels give more thrust to the rocket per pound of rocket fuel but require more complicated pumping and piping systems, whereas a solid rocket engine is simpler to build and operate. On the other hand, since the cool liquid (*liquid oxygen* has a temperature of $-183\,°C$) can be used to cool the rocket engine as it operates, the materials used to construct the nozzle (the exit part) of the rocket engine can be of lower cost and lower strength than for systems using solid propellants. A modern propellant for a solid rocket engine would be a mixture of a chemical which provides oxygen for a combustion process (this component is called an *oxidizer*) and a fuel which reacts with the oxygen to release energy. A typical oxidizer used in modern rocket engines is *ammonium perchlorate*. A common fuel used to react with the perchlorate in solid rocket engines is *aluminum powder*. The use of metal as a fuel will seem odd to anyone not conversant with rocket research, but the heat of combustion released by burning a pound of aluminum is 60% more than that gained from a pound of a hydrocarbon such as kerosene. The binder used to hold the perchlorate and the aluminum powder together is very similar chemically to substances such as gasoline. *Polyurethane* is a binder that is widely used to keep the perchlorate and aluminum powder in an intimate state of mix. A typical proportion by weight of the ingredients in such a fuel would be 65% oxidizer, 15% metal, and 20% polyurethane [9, 10]. The raw ingredients of the solid propellant are usually mixed together in ordinary mixers like those used to mix cakes. During the mixing stage, the chemical which will become polyurethane is present as what is known as the monomer. It is then polymerized to make the complicated compound. (See the discussion in Chapter 6 of cellulose as a polymer of sugar.) At the beginning of the manufacturing process, the mixture of ingredients flows like thick mud. It is poured into a prepared rocket case around a central iron bar which can be removed later to leave a solid cylinder with a hole down the middle. When made in this fashion, the rocket burns from the center of the available fuel outwards to the wall of the rocket. In this way, heat generated during the burning process warms up the next batch of fuel to be burnt, but more importantly, the outside cylinder of the rocket engine is protected from the heat of combustion by the insulating effect of the unburnt propellant. The bar placed down the middle of the cylinder of propellent material when it is cast is described technically as a *mandrel*. Originally this was a French word used to describe an iron bar. Figure 7.5 (a) shows a simple mandrel and the basic structure of a solid fuel rocket. In the words of William Cohen, who has written an interesting review [10] of the design of solid fuel rocket engines,

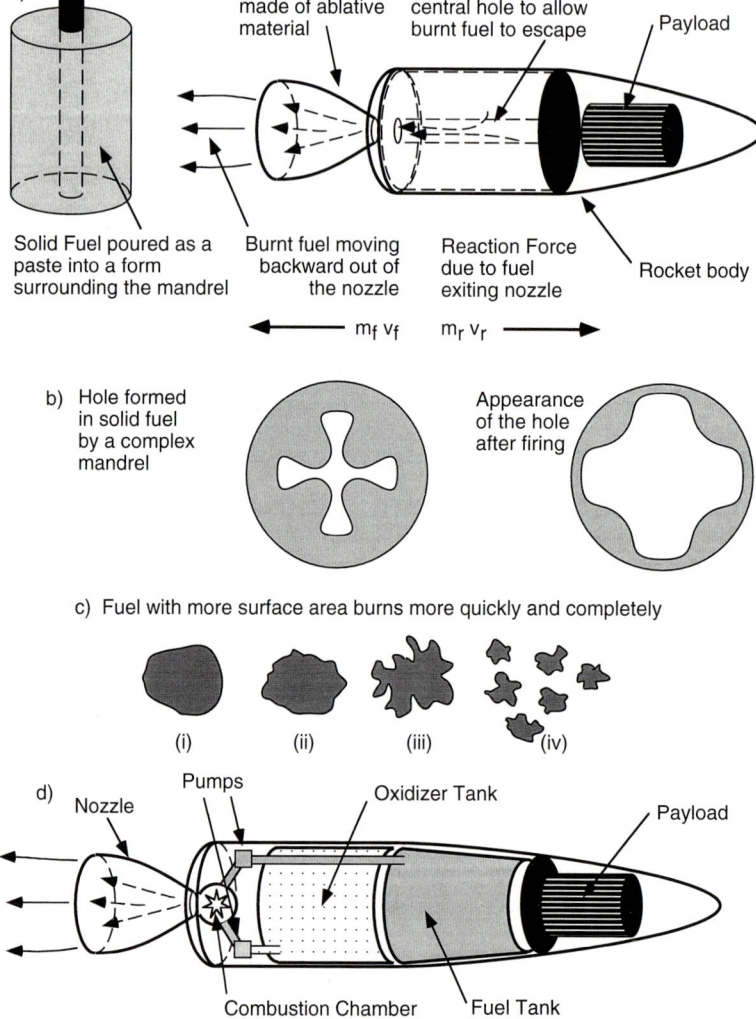

Figure 7.5 The basic structure and operation of solid and liquid fueled rockets. a) A solid fuel rocket is manufactured by pouring a rubbery fuel mixture into a mould around a mandrel. When the fuel hardens, the mandrel is removed and when fired, exhaust gases exit through the hole left by the mandrel. b) A mandrel with a complicated shape is used to form a hole that will allow the fuel to burn evenly. c) The rate at which the fuel burns is dependent on the surface area of the fuel particles. More rugged, or smaller particles will burn more quickly. d) Liquid fueled rockets require more complicated sytems of fuel and oxidizer tanks and pumps.

The cast propellent material can pass for a hard rubber eraser. Pick it up, twist it and it feels like an eraser. It bends, it deforms. Examine it closely and you will see that it is a rubber-like material filled with a crystalline solid. Hit it with your paper weight. Rub it on the desk. Take out a piece with a fingernail. Step on it, nothing happens.

Apply a match and it burns brilliantly with bright streaks shooting out somewhat like a fourth of July sparkler. As it burns it releases a sharp smell of hydrochloric acid.

Cohen explains that the big advantages of a solid rocket fuel are that the structure of the mixture can be closely examined before it is burned. If anything has gone wrong with the mixing process, it can be fixed before the rocket is used. When dealing with liquid fuel rockets, the mixing is done in flight. In this case, if anything goes wrong with the mixing process, it is very difficult to correct the mistake. Another of the main advantages of solid fuel rockets is they can be built bigger without changing the technology.

Cohen points out that it is very important to make sure there are no holes in the mixture of material; if there are open porous pathways within the solid propellant, the burning rate will be much higher than expected creating a distortion in the process of fuel burning and hence in the acceleration of the rocket. This could be fatal for astronauts traveling on top of the rocket. If the rocket is delivering a human payload into space, the acceleration rate must be less than 6g, where g is the acceleration due to gravity. Generally the acceleration rate must be of the order of 5g. The payload may also contain sensitive instruments that can be damaged by excessive acceleration.

Scientists have developed ingenious ways of controlling the burning rate of the solid fuel rocket engine. Figure 7.5 (b) shows the shape of the internal cavity, which, when burning at a uniform rate, will create uniform acceleration in the rocket. To assure this uniformity, the structure of the hollow left by the mandrel is designed to create a complex surface area that burns at a predictable rate.

Two other factors which influence the burning rate of the rocket fuel are the fineness and surface characteristics of the oxidizer and metal powder used in the fuel. A finer powder will mean a higher rate of burning. The surface structure of the metal powder also influences the burning rate; for example, a rougher fine-particle of aluminum will have a much higher surface to mass ratio than a simple blob and will burn much faster. Thus the manufacturer of the aluminum powder and the perchlorate crystals must control the fineness of the powders and the range of powder sizes in any one ingredient as well as the surface structure of the powder. Making very fine aluminum powder can be a dangerous process. The Alcoa company that manufactures aluminum powder for the space shuttle rocket engines had several explosions before the technology was completely mastered. Aluminum powder is somewhat safer than other metal powders in that almost as soon as it is manufactured the surface of the powder becomes covered with a thin layer of oxide. Other metal powders react so violently with the oxygen in the air that they are described as being *pyrophoric*. (See Figure 7.6(b).) In the making of the atomic bomb and nuclear reactors it is often necessary to handle very fine uranium powders. Not only are uranium powders radio-active, but they are also very poisonous and pyrophoric. For these reasons they have to be handled in inert atmospheres in self-contained vessels not only to protect the workers from the radio-active hazard and the poison hazard of any inhaled powders, but also to prevent spontaneous explosions if the powders come into contact with the air.

The ingredients of a solid fuel rocket react to produce high temperature gases of metal oxides — typically at temperatures of 3000–3300 °C. The pressure inside the burning rocket can be anywhere up to 1000 pounds per square inch. This means the propellant must be resistant to cracking under such pressure; otherwise, the burning rate will be uncontrollable. A series of disasters in an army exercise were eventually traced to the fact that the propellant in the tank shells had frozen. Rough handling of the shell caused cracking of the frozen propellant. The cracks created a greater surface area inside of the shell for burning of the propellant. The greatly accelerated burning rate of the cracked, frozen propellant created higher pressures than those that could be sustained by the gun and the base of the gun barrel burst before the shell was able to exit the barrel and relieve the pressure.

The combustion products from solid rocket fuel enter the nozzle of the rocket, traveling at subsonic velocity. Within a distance of a few inches for smaller rockets, and about fifteen feet for very large rockets, the velocity of the gas increases to the local sonic velocity and, in the process of this acceleration, both the temperature and the pressure of the combustion gases drop. The temperature drop is due to the *Joule–Kelvin effect*. Because of the enormous temperatures and pressures in the nozzle, the nozzle must be made of a material which can withstand such conditions. In fact, a typical nozzle of

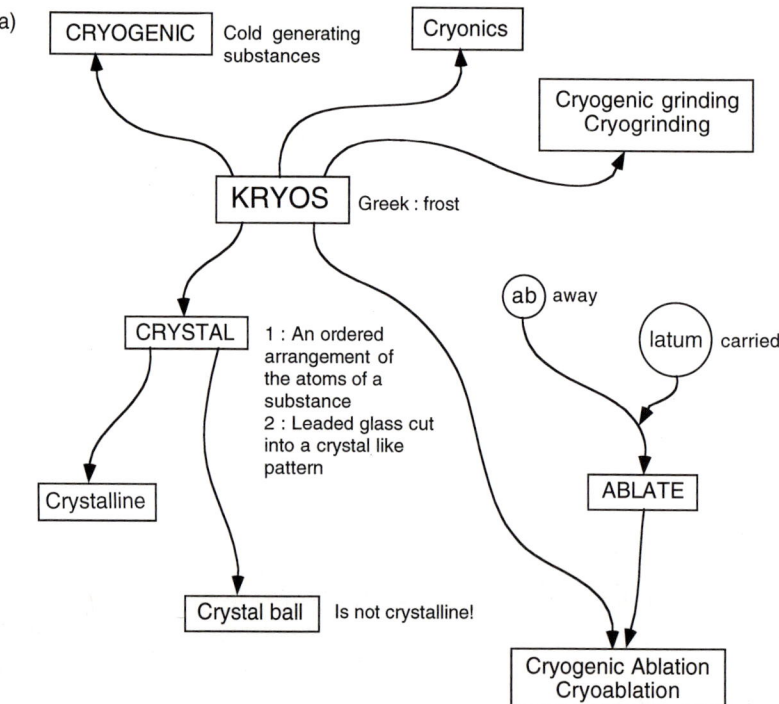

Figure 7.6a

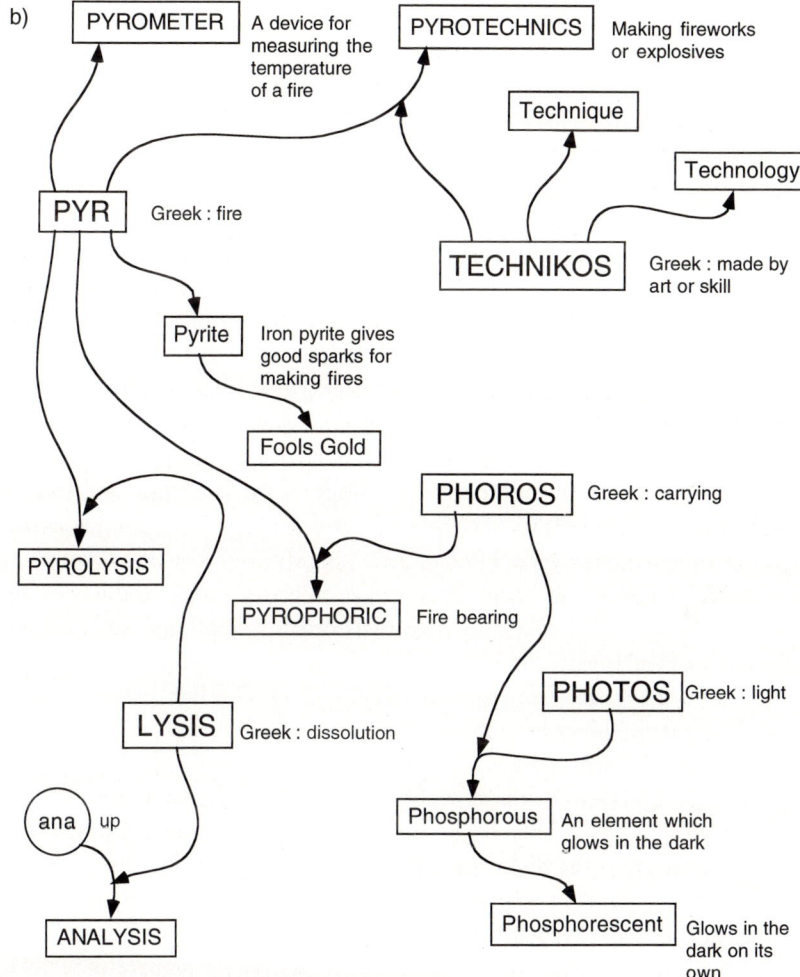

Figure 7.6 The roots of some cold technical terms. a) The Greek word for frost has given us many English words relating to cold or sparkling things. b) The Greek word for fire has given us many words used in rocketry and all applied sciences.

a rocket engine will be made either from a block of graphite (solid carbon) or a graphite laminate. Such materials actually prolong the life of the nozzle by a process which is known as *ablation*. This process will be discussed in greater detail when we look at the problems of protecting items re-entering the atmosphere from great heights.

In the early days of the space shuttle program, some scientists expressed concern about the effect on the Earth's climate of the very finely divided aluminum oxide dumped into the atmosphere during the launching of a space shuttle. As the space shuttle blasts off, its twin booster rockets dump 150 tons of aluminum oxide fine-particles, most of them light enough to remain aloft for a long time drifting in the upper atmosphere. After studying the possible effects of zillions of such fine-particles on the cloud cover of the Earth, Cicerone, an atmospheric chemist at the National Center for Atmospheric Research in Boulder, Colorado, decided that the shuttle exhaust smoke would probably not make much difference to the Earth's climate [11].

An airplane climbing in the atmosphere only needs to carry hydrocarbon fuels such as gasoline which burns with the oxygen of the air. The energy from the burning fuel, in the case of a jet engine, produces a column of hot gas that pushes on the atmosphere to propel the airplane forward. However, outside of the atmosphere, there is no oxygen to be burnt. The acceleration of a rocket comes from the sacrifice of the mass of the fuel which is ejected from the rear of the rocket: the energy of combustion gives the exiting gases high kinetic energy. We have already discovered that in a solid rocket engine the ammonium perchlorate provides the oxygen to oxidize the metal powder and that the forward thrust of the rocket is the reaction to the expulsion of the combustion products out through the nozzle of the rocket. The German V-2 rocket used a mixture of kerosene and oxygen. In the advanced technology of the late 1980s and early 1990s, rockets used a mixture of fluids such as *hydrazine* and what are known as *cryogenic fuels*, liquid hydrogen and liquid oxygen. Hydrazine is a highly corrosive liquid which has the formula H_2NNH_2 or N_2H_4. It burns with oxygen to give gaseous products of high energy. The equipment needed to handle such fuels before takeoff and during the operation of the engine is relatively complex because the normal boiling point of liquid oxygen is $-183\,°C$, while that of liquid hydrogen is $-253\,°C$. High pressure equipment and thermally insulated pipes have to be used throughout the fueling equipment until the rocket is above the atmosphere.

Hydrazine is about 12–15% less efficient than cryogenic fluids as a rocket fuel but is simpler to use and more reliable. In the late 1980s cryogenic fuels were used to launch large rockets, but to adjust the flight patterns of satellites in orbit they are equipped with smaller and simpler hydrazine engines.

The origin of the term cryogenic is illustrated in the word web of Figure 7.6(a). The Greek word for "frost" is "kryos". When chemists started to prepare other materials in a form that looked like little pieces of frost, they called this type of material *crystalline*. Today the specialist who studies the way in which materials form crystals is called a *crystallographer*. In everyday speech in North America people refer to good quality cut glass as *crystal*. If a good quality glass is made with a high content of dissolved lead, the optical qualities are such that when cuts are made in the glass the reflected light seems to sparkle with high intensity. *Crystal chandeliers* are made of this type of glass, not of real crystals. The *crystal balls* in which fortune tellers are supposed to be able to see future events, again have no crystals in them, but are just round spheres of very pure, flawless glass. When scientists started to be interested in very cold conditions and in the properties of very cold materials, they used the same Greek word but with a slightly different spelling to create the term *cryogenic* which means "cold creating." Cryogenics became a specialized area of physics. One of the popular demonstrations that students often see in their early introduction to physics is how a rubber ball, when cooled to the temperature of liquid nitrogen ($-196\,°C$), shatters if dropped onto a hard surface. This illustrates how a material that is flexible at room temperature becomes brittle at very low temperatures. This principle is also used in what is unfortunately known as *cryogenic grinding*, or *cryogenic pulverization*. The grinding and the pulverization does not produce the cold condition. Cryogenic fluids are used to cool the materials before they are pulverized. It would be more correct to call the process cryogrinding or cryopulverization, but unfortunately the earlier terms are firmly entrenched in the vocabulary of the engineer. In Chapter 11 we will look at the way in which powders of spices such as nutmeg are prepared. Cryogrinding is used to make the nutmeg brittle before pulverization and also to prevent the loss of taste-bearing oils from the nutmeg during the grinding process.

Liquefied natural gas which is a mixture of hydrocarbon gases, the chief constituent being methane (CH_4) typically found in oil-bearing rocks, is shipped across the oceans to Japan. It then has to be turned into a gas by warming it up to ordinary temperatures. The recycling of plastics is difficult because it is not easy to pulverize waste plastic. Japanese scientists are thus using large containers of shredded plastic to warm up the liquid natural gas and then they grind the chilled plastic with their pulverizing equipment. The plastic starts off very cold and the pulverizing process does not create enough heat to hold the powder grains together as they are created. The word web of Figure 7.6(a) includes the term *cryonics*, which describes a pseudo-science. (*Pseudo* in Greek meant false or something that pretends to be something else. Thus a *pseudonym* is a fictitious name assumed by an author who does not wish his real name to be known.) A pseudo-science appears to be a scientific pursuit, but under closer inspection is seen to be based on very shaky grounds. "Cryonics experts" claim that a human, e.g., someone suffering from a terminal disease, can be frozen solid using a cryogenic fluid, such as liquid nitrogen, and preserved until the day when a cure has been found. When the cure is found for the terminal illness, then the individual can be thawed out and, hopefully, cured. The immortality promised by such pseudo-scientific ideas tempts the critically ill or their relatives into parting with large sums of money for a procedure which ranks with searching the pot of gold at the end of the rainbow.

A term that we will meet later in this chapter is *ablation*. To *ablate* something literally means "to have it carried away." We have already met the present tense of the verb "to carry" in Latin, "ferre", and "latum" is the past tense describing something which has been carried (why these two words are dissimilar is hard to explain, but then language is full of illogical developments). Modern medical practice uses cryogenic liquids such as liquid nitrogen to carry out surgery by freezing something which is then broken away. Such surgery is called *cryoablation*. Again, unfortunately, some people say cryogenic ablation, but the correct term is cryoablation. Small areas of a skin cancer can be removed using cryoablation techniques. One of the modern surgical techniques for removing the clouded eye lens, the condition known as a *cataract*, is to use a small probe filled with liquid nitrogen to freeze the lens of the eye and break it into pieces and pull it away before replacing it with a plastic lens insert.

Section 7.3 What goes up must come down … Most of the Time

A 5000 mile, or 8000 kilometer range intercontinental ballistic missile (ICBM) reaches a maximum speed of the order of 6–7 kilometers per second (15 000 mph). Its trajectory reaches a maximum height of 1100 kilometers (700 miles) and returns to Earth along a path inclined at about 30° to the horizontal. Figure 7.7 shows the physical processes occurring around a blunt-nosed warhead of an ICBM coming in at an angle to the horizontal at a height of 60km. It can be seen that there is a turbulent wake behind the warhead and shock waves. Scientists report that at the center of the nose cone the temperature reaches 6000 °C. This is of the same order as the temperature of the outside of the Sun. The nose cone is said to be *white hot* and *incandescent*.

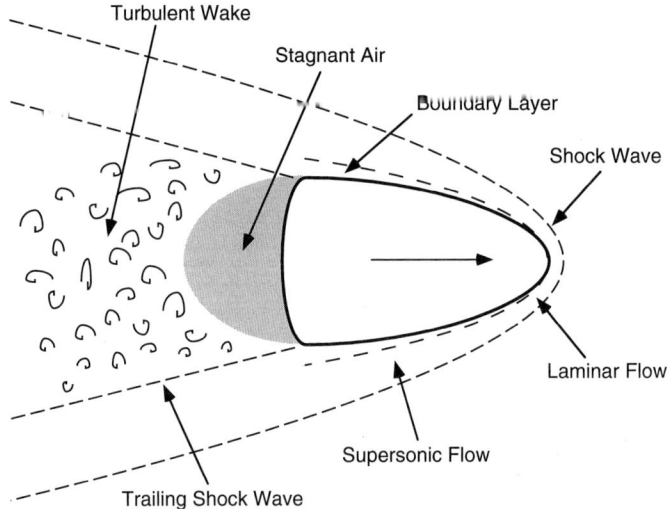

Figure 7.7 Conditions on the outside of a missile re-entering the Earth's atmosphere can be severe and depend upon the angle of entry into the atmosphere with respect to the Earth's surface. The surface temperature can reach 6000 K for a missile moving at 6.4 km/s.

In Latin the word "candescere" meant "to shine with light." This word has also given us the word *candle* for a small light. Anything that was white was called "candidus". When Romans ran for office they used to wear white togas to symbolize their purity. Cynics would say that not many modern politicians (candidates) would be able to wear white robes! In general, people who wore white were obviously seeking political office and this gave us the word *candidate*.

The track of a manned orbital vehicle returning through the atmosphere follows a lower speed trajectory and does not attain the higher temperatures which are possible

7.3 What goes up must come down ... Most of the Time

for the ICBM. The difference between the heating in the two trajectories comes about because of the angle of inclination with which they enter the atmosphere. Space capsules carrying humans travel at an inclination to the horizontal of only 1° and the capsule is shaped so that it gets some lift from Bernoulli's principle to prolong its path and to change its deceleration as it moves through the atmosphere.

Manned capsules are protected by an ablative shield. In terms of space technology ablation means the erosion or vaporization of material. In this chapter we will review the development of technology aimed at protecting the inside of space capsules and nose cones of ballistic missiles from the severe conditions that they encounter during re-entry. To be able to appreciate this technology we must develop several concepts associated with heat transfer and also be very clear about the terms we use to describe the atmosphere and the directions of re-entry.

When we look at the distant sky and Earth they appear to meet at a boundary. That boundary was called "*horizon*" in Greek and has become our word for the boundary between Earth and sky. A line which lies flat like the horizon is said to be *horizontal*. At right angles to the horizontal we have what is known as the *vertical*. This word has a rather strange origin in that it comes from a word "vertere" meaning "to turn." A whirlpool, the swirling water moving round the central point is described as a *vortex*. When Greeks used to look into the night sky there was one point up in the sky where the whole panorama of the stars seemed to turn around that point. Thus the stars seemed to turn as one, so they were called the *universe* from the Greek word for one. The straight line going up to the point about which the sky turned became known as the vertical line because it went from Earth up to the turning point. As the word web of Figure 7.8 shows, other words related to vertical are *invert*, to turn upside down, *divert* meaning to turn away, *revert* meaning to turn back, *pervert* to turn away from a normal pattern of behavior, and *convert* turned into something. An advertisement was originally a notice which turned your attention to a product. But some adverts are counterproductive, creating *aversion* (a turning away from the product). Originally a *university* was a group of scholars who all worked together to turn the world into a better place.

The angle of *inclination* of the trajectory of a re-entering space capsule is the angle it makes with the horizontal. The word *incline* comes from the Greek word to bend. A hill which slopes up from the horizontal is described as an incline. If we lean back from the vertical, we have *decline*. In everyday speech the meaning of the word decline has also come to mean downhill in the sense of "degraded" or "decaying away." If we are really laid back then we are *reclining*. The word "klinein" meaning "to slope" is closely related to the Greek word "to bend" and this has given us the word *clinic* where people are put into beds and looked after. It has also given us the word for a ladder so that a *climax*, meaning highpoint of a series of experiences, comes when we reach the top of the ladder. Curiously, the word *climate* comes from the same root word since the early Greeks used to think the world was flat and sloped upwards to the North and that was why the weather was different in the north than around the *Mediterranean*.

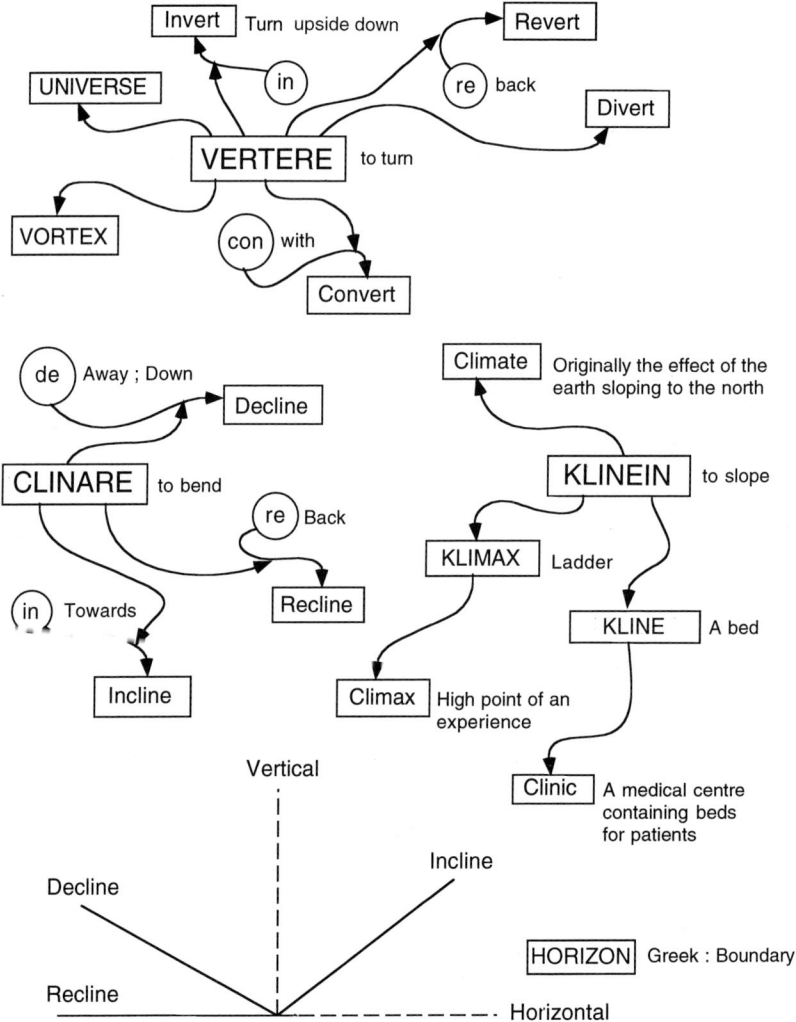

Figure 7.8 Students often show little inclination towards learning declination (or declension) in language studies. They are also often confused by words showing direction.

We now know that climate is caused by the inclination of the Earth as it spins on its axis with respect to the Sun; it is the angle of the axis of the Earth in its orbit that determines the climate.

Many of the problems associated with the design of safe space capsules were too difficult to solve from theoretical considerations alone and many tests had to be made with models of various designs of space capsules in specialized wind tunnels (known as a *hypersonic* or *supersonic wind tunnels*). A detailed discussion of such a system is beyond the scope of this text, but we will discuss briefly the construction of a supersonic wind tunnel used to study the design of various types of aircraft. In the supersonic closed-circuit wind tunnel of Figure 7.9(a), the shape of the part of the tunnel where models are placed can be altered with the flexible liners to change the flow conditions around the model. A delta wing aircraft mounted on a support is shown in the working section of the wind tunnel. The very high velocity air needed to flow around the model is passed through a multi-staged compressor. It will be noticed that a cooling system is installed immediately after the air compressor. This is because the compression of the gas raises its temperature in the same way that compression of air by a bicycle pump makes the pump become hot. After compression, there is not sufficient time for the gas to cool down by natural means and hence the cooling system [12].

The supersonic wind tunnel of Figure 7.9 can be used to study the optimum design of aircraft. Using such tunnels, scientists have discovered the best shapes for different types of airplanes intended to fly at different Mach numbers as shown in Figure 7.9(b). The reader should note that for the first time in these studies we have used a specialized scale to compress the information onto the graphical display. Before you study Figure 7.9(b), it is important you understand how this figure was scaled. The type of scale used is known as a *logarithmic scale*. Equal intervals on such a scale represent equal increases of the ratio of different numbers. The word logarithm comes from two Greek root words: LOGOS which can mean "proportion" or ratio and "arithmos" meaning "number." *Logarithms* as a number system were invented by a Scottish mathematician John *Napier* who lived from 1550 to 1617. On a logarithmic scale, equal changes in a ratio occupy equal space on the axis. Thus the points 10, 100, 1000, 10 000, ... would be equally spaced, and, on our graph, the distance representing the difference in aircraft range between 10 000 and 20 000 units is the same as the distance separating 1000 and 2000. Logarithmic scales are widely used in graphical displays to compress data when the quantities being studied vary over large ranges. Thus if we were to draw a graph such as that of Figure 7.9(b) so that the distance between 1000 and 2000 was the same as between 2000 and 3000, etc., right up to 20 000, we would need a graph ten times as tall. Before the invention of high speed modern computers, science students needed to know how to use logarithmic numbers to do mathematical calculations. This is why older textbooks had tables of logarithms at the back. Fortunately this type of tedious calculation is no longer necessary but it is still very useful to use logarithmic scales on graph paper. The reader should always check the type of scale being used on a graph. Very strange conclusions can be reached if one fails to recognize that a scale is logarithmic rather than linear. Take time now to study the structure of Figure 7.9(b) very carefully because we will be using logarithmic scales at various times later on in this book.

As we discussed in an earlier section, the use of the freezing point of water as a convenient reference point for temperature measurements was established by Celsius; however, when we are dealing with cryogenic technology it sometimes becomes quite confusing to work with negative temperatures such as the boiling point of oxygen which is $-183\,°C$. To avoid this problem, and to express some of the important relationships discovered by exploring the properties of materials at low temperatures, scientists often

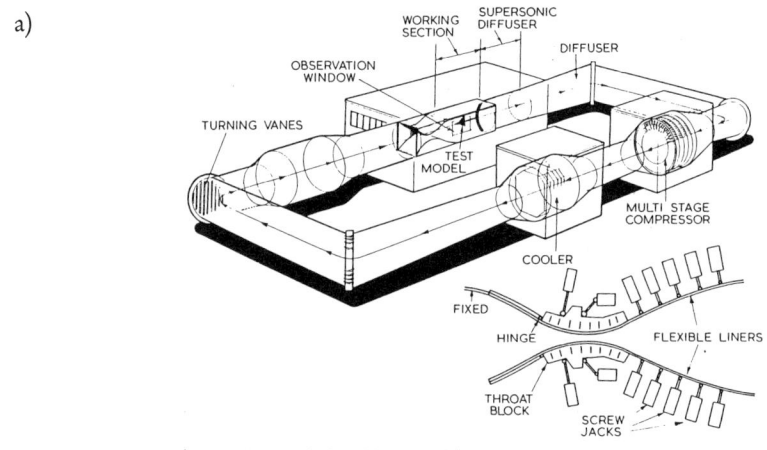

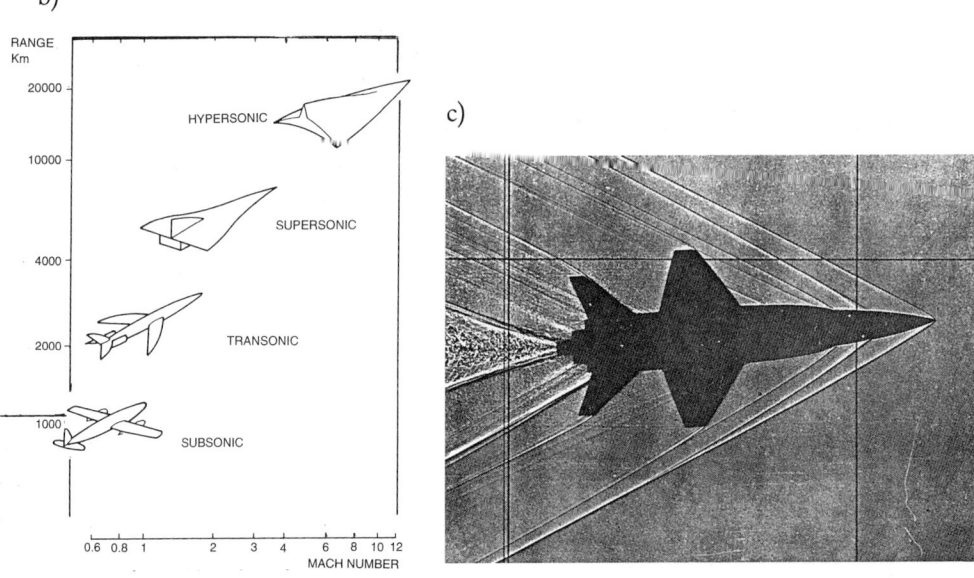

Figure 7.9 In a supersonic wind tunnel, compressed air is driven at high velocity to simulate the conditions encountered by an aircraft in supersonic flight. Reproduced from Aerodynamics: The Science of Air in Motion by John E. Allen, Second Edition, Granada Publishing Limited-Technical Books Division, 1982.

se a temperature scale known as the *absolute temperature scale*, also known as the *Kelvin scale*. To understand the concepts embodied in the use of an absolute temperature scale, it is necessary to backtrack in history and consider the experimental work done by a French scientist on the expansion of gases when they are heated.

The French physicist who carried out this work was Jacques *Charles* (1746–1823). Charles was an interesting person. He was the first to use hydrogen balloons to explore

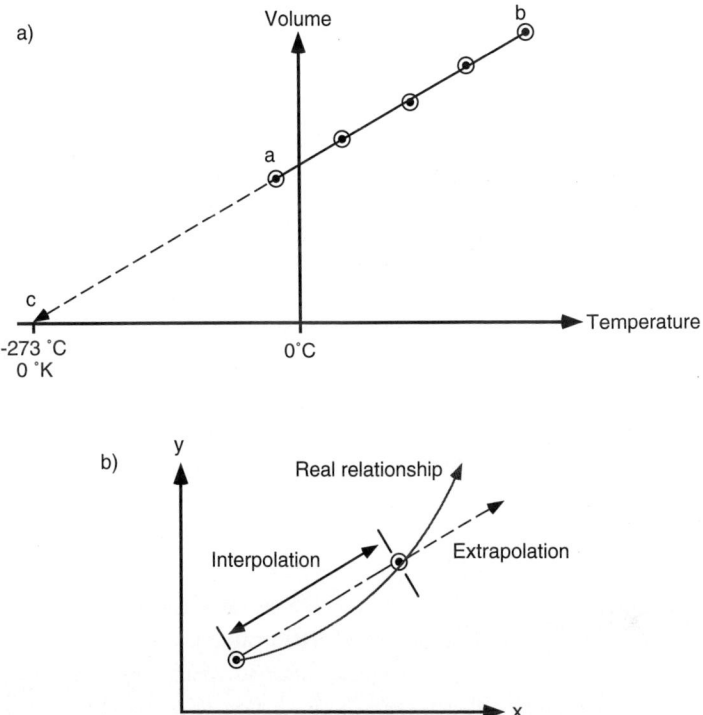

Figure 7.10 Extrapolation and interpolation are "data-extending" techniques that scientists should use with caution. a) A daring extrapolation of some limited data on the expansion of gases at constant pressure was the initial stimulus of the search for "absolute zero." b) The real relationship between data points may be very different from the interpolated or extrapolated relationship.

the atmosphere. He built the first of these in 1783 and went up in such balloons several times. He once reached a height of over one mile. His interest in ballooning inspired him to study the way in which gases expand when they are heated. In the course of his experiments he discovered that if he plotted the volume of a gas against its temperature, he obtained data points such as those shown in Figure 7.10(a). As he studied the effect of temperature on the volume of the gas at constant pressure, Charles generated the experimental data over the region *a* to *b* in this figure. He noted that he obtained a straight line relationship in which the volume of the gas increased by approximately 1/273 for every increase of one degree centigrade. Sometimes when scientists have a limited set of data such as that from *a* to *b* they indulge in what is known as *extrapolation* of their data. This word comes from the Latin word "extra" meaning "outside of" or "something added" and the word "to plot," as in to put points on a graph. The word plot in this sense is another of the very ancient words which originated in science from the surveying of land. It comes from an old English word meaning a piece of land. When marking the boundaries of a piece of land, the surveyer put in a series of posts which became points of reference along the line of the boundary. The mathematicians took this imagery to describe the creation of a pattern of data points that defined the boundaries of their mathematical curves. By transfer of meaning, the reference points came to be described as plotting the boundary. Sometimes the data points on an experimentally generated graph are widely spaced. When the scientist makes an educated guess as to what happens between two data points, the process is described as *interpolation*. Interpolation can sometimes mislead scientists if the properties of what they are studying change radically between the two data points they have established. This is suggested by the solid curve labeled real relationship in Figure 7.10(b). When mathematicians venture outside of their data points, the extended curves that they draw are said to be extrapolated to generate anticipated relationships. Again this can be a misleading exercise in data interpretation (See the discussion of elephants in face powder in Ref. [13]).

When Charles looked at his data points between *a* and *b*, he made the daring extrapolation that if he extended his line to *c*, he would find a temperature at which the volume of the gas would shrink to zero and that it would thus be impossible to have a temperature lower than that. It must be remembered that at the time Charles carried out his work, scientists were unaware of the fact that gases could eventually be liquified and solidified. It was in fact a much more reasonable extrapolation, given the concepts of the time, than we would judge such an extrapolation today. As it turns out, the prediction of $-273\,°C$ C as the lowest temperature attainable turned out to be surprisingly near the truth. The accepted value today for the lowest possible temperature is minus $273.18\,°C$. Students often mistakenly state, on the basis of Charles work, that at absolute zero the volume of gas is zero. In fact, the correct statement, based on our current knowledge should be that, the free energy of a gas molecule is zero at absolute zero; it is a solid; it still has volume but cannot move!

Charles did not publish his experiments. In 1802 the French Chemist Gay-Lussac (1778–1850) reached the same conclusions and so the law is often called *Gay-Lussac's law* as well as *Charles' law*. Gay-Lussac was also one of the early pioneers of ballooning for scientific purposes. In one of his flights, Gay-Lussac reached a height of 4 miles which is higher than the tallest peak of the Alps. He found no change either in the composition of the air or in the Earth's magnetic field at this height.

The effort to generate very low temperatures in the laboratory was spearheaded by *William Thompson* (1824–1907), a Scottish mathematician and physicist who later became *Lord Kelvin*. Lord Kelvin was the son of an eminent mathematician. We are told that as a child of eight "William attended his father's lectures with delight." He entered the University of Glasgow at eleven years of age and wrote a scientific paper while still in his teens. In 1848, after studying the predictions made by Charles and based upon further experimental work, he proposed that at absolute zero the kinetic energy of the molecules was zero. Thompson also suggested there were advantages to measuring all temperatures in the universe from this absolute zero. From the perspective of absolute zero, the freezing point of water is 273.18 Kelvin. The temperature scale based on absolute zero as a reference point is called "the absolute temperature scale" or "the Kelvin scale" in honor of William Thompson. In 1892 Thompson was made Baron Kelvin of Larges; a title acknowledging that he worked near the Kelvin River which flows near Glasgow in Scotland [14].

Thompson worked for many years on developing instruments that made it possible to have submarine telephone cables across the Atlantic. However, the end of Thompson's life was not quite as illustrious as its beginning. Asimov [14] has the following comments.

> *It is sometimes the fate of scientists who in their youth forge new trails and who lead the way towards new concepts to pass their last days bewildered by still newer developments they cannot accept.*

In Asimov's words

> *With almost his last breath, as an old man in his 80s he who had been so brilliantly a revolutionary in his youth set his face against novelty and bitterly opposed the notion that radioactive atoms were disintegrating or that the energy that they released came from within the atom.*

Lord Kelvin's view of new theories when he was old is a good example of the cynical saying of some young innovative scientists that old scientists do not change their minds, they just die off. If our present system of granting research funds had been in place in the late 1890s, Lord Kelvin, as a reviewing expert, would probably have denied funding to those seeking to study the phenomenon of radioactivity. Peer evaluation has its limits!

Section 7.4   Heat Transfer Mechanisms

To understand how we are able to protect missiles and space capsules from the extreme heat generated by atmospheric friction upon re-entry it is necessary to look at the various ways in which thermal energy can be transferred from one body to another. It is useful to describe heat transfer in terms of three basic mechanisms: conduction, convection, and radiation. The thermal energy of a body is the kinetic energy of the vibrating molecules in the body. As the energy of the molecules increases, the temperature rises. When a point is reached at which the vibrational energy of the molecules is so large that they cannot be held together as a rigid body, they break apart to form a liquid and, from the outside perspective, we say that the body has melted. Transfer of vibrational and kinetic energy from molecule to molecule is described as *heat conduction*. The origin of this word and some related words is shown in the word web of Figure 7.11. Metals are generally good conductors of heat whereas a material such a cork is a poor conductor of heat and is called a thermal *insulator*. The word *insulate* comes from

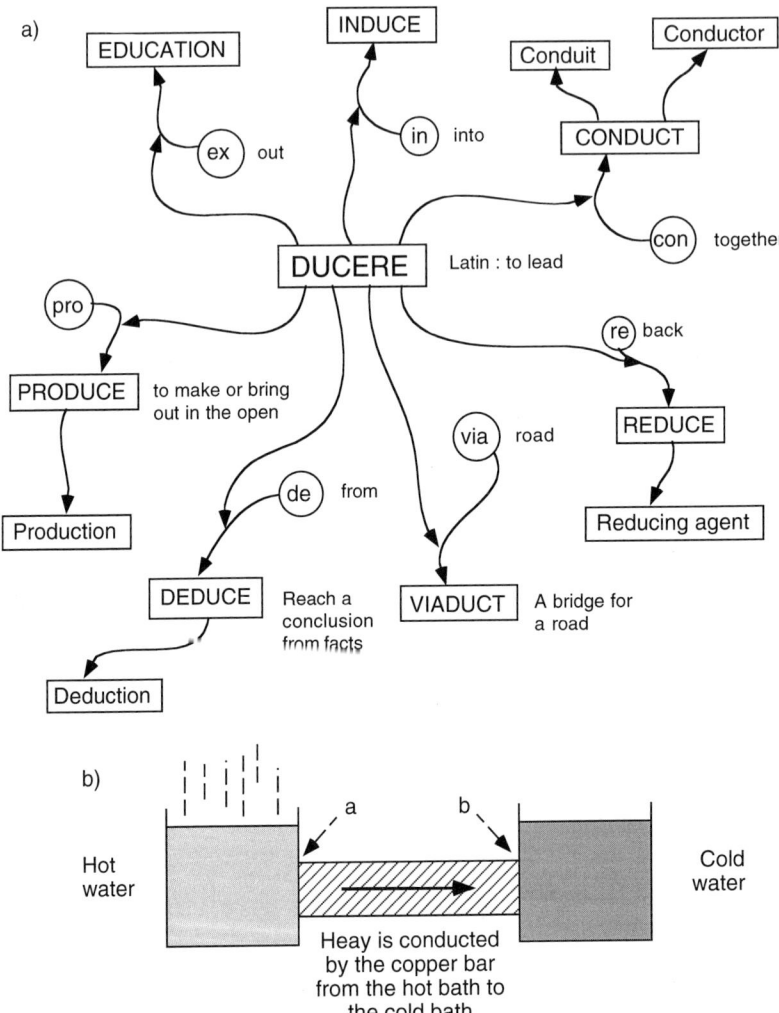

Figure 7.11 Conduction is a major heat transfer mechanism.

the Latin word for an island (Figure 7.12). Thus if we surround a hot body with an insulator, we turn it into an environmental island which cannot exchange heat with its environment. There are no perfect insulators, but for many purposes, substances such as porous plastic can be regarded as very good insulators. In substances such as fiberglass, cork, and expanded styrene, the actual insulators are the tiny pockets of gas trapped by the thin network of fibers or cork. Fur is a good insulator because if traps air in its structure and insulates the wearer from the cold air outside. Birds and fur-coated animals are killed by oil contamination because the oil causes the feathers or fur to cling together eliminating the air pockets which enable the animals stay dry and warm.

7.4 Heat Transfer Mechanisms

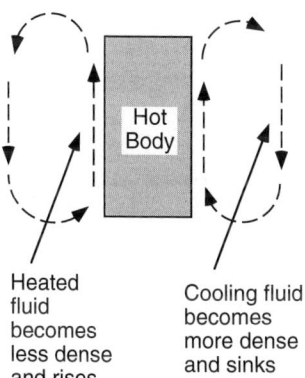

Figure 7.12 Convection drives the motion of boiling water in a pot, as well as major weather systems in the atmosphere.

A word which is similar to insulation but which has a complicated historical derivation is insulin. The disease diabetes is due to a deficiency of insulin in the body. The chemical we now call insulin is made in the body by some special cells in the pancreas. These special cells were discovered in 1869 by Paul *Langerhans*. and, in his honor, were named the *Islands of Langerhans*. As Asimov comments, "perhaps the most romantic sounding name in the body." In 1916 a British physiologist suggested that the Islands of Langerhans produced a hormone that controlled the manner in which the body handled sugar. He gave this hormone the name *insulin* from the Latin word for island: a name which is universally used throughout the medical profession [15].

In experimental studies of heat conduction, errors can easily arise if the experimenter fails to establish a good contact between a hot and a cold body. Thus in our sketch in Figure 7.11(b), the bar bearing the arrow is considered to be a copper rod which is intended to conduct heat from a hot body to a cold body. Air gaps at the points shown by the letters *a* and *b* can interfere with the efficient transfer of heat energy along the bar. When carrying out this type of experiment, the application of a liquid such as glycerine, a good conductor of heat, at the junctions *a* and *b* eliminates the air gaps. In this case I speak from experience because as a physics student at university I had to repeat this particular experiment because in my first attempt I had bad thermal contacts!

As seen in our wordweb of Figure 7.11(a) several important words in the vocabulary of science are related to conduction. Thus the word *education*, used to describe the way in which one person develops the knowledge of another person, comes from an old idea that everything that we know is already inside us and that the process of teaching leads us to discover and bring out what is already in ourselves. The process of thinking which we call *induction* is the reasoning from particular cases to general conclusions. It is the opposite of *deduction* which is described as the achieving of a particular truth from general information. Science is generally considered to be a field of knowledge where deductive reasoning is the main intellectual process used when interpreting experimental data.

In chemistry, the word *reduction* has a very special meaning and describes the elimination of oxygen from a substance or the conversion of something into an elementary form. The chemical that achieves this reduction of the compound is called the *reducing agent*. In making iron from iron oxide, coke, which is a form of carbon, is used as a reducing agent in the process. This process produces carbon dioxide and leaves the iron free of oxygen.

A gas is a poor conductor of heat, but gases and liquids can carry heat away from a body by a process known as *convection*. Consider the hot body surrounded by fluid shown in Figure 7.12. Heat transfer from the body to the fluid causes the fluid to expand and become less dense. This in turn causes the fluid to move away, upwards from the body, taking away the heat energy it has acquired from the hot body. As this hot liquid moves upwards, cooler fluid moves into contact with the body. In turn, this cool fluid is heated and subsequently moves up.

The word convection comes from root words meaning "to carry away". The resultant movement of fluid is described as a *convection current*. The rising convection currents from objects, such as radiators in the living room, tends to deposit dust on the ceiling where the rising air current is deflected by the ceiling. The dust particles in the rising air have mass and their inertia, or kinetic energy depending on how you view the system, carries the dust on a straight path to the ceiling as the rising current of air is deflected by the ceiling and thus the dust deposits at that point. For the same reason one tends to get a dirt pattern on the ceiling above an incandescent light bulb. In science the impact of dust onto a surface by the hot convection current is described as *thermophoresis*. This word is created from the term that gave us Christopher and pyrophoric (Figure 7.6(b)).

In the design of thermally insulated windows (systems widely used in cold climates), the air gap between the two panes of glass must be free of moisture to improve the insulating properties of the air. The air gap must also be small so that convection currents, which would more rapidly transfer heat from the inner pane to the outer pane, are suppressed. In the early 1990s, as society became more aware of the need to build thermally efficient windows, scientists were beginning to develop systems which were extremely fine sponges of silica powder, invisible to the naked eye; when placed between the two panes of glass these sponges completely suppress any air movement and improve the insulating properties of the air gap by a factor of seven. This very fine sponge of silica is known as *silica gel* or *aerogel* [16, 17].

A very important source of cooling for space capsules re-entering the Earth's atmosphere is radiation of heat energy from the hot body. The radiant energy leaving a hot body is of the same type as light energy except that most of the heat energy is of longer wavelength than visible light and is generally referred to as *infrared radiation*. The term infrared developed historically from studies of the radiant energy from the Sun. When the energy from the Sun was dispersed into its different wavelengths using a glass prism, scientists discovered that there was energy coming from the Sun at shorter wavelengths than those that were visible. This was called *ultraviolet* radiation, meaning "beyond the violet", violet being the shortest wavelength of visible light. More energy was found beyond the red at the other end of the visible spectrum. This type of radiation was called

7.4 Heat Transfer Mechanisms

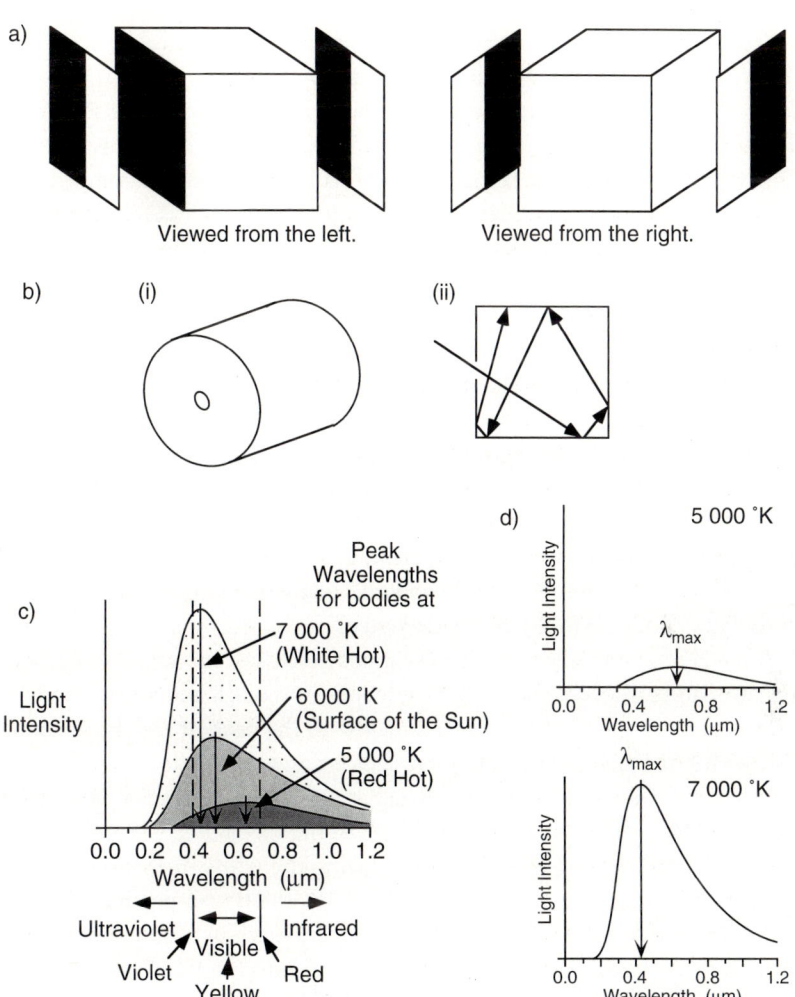

Figure 7.13 Radiant energy relationships discovered by Stefan and Wien are very useful when studying how energy is absorbed, reflected, or radiated from bodies at various temperatures. a) Leslie's cube is a device which is used to study how heat is radiated or absorbed by black and white surfaces. b) A simple representation of a black body would be the way in which a can with a small hole in the top would trap light entering via the hole. c) As the temperature of a black body increases, the wavelength distribution and total radiation output changes. d) The difference in output between a red-hot body and a white-hot body is shown by the movement of the peak wavelength toward the blue end of the spectrum, and the much larger area under the graph indicating greater overall energy output.

infrared radiation from the Latin word "infra" meaning below. The word *spectrum* itself means "that which can be seen," from the Latin word "specere" meaning "to look at." This word has also given us the word *spectacle* for something worth looking at. A *spec-*

tator is someone who looks at something and *spectacles* are optical devices that enable us to look at things more clearly by correcting eye defects.

When a hot body radiates heat it is because the molecules become so agitated with their thermal energy that they begin to act like miniature radio transmitters. In fact there is always some radiation from any body as long as it is at a temperature above absolute zero. The amount of energy radiated by a body increases enormously as the temperature increases. The wavelengths of the radiation emitted by a warm body change with temperature as shown in the diagrams of Figure 7.13(c).

Before we can discuss the structure of the radiant energy temperature curves of Figure 7.13(c), we must develop the concept of a *black-body radiator*. This concept is very important in the history of science since a study of the behavior of black-body radiators led to the development of what we now know as the *quantum theory* of physics. Fifty years ago students used to be taught physics via experiments rather than by the lecturer filling the board with a confusing array of mathematical symbols. Thus when I was introduced to the concept of radiation, our teacher didn't use mathematics but demonstrated radiative heat transfer with an experiment known as *Leslie's cube* [18]. (This simple experiment has disappeared from "modern" textbooks, but it is well worth reviewing as a teaching/learning tool.) The basic system of this experiment is shown in Figure 7.13(a). A metal box, one side of which is painted black and the other white, was placed between two panels. Each of these panels was split into black and white areas as shown. Thermometers were placed in intimate contact with the back of each panel and the box was filled with hot water. This experiment demonstrated very effectively that black surfaces were good emitters and receivers of radiation. Thus the highest temperature was reached in the black panel facing the black side of the metal box. Conversely the lowest temperature was achieved by the white panel facing the white surface of the metal box. This experiment demonstrates why both Eskimos in the frozen north and Arabs in the hot Sahara desert wear white clothes. The Eskimos use white to stop radiation of body heat out into the surroundings whereas the Arabs wish to keep the hot Sun from warming up the space inside their clothes. Since hot air rises, the Eskimos also prefer loose clothes so that the hot air can collect underneath the shoulder region and further insulate the body from heat loss. It is for the same reasons that polar bears are large and white. The amount of energy lost from the surface of the body is a function of the ratio of surface area to volume of the body. This relationship can be calculated as follows. First, we consider the volume of a sphere given by

$$V_s = \tfrac{4}{3} \pi r_s^3$$

where r_s is the radius of the sphere. Then we consider the surface area of the sphere given by

$$A_s = 4 \pi r_s^2$$

From these two terms we can see that the ratio of the surface area to the volume is

surface area/volume = $A_s/V_s = 3/r_s$

Therefore the bigger the body, r_s, the less the surface area to volume ratio.

This is the reason why children lose body heat more quickly than adults in a cold climate and must be given more protection against the cold. Chubby people can claim that they are not overweight but merely better adapted to living in the north.

In Europe it was common practice to whitewash the outside of a house to keep it both cool in the summer and warm in the winter. Thus whitewashed cottages are a feature of the landscape from northern Scotland to southern Greece. Ordinary folk discovered for themselves, long before radiation physicists worked out the appropriate theory, that it was an advantage to keep the outside of the house as white as possible. Therefore they developed a kind of non-permanent coating which tended to shed a dirty outside layer every time it rained. Each year they renewed the *"whitewash"* to keep the outside as white as possible.

In the late 1800s, as scientists started to study the physics of radiation, they conceived of what is known as a *perfectly black body*. Surprisingly enough this did not consist of a painted surface but of a hollow cylinder with a small hole at the top. This hole absorbed all radiation that entered it because of the multiple reflections that took place inside of the body, with some energy being absorbed by the surface at each reflection. This effect can be seen by looking at the small hole in the top of a pop can. The inside of the can is shiny but looks dark when viewed through the small hole because light entering the hole has little chance of being emitted because of the multiple reflections inside of the cylinder. For the same reason, the windows of a house look black from the outside unless a light is on or there is a window on the other side of the house. People are often surprised to discover what looked like a dark room as they walked past a house turned out to be quite light when they themselves come into the room. The reason for this effect is that the multiple reflection of light entering the window bounces off the walls and does not escape through a window to allow people outside to see what is inside the room. Because of the multiple reflections that take place in a black body, it also behaves as a perfect radiator when it is heated. The *black body* used by scientists in radiation studies was a hollow ceramic body with a small entrance hole. The ceramic body was heated by an electric furnace. Using this type of device scientists generated curves such as those shown in Figure 7.13(c). As the temperature of the black body increased, the total amount of energy radiated increased as illustrated by the increasing area under the radiation-distribution curves. This aspect of the structure of the curve is shown in Figure 7.13(d). At the same time, the distribution of energy at various wavelengths of energy shifted, with the peak emission wavelength, λ_{max}, shifting towards shorter and shorter wavelengths with increasing temperature. Only a small amount of the radiation given out by a hot body is visible light, as illustrated in Figure 7.13(c). If a piece of iron such as a horseshoe is heated to a temperature of approximately 1000 K it appears red hot. As it is heated further it gives out more and more radiant energy visible to the eye, and it is described as becoming yellow hot, blue hot, and finally white hot.

Experimental scientists investigating the structure of the radiation curves established two important relationships, both of which are still used by modern industrial scientists. One relationship was discovered by Josef *Stefan*, an Austrian physicist (1835–1893), in 1879. This relationship, known as *Stefan's law*, states that

> *The total energy emitted by a black-body radiator is equal to a constant times the fourth power of the absolute temperature of the body.*

This can be written symbolically in the form

$$E = \sigma T^4$$

where T is the temperature on the Kelvin scale and σ is the Stefan–Boltzmann constant which has a value of $\sigma = 5.6697 \times 10^{-8}$ W m^{-2} K^{-4}.

This important relationship enabled scientists to measure the Sun's temperature and that of other stars in the universe. The technical definition of a *star* is that it is a self-luminous body. (In popular speech, some refer to the *planets* as seen in the night sky as stars but they are only visible to Earth-bound observers by reflected light and are not stars according to the technical definitions of the astronomer.) Using Stefan's law scientists determined that the outside of the Sun was at a temperature of 6000 K. The peak wavelength of the energy emitted at this temperature is 0.5 micrometers or 500 nanometers as shown in Figure 7.13(c). Scientists find it convenient to measure infrared wavelengths in micrometers whereas they switch to nanometers when describing the wavelength of visible light. In Figure 7.14(a) the wavelengths of the various components of the visible light are shown. White light contains many different wavelengths. A black surface appears black because all of the light energy falling on it is absorbed.

In Figure 7.14(b) the range of wavelengths encountered in the various parts of what is known as the electromagnetic spectrum are summarized. Just what is meant by describing light and other radiant waves as being electromagnetic is very difficult to explain. The student is advised simply to accept the fact that these types of waves have energy that scientists describe as associated with electricity and magnetism. The student should grasp firmly that, to the scientist, what is known as electromagnetic radiation contains energy which is described mathematically using the concepts of wave theory and that the distance between one wave crest and the next can be described as a wavelength.

Since humans evolved on a planet lit by sunlight, it is not surprising that the human eye is most sensitive to the yellow light whose wavelength corresponds to the peak intensity emitted by the Sun. A psychological study of devices used in the classroom established that the use of yellow chalk on green backgrounds is much more restful to the eye than the traditional white chalk on black backgrounds. Accountants discovered experimentally a long time ago that when you had to do a lot of calculations on paper, *green paper* was more restful than white paper especially when your bank balance is

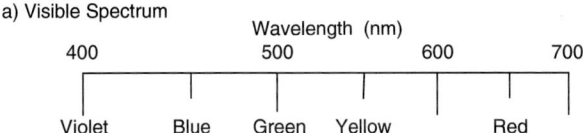
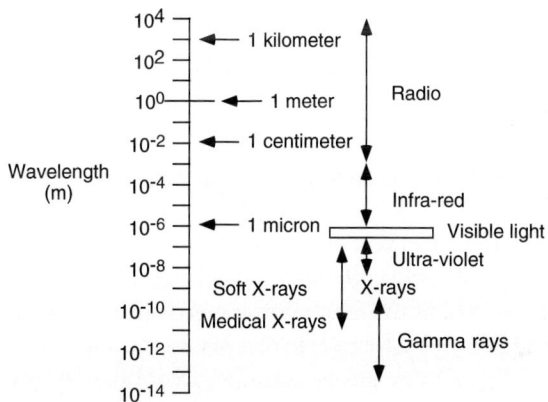

Figure 7.14 Visible light makes up only a small portion of the full spectrum of electromagnetic radiation. a) The wavelengths of visible light range from about 400 to about 700 nanometers (0.4–0.7 micrometers). b) Different names are given to various ranges of wavelengths within the full electromagnetic spectrum extending from gamma rays at the shortest wavelengths to radio waves at the longest.

written in red ink! For the same reason, notebooks made from yellow and green paper achieve significant sales.

Scientists can use Stefan's law to measure the temperature of incandescent bodies such as returning spacecraft and meteorites descending through the atmosphere.

The term *meteor* comes from the Greek term "ta meteora" which literally meant "things on high". The Greek word is formed from the root words "meta" meaning "beyond" and "aeirein" meaning "to lift," thus although a meteor is something that is coming down, the Greeks regarded it as something that was first put up in the atmosphere by the gods. The popular name for a meteor is a *shooting star*. Because of air friction, meteors become white hot and tend to burn up as they fall towards the Earth. A meteor which survives its passage through the Earth's atmosphere to hit the Earth is called a *meteorite*. The technical specialist who studies the weather is called a *meteorologist* since he is the one who studies things that have been lifted up into the sky. [Generations of students have been confused by the fact that the specialist in engineering measurements is called a *metrologist* (from meter "to measure").] As we will discuss in greater detail later in this chapter, some of today's shooting stars are debris from spaceshots and satellites which fall back to Earth through the atmosphere.

The other important empirical law deduced from experimental investigation of black-body radiators is known as *Wien's law* after its discoverer Wilhelm *Wien*, a German physicist (1864–1928). Wien showed in 1893 that the wavelength of the maximum energy radiation in the black-body spectrum times the absolute temperature of the black body is a constant. This relationship is usually written in the form

$$k = \lambda_{max} T$$

where λ_{max} is the wavelength of the peak of the energy curve, T is the temperature in Kelvins and k is a constant equal to 2.8970×10^{-3} mK.

Another, equivalent, way of expressing this rule is that the wavelength of the maximum energy emission from a black body is inversely proportional to the absolute temperature of the body:

$$\lambda_{max} = k/T$$

The scientists of the late 1800s were both fascinated and frustrated by the shape of these radiation curves. A more colloquial statement of Wien's law is that as the temperature rises the wavelengths of the maximum energy (the peak) part of the radiation curve shifts to shorter frequencies. Attempts to derive a formula to describe the emission curve of a black-body radiator from theory were unsuccessful. Towards the end of the nineteenth century Lord *Rayleigh*, an English physicist (1842–1919), put forward a formula which described the emission of the black-body at long wavelengths but also led to the impossible conclusion that at short wavelengths (ultraviolet light) the body would radiate infinite amounts of energy. One of the aspects of the black-body radiation curve that baffled physicists was the fact that although the energy given out at long wavelengths trailed off to very low values, the energy spectrum at the higher frequencies had an abrupt cutoff dependent upon the absolute temperature of the black body. Since Lord Rayleigh's formula predicted infinite energy emission at short wavelength, the failure of the physicists to deduce a valid equation from theory became known as the *ultraviolet catastrophe* [2, 18].

The problem of emission curve of black-body radiators was finally solved by Max *Planck*, a German physicist (1858–1947). Planck noticed that if he subtracted one from one of the relationships in Lord Rayleigh's equation, he could obtain a modified formula that accurately described the emission curve. One commentator points out that this meant that Planck was in the situation faced by many students who have to solve a problem set out in a textbook and have already taken a peek at the answer at the back of the book. The individual then faces the situation of having to work from the problem to the known answer in an intelligent manner. Whatever the truth of the matter, Planck was able to derive the correct equation by throwing away an *implicit* assumption that scientists had been making ever since modern science evolved from the work of people such as Galileo and Newton. In a scientific theory an *explicit* assumption is one that is stated openly at the beginning of the derivation of a theory. An implicit assumption is one that everybody assumes but often fails to recognize. Scientists, like everyone else, often overlook things that are so obvious that they have never aroused discussion. Thus, the scientists who developed the laws of thermodynamics proposed the first, second and third laws of thermodynamics. Only afterwards did they realize that they had been assuming all along another truth which is now known as the "*zeroth law* of thermodynamics." It had to be given this designation since it was too late to alter the numbers of the other laws.

The implicit assumption that Planck had to discard en route to his solution of the black-body radiation formula was the infinite divisibility of energy. Until faced by the ultraviolet catastrophe, scientists had for three centuries implicitly assumed that if one had a small amount of energy one could always split it into two parts. However, to reach the correct formula describing the radiation from a black-body, Planck had to assume that there was an ultimate small packet of energy that could not be split up into subunits. This small amount was called a *quantum* of energy. Scientists now know that the smallest package of energy that can exist for an electromagnetic wave is given by the following formula.

$$E = h\nu$$

where ν is the frequency of the radiation and h is known as Planck's constant and has a value of $h = 6.6262 \times 10^{-34}$ J · s $= 6.625 \times 10^{-27}$ erg · s An erg is a unit of energy. Planck's constant is one of those fundamental constants of the universe the magnitude of which cannot be deduced from any theory: it has to be measured experimentally. Planck's *quantum theory* explained why the high energy part of the black-body radiation curve had a sharp cutoff. Concerning the impact of quantum theory Asimov [14] comments that,

> *The quantum theory was so revolutionary that it was not accepted by physicists at once. In fact, Planck himself did not quite believe it. He half suspected it might be only a piece of mathematical jugglery without any correspondence to anything real in nature. He struggled for years to find a way around his own discovery and would not accept the statistical interpretation of thermodynamics based upon his work.*

In 1918 Planck received the Nobel Prize in Physics for the development of the quantum theory. Essentially the quantum theory of the universe says that the world we see around us is not a painting but a mosaic made up of very tiny elements of energy. We can regard the smallest packet of energy as being a small tile or pixel in the mosaic of reality.

If you find it difficult to understand the quantum theory you are not alone. A generation and more of scientists fought fierce intellectual battles over its concepts. It is one of those facts of science that has to be labelled "grasp me even if you don't understand me!"

Wien's law is used in industry to measure the temperature of systems such as furnaces in which metal is melted or ceramics treated. The instrument used to measure the temperature of such systems is known as the *disappearing filament pyrometer*. In this instrument a glowing filament of a lamp is viewed against the background of a small area of the furnace. The temperature of the glowing filament can be raised or lowered by altering the amount of electrical current flowing through the wire. The electrical current in the wire is adjusted until the filament is the same color as the furnace at which point it can no longer be seen as a separate item in the field of view. From the known electrical resistance of the wire and the amount of electrical current flowing through the wire, the temperature of the wire can be deduced and hence the temperature of the furnace can be evaluated.

As we can see from the graph of Figure 7.13(c), a great deal of the energy leaving the Sun is in the shorter wavelengths. Although the Earth as a whole is not a black body for most purposes, it can be regarded as a black-body radiator at a temperature of 300 K where K denotes absolute temperature.

The *transparency* of a body is a function of the molecular structure of the substance and the wavelength of the radiation. (Transparent means something that can be seen through from the Latin words "trans" meaning "through" and "parere" meaning "to appear"; this means when we look through a transparent body things appear as if there were nothing between us and the object we are looking at.) Glass is transparent to the wavelengths of visible light, but is opaque to infrared radiation. As illustrated in the sketches of Figure 7.15 a house made of glass warms up in sunlight because the incoming radiation can pass through the glass whereas the infrared radiation re-emitted from the soil is trapped by the walls of glass. Therefore people can grow plants in cold climates using such glass houses or *greenhouses*. When thin plastic sheets became available in the late 1950s, people thought they had a cheap way of building an unbreakable greenhouse only to find that a single-walled plastic building was not very effective as a greenhouse. Although the average gardener did not know the reason for this, it was because the plastic was transparent to infrared radiation as well as light. However, gardeners soon discovered experimentally that if they used double-walled plastic structures, the plastic made a very good greenhouse. The double-walled plastic worked to trap heat because water condensed in the gap between the two plastic films and this water absorbed the infrared radiation to give the same effect as a glass greenhouse.

The gas carbon dioxide is a strong absorber of infrared radiation. Towards the end of the 1980s some scientists became concerned that rising levels of carbon dioxide in the Earth's atmosphere from the combustion of carbon-based fuel was creating what became known as the *global greenhouse effect* [19]. Essentially this effect arises from the fact that short wavelength radiation is absorbed by the ground and then the re-radiated from the surface of the Earth as infrared radiation. The infrared radiation is absorbed by the carbon dioxide instead of traveling back into outer space. Therefore the net effect of increasing carbon dioxide, it is claimed, will be a warming of the Earth; however, a full discussion of this greenhouse effect is beyond the scope of our presentation here.

The idea that a black body cannot be seen since it does not radiate any light has been used in fiction to create the idea of an invisible man who can operate without being

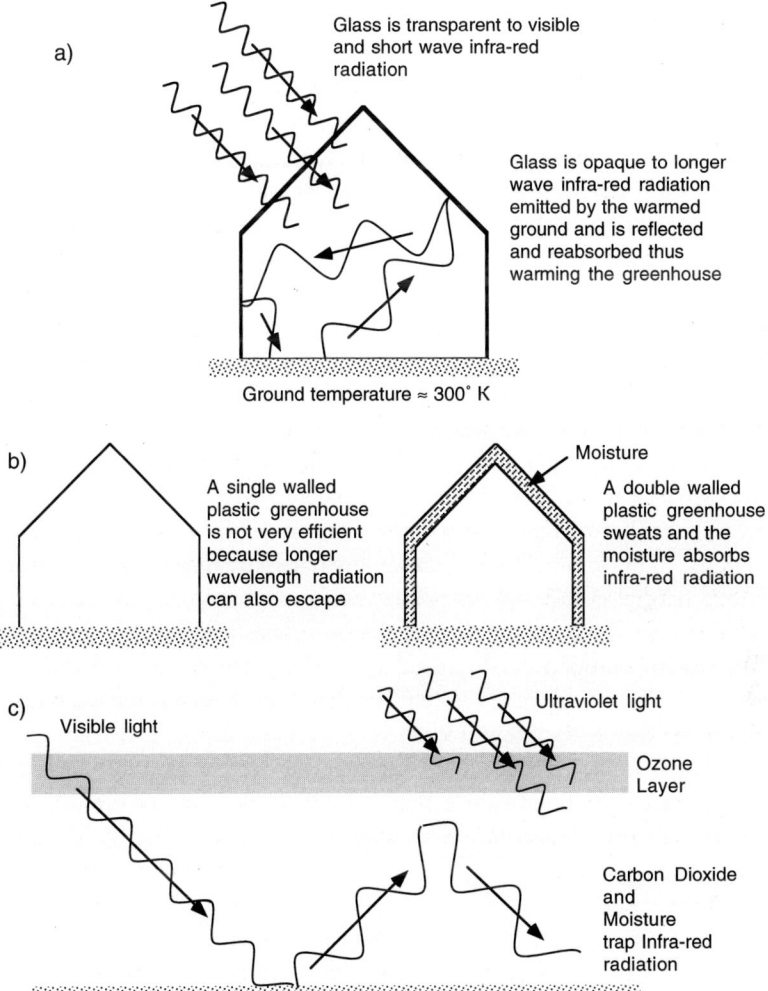

Figure 7.15 The absorption of shorter wavelength radiation by the surface of the earth and the re-emission of longer wavelength, infrared (heat), energy is an important aspect of life on earth. This effect can be used to construct a greenhouse which will remain warm on a sunny day even when temperatures outside the greenhouse are low. (Note that the wavelengths shown are not to scale.)

seen when he is covered with material that completely absorbs light. Unfortunately, this concept would not work. Although in theory the person could not be seen by re-radiated light, a black silhouette would be created against a viewing background and would be all too obvious to any observer. The fact that black material absorbs light is used to create "magical" effects in puppet shows and magic demonstrations. When a person dressed in black is viewed against a black background, the person is not visible to the naked eyes of the audience. Therefore unseen "blacked out" actors can manipulate puppets or bring about mysterious transformations of objects on a stage.

In the 1980s scientists started to look for objects in space that would absorb everything that came near them. These entities are known as *black holes* [20]. Science fiction writers were quick to seize upon the concept that a black hole might be a tunnel to the other side of the universe and a movie was made called *The Black Hole*. In this movie the starship adventurers travel into a black hole and out into the other side of the universe (whatever or wherever that is!). It is interesting to note that the makers of the film, science fiction buffs, did a market study before releasing the film and were dismayed to find out that the public had no idea what a black hole is. Advance publicity for the movie had to be changed to do some quick concept-education.

Systems for transporting rocket fuels such as liquid oxygen make use of the fact that a sphere has the smallest surface area for a given volume. A truck with a spherical vessel on the back is used for shipping liquid oxygen. This container is painted white to minimize heat exchange with the atmosphere, but there is always an opening at the top of the vessel so that the small amount of evaporation of the oxygen that does take place does not build up unsustainable pressures in the transport vessel. Tanks for storing gasoline, oil, and natural gas are always painted white to minimize heat transfer from the atmosphere. In a similar way, the *thermal design of flowers* exploits the properties of hot air and the radiative properties of various colors. Thus one of the first flowers of spring is the snowdrop shown in Figure 7.16(c). The flower of the *snowdrop* is a white inverted bell. During the day as the ground warms up with the sunlight, the hot air rises into the bell. At night the bell closes down, trapping the warm air around the sensitive part of the flower. During the night the white exterior of the petals minimizes heat loss by radiation to the surrounding atmosphere. Snowdrops are so thermally efficient that they can sometimes be found blooming below the snow as the snow melts away under the spring sunshine.

The first laboratory device developed for storing cryogenic fluids, known as a *Dewar flask*, attempted to minimize heat losses from the vessel by several mechanisms as shown in Figure 7.16(a). First of all, the actual container is a double-walled glass vessel and the space between the walls is made into a vacuum by pumping out the air. The internal surfaces of this double-walled vessel are covered with a thin film of silver to reduce radiant heat transfer through the vacuum. The vessel is supported on an insulator and the neck of the vessel again supported with an insulating material; the stopper is made out of a substance such as cork which is a poor conductor. These containers are called Dewar flasks after their inventor Sir *James Dewar*, a Scottish chemist, (1842–1923) [14]. (See discussion of the inventors of the explosive cordite in Chapter 6). Today such vessels are available commercially at quite low prices; however, they are somewhat fragile and have to be handled carefully. Lower cost thermally insulated storage vessels are made entirely out of insulating expanded plastic. These are more robust but less thermally efficient than a Dewar flask.

7.4 Heat Transfer Mechanisms

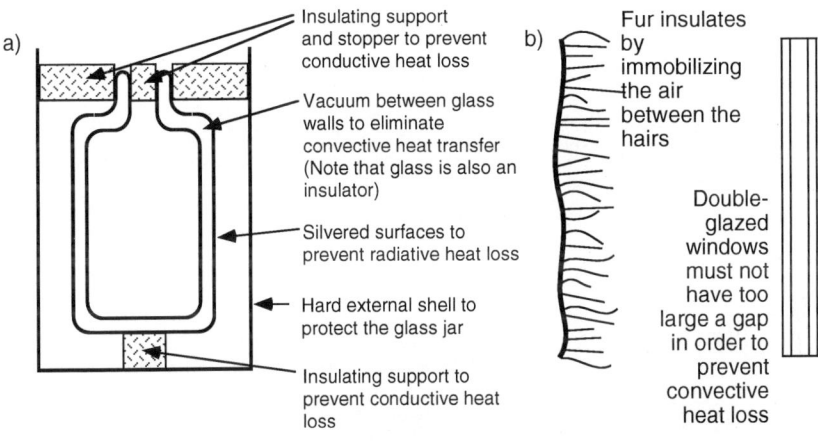

Figure 7.16 Dewar flasks, fur, double-glazed windows and snowdrops are thermally efficient systems. a) A Dewar flask employs several devices to keep its contents warm (or cold). b) Fur preserves a warm layer of air close to the body. Double glazed windows use air, another gas, or a partial vacuum as an insulating layer. c) Snowdrops are a thermally efficient flower. Their white color helps prevent heat loss.

Before we can use our knowledge of the mechanisms of heat transfer in natural systems to study the problems of making systems designed for the safe return of missiles, it is necessary to develop some vocabulary associated with the description of the atmosphere surrounding the Earth. In Figure 7.17(a) the various layers of material forming the Earth are shown in a cross-sectional diagram. In the words of Richard Fifield who has written an interesting article [21] on the structure of the Earth,

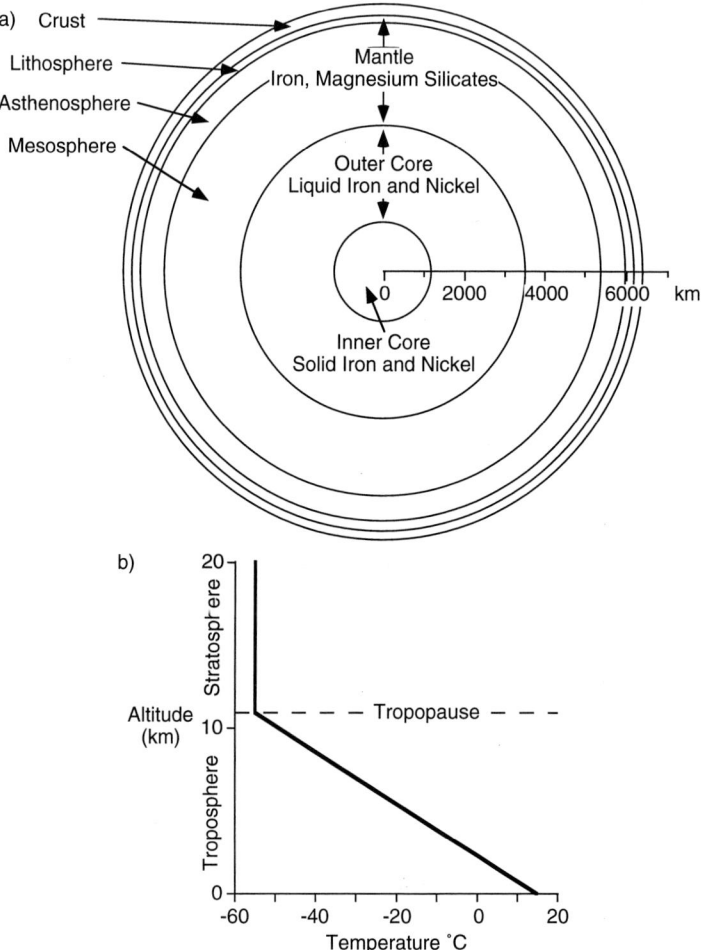

Figure 7.17 Scientists find it very useful to give different names to the various layers of the earth and the atmosphere surrounding it. a) The internal structure of the earth. b) Theoretical temperature profile for the various layers of the atmosphere.

The thin crust on which we live represents a mere one percent of the volume of the Earth. It is part of the rigid outer shell called the lithosphere.

The Greek word *"lithos"* meaning "stone" has given us several technical words in English such as *monolith* meaning "a large single stone," and *lithography*, a form of printing in which originally a special type of stone was used to make the printing plates. In modern lithography other forms of material are used but the name has remained. Below the lithosphere is a layer of material known as the *asthenosphere*. Again in the words of Richard Fifield [21]

The asthenosphere is tough and elastic and extends for a depth of 200 kilometers.

The crust of the Earth has the same relative dimensions to the diameter as a sheet of newspaper wrapped around a soccer ball. The final layer of the Earth is known as the *mesosphere* which together with the other layers are collectively known as the *mantle*. Much of the inner Earth is white hot at a probable temperature of 3000 °C which falls to 275 °C at the mantle/crust boundary. The waters of the surface of the Earth are sometimes collectively referred to as the *hydrosphere*. "Hydro" means "water"; however, visitors to Canada, when informed that the cottage they are about to rent is equipped with hydro, would find out that it has electricity. This use of the word hydro to describe electricity in Canada grew from the fact that, in the original development of electrical services in Canada, the bulk of the supply of electricity came from hydroelectric power stations in which the power of falling water was harnessed through turbines to generate electricity. Inevitably in everyday speech hydroelectric power became shortened to hydro.

The Greek word "atmos" meant "vapor," and the *atmosphere* refers collectively to all of the gaseous envelope that stretches out hundreds of miles above the surface of the Earth. Seventy-five percent of the mass of the atmosphere lies within the first 11 kilometers of the Earth. It is this layer of the atmosphere in which the clouds and storms occur. It is known as the *troposphere* from the Greek word "tropos" meaning "change." Above the turbulent troposphere there is a more serene, spread out layer of gas which is known as the *stratosphere*. This name was created from the Latin word "sternere" meaning "to spread out." Stretching several hundred miles above the stratosphere is the ionosphere which is so called because the molecules of gas in this layer have had outer electrons stripped off by violent collisions with short wave radiation from the Sun. All molecules which have had electrons stripped off or added are referred to as *ions* from the Greek "ienai", to go. Molecular ions are able to move under the influence of electric voltages and hence the name given to them as they travel from one place to another. Between approximately 15 and 30 kilometers above the surface is a layer of gas that is rich in ozone, a very chemically active form of oxygen. Ordinary oxygen molecules consist of two oxygen atoms joined together, O_2, whereas *ozone* is three atoms of oxygen linked together, O_3. The ozone layer absorbs high-energy ultraviolet light which, if it reached the Earth, would cause damage in the form of skin cancer to human beings exposed to the radiation. In the late 1980s and early 1990s there was evidence that chemicals used by man were causing a depletion of the ozone layer which could lead to higher incidences of skin cancer. Action has been taken to ban the aerosol gases that cause the depletion of the ozone but, due to the nature of the chemicals involved, it will be several years before we can expect to see any improvement in the situation.

High-flying transatlantic jet airliners fly on the upper edge of the stratosphere. When they were first introduced into international travel, there was some concern that the crew and passengers would be exposed to higher levels of ozone than experienced in lower flying aircraft. In the mid 1960s I was involved in a minor way in a program to monitor the ozone levels in high-flying jet aircraft. The ozone layer tends to move lower during the winter and there have been cases reported of passengers suffering *ozone sick-*

ness, a form of nausea, on high-flying aircraft crossing the Atlantic during the winter. On one particular occasion when I was a passenger flying from London, England to Toronto, I could smell the pungent odor of ozone in the aircraft, but the Captain was unwilling to answer my questions about the smell.

As seen from Figure 7.17(b), the temperature of the air falls as one leaves the ground and travels up to the stratosphere. For this reason high-flying aircraft must be temperature controlled and, due to the lower atmospheric pressure at cruising altitudes, the cabin air must be pressurized to assure the comfort of the passengers. Such aircraft are equipped with oxygen masks in case the skin of the aircraft is somehow punctured during the flight since, at high altitudes, this would result in a drop in cabin pressure to the point where it would be difficult to breathe and could result in passengers and crew becoming unconcious.

Section 7.5   Designing Heat Protection Shields for Returning Space Missiles and Capsules

As mentioned earlier, the nose cones of intermediate range missiles can be allowed to enter the atmosphere at high speeds and decelerate quickly in the final forty kilometers of the atmosphere. The same is true of the intercontinental-range nose cones; however, the frictional forces involved in sharp entry trajectories cause very high temperatures on the outside of the nose cones. The temperature can rise to as high as 600 °C on the outside of the intercontinental missiles. The very high rates of deceleration of such nose cones in the final part of the atmosphere would be intolerable for astronauts in space capsules. Therefore space capsules enter at a much shallower angle and decelerate more uniformly over a longer path. The outside of manned capsules does not reach such high temperatures as the outside of the nose cone of ballistic missiles.

If asked to design a protective nose cone for a re-entering missile the instinctive choice would probably be to select a ceramic outer shield which could withstand the high temperatures created by the friction of the atmosphere. It could also be anticipated that since ceramics are generally not good conductors of heat that the inside of the missile nose cone and capsule would be protected from the effects of the severe thermal conditions on the outside of the cone. To minimize the weight of the space capsule, the protective coating is only placed on the leading surface of the capsule.

The first type of re-entry protection systems were indeed ceramics, but later it was discovered that, in fact, substances such as glass-fiber-filled plastic made better shields. They not only became incandescent, thus radiating heat away, but also dissipated thermal energy from the surface of the protective coating by shedding mass. Since they provided thermal protection by sacrificing mass carried away by the flow of gases around the system, they were described as *ablative shields* [22]. Theoretical studies showed that the ability of a material to absorb heat, called the *specific heat* of the substance, increased as the number of hydrogen atoms inside the material increased. This is the reason why scientists turned to hydrocarbon type plastics such as nylon fibers in phenolic resins. The idea that such materials would make good protective covers for re-entering missiles and capsules was not readily accepted by the scientific community. Thus Gruntfest and Shenker tell us in an article on space age ablation [22],

> *We recall all too clearly the amused contempt which greeted early requests to test reinforced plastics in ablative environments.*

The heat protection given by ablative coatings takes place in two stages. First, the outside of the surface begins to heat up and the hydrocarbon material starts to degenerate by giving off gases which cool the surface and carry energy away. An almost pure *carbon char*, similar to the charcoal used for barbecuing, is left by the exiting gases. The interface between this char and the virgin material advances into the protective coating as the space capsule descends into the atmosphere. The carbon char is a very good insulator and it becomes incandescent, radiating away heat. Because it is now carbon, the outer coating can be heated to a very high temperature and, as predicted by Stefan's law, it becomes a very effective dissipator of energy. The final stage of protection comes when the incandescent coating starts to flake off, carrying away large quantities of heat. Although the pioneers of this technology had a difficult time persuading their colleagues that this type of material was a very effective defense against heat, they could have discovered that nature already used ablative techniques to protect plants. A California Redwood tree has very thick bark which ablates during a forest fire to protect the tree. Allen tells us that the Russians are rumored to have used the naturally occurring hard wood *lignum vitae* as a spacecraft re-entry heat shield [12]. Pure graphite makes a good heat shield but is difficult to make in large uncracked assemblies. Graphite fins are used in the construction of rocket engine nozzles to guide the rocket during its early flight path. During the operation of the rocket engine, the graphite ablates to protect the overall construction of the rocket engine.

Section 7.6 Did Astronauts see Shooting Stars in their Eyes?

The first astronauts in space reported to ground control that as they were moving through space they saw flashes of light which appeared to be inside their eyes. This was one more phenomenon which scientists should have predicted before the astronauts took off from the Earth but which had to be experienced before it was explained. To understand the origin of these flashes of light, we must look at the history of scientific attempts to measure the speed of light. The first successful attempt to measure the speed of light on the surface of the Earth was the experiment carried out in 1849 by the French physicist Armand *Fizeau* (1819–1896) [2, 14]. The basic principles employed in his experiments are outlined in the sketch shown in Figure 7.18(a). Light was focused with a lens to a point between the teeth of a wheel. The surfaces of the teeth were blackened to avoid light reflection. A half-silvered mirror was used to deflect the light from the source off line as indicated in the diagram. When the light had passed through the teeth it went to another lens which made it into a parallel beam to transmit it over a distance of eight kilometers. At that distance was another mirror which returned the light along its path. At the beginning of the experiment, the observer looked through a telescope to see the light being returned to the point between the teeth of the wheel. Then the wheel was spun at a known speed. Only if the light traveled the distance between the teeth to the distant mirror and back again in the same time that it took for a tooth to move exactly one tooth position could the eye observe light through the instrument. The wheel was spun slowly and then the speed of rotation gradually increased until a speed was reached at which a tooth moved into the path of the beam in the time it took the light to travel to the distant mirror and back. For such a position, darkness was experienced by the operator. Using this equipment Fizeau estimated the speed of light to be $3.13300 [ft56] 10^8$ meters per second. The modern value for the speed of light in a vacuum is $2.99792 [ft56] 10^8$ meters per second, and this is held to be accurate to three parts in a million. Considering the difficulties tackled by Fizeau in his experiments, his value was remarkably close to the modern value. A year after Fizeau completed his experiment Foucault, who we met in our earlier discussion of the pendulum, carried out another set of experiments. Foucault's equipment was more compact than that of Fizeau and he was able to carry out an experiment to measure the speed of light in water.

Readers who find it difficult to understand the nature of light are in good company. For centuries scientists have debated whether light was made up of small ball-like bundles of energy called *corpuscles* (Latin for small bodies) that bounced around or was a wave motion like sound. The English physicist Thomas *Young* (1773–1829) developed the wave theory of light. The supporters of both theories knew that the wave theory of light predicted that light moved more slowly through water than through a vacuum.

Foucault's experiments demonstrated that the speed of light in water was less than in a vacuum, thus helping to establish the wave theory of light [14]. Physicists of the time, however, were still baffled by the nature of light. They argued that if sound waves were carried by air, then light waves must be carried by something. They therefore put forward the idea that the space between the stars was filled with some mysterious material that carried the light energy. They gave this material the name *luminiferous ether*, or simply, the ether. This grand sounding name came from root words "luminiferous" meaning "light carrying" and the word "ether" which was a term used by Aristotle to describe the components of all objects outside the Earth's atmosphere. Thus things observed in the sky were sometimes referred to as being *ethereal*. An older dictionary [15] defines *ether* as

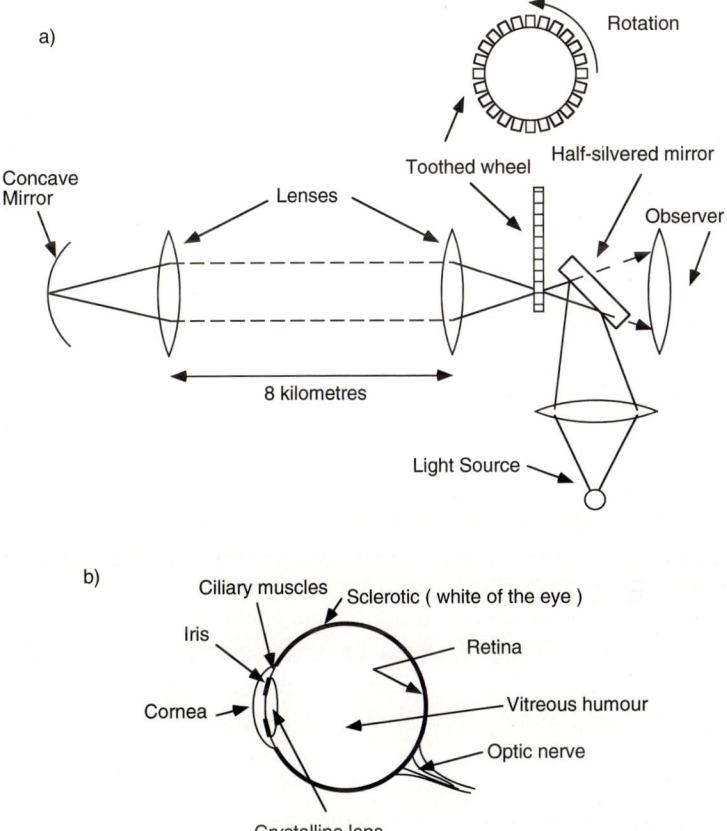

Figure 7.18 Light has been studied for centuries. Many experiments have been conducted to determine the nature and speed of light as well as how it travels. a) Basic equipment used by Fizeau to measure the velocity of light. b) The human eye is filled with a fluid described as the vitreous humor. Images are focussed on the retina by the cornea and carried to the brain by the optic nerve. The iris controls the amount of light reaching the retina.

the clear upper air, the material medium formerly supposed to fill all space and transmit electromagnetic waves.

When students are told that the space between the stars was once thought to be filled with ether, they are thoroughly confused because they know of the chemical substance, ether, which is used as an organic solvent and as an anesthetic. This unfortunate double use of the term ether can be traced back to the chemist *Frobenius* who gave the term "spiritus aethereus" to a new liquid he discovered in 1730. This liquid evaporated so quickly that Frobenius thought it must be the very essence of the upper air and hence its name. We now know that it is a chemical we call methyl ether.

In the late 1880s scientists argued that if electromagnetic waves such as light were indeed carried through space by the luminiferous ether, then the speed of light should be different depending on whether a star was moving towards us or away from us. The American scientist Albert Abraham *Michelson* (born in Prussia in 1852, died in the United States in 1931) made a lifetime study of the speed of light and was awarded the Nobel Prize for his optical experiments in 1907 [2, 14]. One of the main thrusts of the experimental investigation carried out by Michelson was to detect the presence of the ether. By 1887 it became obvious from his experiments that there was no such thing as the ether. The crucial experiment is often referred to as the *Michelson–Morley experiment* and it caused an intellectual crisis in physics comparable to the one generated by the black-body experiments that we discussed earlier. Edward William Morley (1838–1923) was an American chemist who trained originally to become a congregational minister. In the words of Asimov [14],

> *The Michelson–Morley experiment is undoubtedly the most famous experiment that failed in the history of science.* (That is, it failed to measure a change in speed of light from a retreating star).

The implication of the Michelson–Morley experiment was that the speed of light in a vacuum could not be exceeded anywhere in the universe. In 1905 Einstein announced his *special theory of relativity* which began by assuming the speed of light in a vacuum to be an unvarying and fundamental constant that could only be known by experiment. Michelson himself could never bring himself to accept relativity. Understanding the empirical fact that the speed of light in a vacuum is the highest speed that can be achieved is not something that is understood like two oranges plus two oranges equals four oranges; it is rather a difficult fact that has to be grasped on the basis of "that's the way the universe works." Ninety years after the launching of the theory of relativity, some scientists teach relativity as a self-obvious truth when, in fact, it was considered outrageous and incomprehensible at the time it was put forward.

Another mysterious prediction from the theories of Planck and Einstein is that as an object travels faster and faster, its mass increases according to the following formula

$$m_v \frac{m_R}{\sqrt{1 - v^2/c^2}}$$

where v is the velocity of the object, c is the speed of light, m_R is the rest mass of the object and m_v is the mass of the object at velocity v.

This equation was first developed by Einstein who also put forward the now very famous equation relating the energy of an object to its mass and the speed of light

$$E = m c^2$$

where E is the energy released when a mass m is completely transformed into energy and c is the speed of light.

Before we can use the information we have discussed concerning the speed of light to explain the flashes seen by the astronauts, we first need to explore what is meant by *cosmic rays*. Beginning in 1911, the Austrian physicist Victor Francis *Hess* (1883–1964) started to measure radiation in the upper atmosphere by sending instruments up in balloons to heights of up to six miles. He expected that the radiation levels at the surface of the Earth, which he thought were only due to radioactive material in the ground, would be lower in the upper atmosphere. To his surprise, the radiation he measured became more intense the higher he went. The American physicist Robert Andrew *Millikan* (1868-1953) became very interested in the work of Hess and gave the name cosmic rays to this radiation. He believed it was high energy radiation coming from the edge of the universe.

The Greeks used to believe that, before the gods created the universe we know, unorganized matter described as *chaos* filled the universe. The word cosmos can be traced back to the fact that the Greeks believed that when the first gods organized the chaotic material of the primeval universe into our present universe, they described the new situation as the *cosmos*. Chaos and cosmos have given us several words in the English language as illustrated in Figure 7.19. In a picturesque statement, Millikan called the energy present in cosmic rays "the birth cry of matter" since he thought it came from matter being created at the edge of the universe. Millikan, who himself was the son of a Congregational Protestant minister, was deeply religious and used to say that the presence of cosmic rays showed that "the creator is still on the job." Cosmology is the study of theories of how the universe was formed, continuous creation is one cosmological theory [20]. Later it was found that cosmic rays are essentially parts of atoms traveling at high speeds across the universe. Some of these highly energetic fragments of the atom are now known to be traveling close to the speed of light in a vacuum.

The final piece of information that we require in our search for an explanation of the flashes seen by the astronauts is the theory of *Cherenkov radiation*. This type of radiation is named after the scientist who studied it, a Russian physicist Paval Alekseyevich Cherenkov who was born in 1904 [14]. In 1934 Cherenkov observed the radiation given out when a highly energetic atomic fragment passes through a material. Two other Russian physicists *Ilya Franc* (1908-) and *Igor Tam* (1895-1971) worked out the theory behind the emission of Cherenkov radiation. They pointed out that the cosmic ray entering a material was traveling at a speed in excess of the speed of light in that medium and that the ray had to slow down extremely quickly. As it slowed down, the cosmic ray had to get rid of this excess energy and it was given out as light. Thus Cherenkov radiation is light energy given out by a cosmic ray slowing down to the speed of light in a medium. Figure 7.18(b) shows the structure of the human eye. The *ciliary muscles* can pull on the lens of the eye to make it bulge or stretch, thus enabling us to see close and distant objects. As a human being ages, these muscles lose their flexibility and we need glasses to enable us to see things very close or far away. The *iris* is a screen that opens up or closes according to the amount of light that falls upon the eye. Iris was originally the name given to the Greek goddess of the rainbow. Iris was used to describe the eye because this part of the eye came in a rainbow of colors. The iris of the camera is named after this part of the eye and functions in the same way. The center of the eyeball is filled with a fluid described as *vitreous humor*. The Latin word

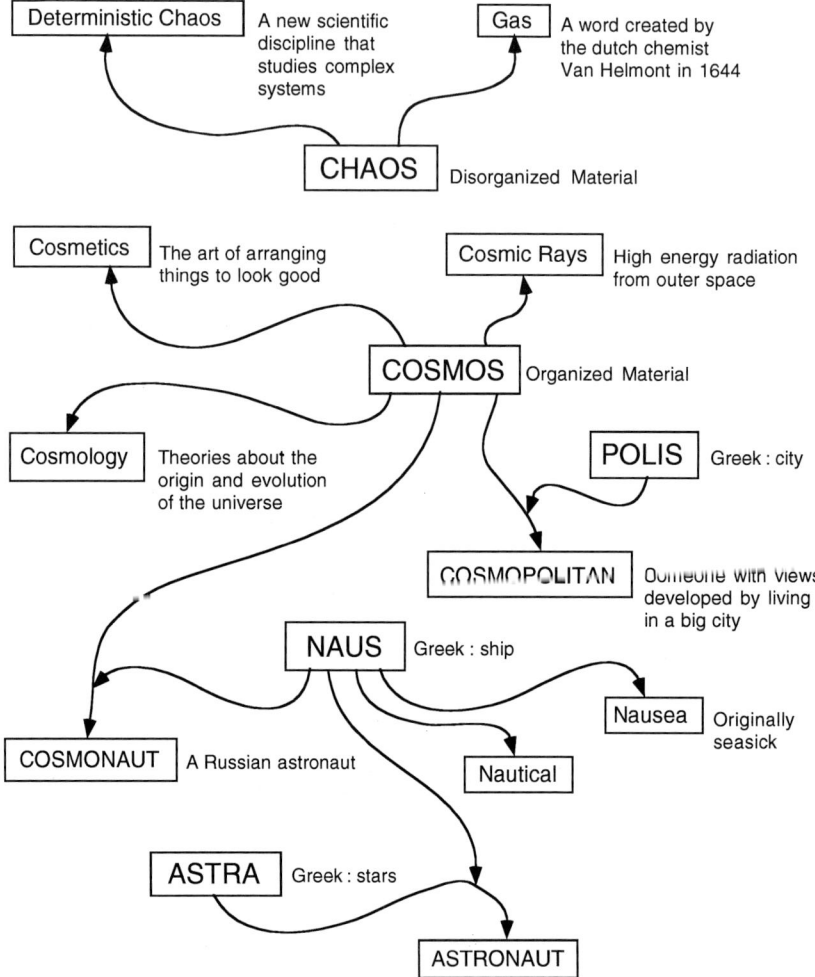

Figure 7.19 The Greeks used to teach that the gods created the organized universe, the "cosmos", out of unorganized material, "chaos".

for "glass" was "vitrum" and "humor" was a word meaning "fluid"; therefore, the term vitreous humor simply means a transparent, glassy liquid. People used to believe that body fluids were of four different types and that the balance of these types of fluids determined the emotional state of a person. When the fluids were in a good balance, people were said to be in a *good humor*. Because people in a good humor were often funny, *humorous* came to describe something causing laughter.

Now we can finally put all the pieces of information together and realize that the astronauts, when they saw flashes of light, were seeing Cherenkov radiation caused by the bombardment of their eyes by cosmic rays traveling close to the speed of light in a vacuum. When they enter the vitreous humor they are moving faster than the speed of light in this medium. Therefore they must slow down very quickly and as they do so they give off Cherenkov radiation, and hence the flashes of light.

Section 7.7   Circulating Missile Messengers

As early as 1687, Sir Isaac Newton discussed what would happen if missiles were fired into the air with higher and higher velocity. The type of sketch that he gave in his famous book *Principia* is shown in Figure 7.20. (This diagram is not to scale; it is meant to illustrate principles rather than exact dimensions.) At first, as the energy of a cannon shell is increased, it travels farther around the circumference of the Earth. However, as the velocity continues to increase, finally a situation is reached in which the trajectory of the shell is as shown by the pathway labelled orbital trajectory in Figure 7.20. In other words, at a certain speed a shell goes into orbit. One of the Latin words for "a circle" was "orbis" and the word "orbita" could mean "a wheel." Thus an *orbit* is the circular path followed by a projectile, and this path looks like a wheel drawn around the axis of the body being orbited.) The *orbital trajectory* is a circle around the center of the Earth. From the point *a* in the orbital trajectory a tangent has been drawn to this circle. By the time the shell reaches the position *c'* it has dropped a distance δ away from the tangent. In the same time, if a tangent was drawn to the Earth's surface at the point *b*, immediately below *a*, the Earth's surface is now almost the same distance δ from that tangent. From one perspective, the shell can be regarded as moving the distance *a* to *c* and falling the amount δ toward the Earth; but because of the shape of the Earth, the surface of the Earth is still almost the same distance, δ, below the tangent at *b*: In effect, when the shell is at the point *c'* it is still the same distance from the surface of the Earth as it was at *a*. Since the air above the Earth is so thin at the height of the orbit, there is no resistance to the movement of the shell, and it keeps on moving around the surface of the Earth. Although it is always falling toward the center of the Earth, it never manages to get any closer to the Earth's surface. The velocity of the shell going into the orbital trajectory is represented by the symbol v_0. If fired with an even higher energy, the shell would go into orbit at a greater distance from the Earth. Sir Isaac Newton was able to show that if the shell could be given a velocity of 11.2 kilometers per second, then it would escape completely from the gravitational pull of the Earth. In the next section we will discuss the fate of such a shell, but here we are more concerned with orbiting missiles.

Although Newton and his contemporaries knew that in theory they could fire a missile out to infinity, they also knew that they lacked the technology to even begin that type of missile research. Interest in making orbiting missiles lay dormant until the rocket research carried out during World War II made scientists believe that such orbit-

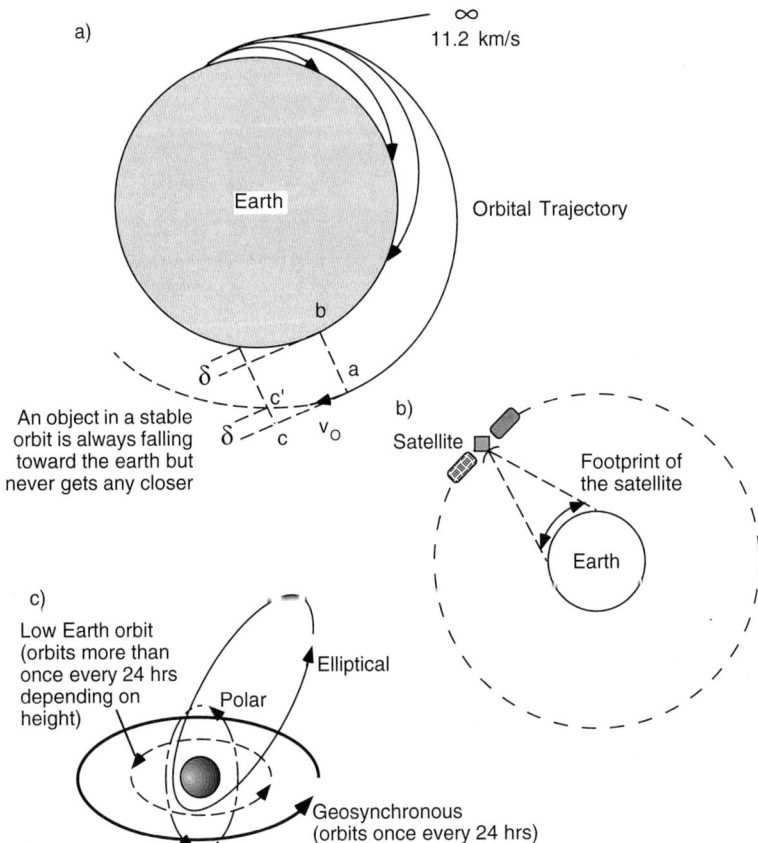

Figure 7.20 As early as 1687 Sir Isaac Newton discussed what would happen if cannonballs were fired into the air with more and more energy. a) Newton showed that eventually a high energy cannonball could circle the earth indefinitely (go into orbit). He also showed that a cannonball travelling at 11.2 kilometers per second can escape the pull of Earth's gravity and continue indefinitely. b) The footprint of a satellite is the area of the Earth's surface from which the satellite can be seen. Several satellites can relay information anywhere on the Earth if their footprints touch or overlap. c) Satellites can be placed in various orbits depending on their purpose and country of origin. Geosynchronous satellites remain stationary above one point on the Earth's surface.

ing missiles were within their grasp. The first orbiting missile created by man was known as Sputnik 1. It was launched by the Russians on October 4, 1957. At that time the name *satellite* was used to describe such orbiting missiles. The word satellite comes from the Latin word "satelles" used to describe a servant who traveled with his master (see Figure 7.22). Today satellites are used extensively to relay messages from one part of the world to another. In a review article on communication satellites Helen Gavaghan [9] makes the following statement:

7.7 Circulating Missile Messengers

Since 1957 when the Soviet Union launched Sputnik 1 the world has put more that 3000 satellites into space. Today satellites relay telephone calls, monitor the weather, survey the Earth's surface for minerals and gather sensitive military information. Satellites fill the space around the Earth like a swarm of bees kept there by the gravitational pull of our planet. Their orbits can vary from a few 100 to about 200 000 kilometers above the surface of the Earth.

In Figure 7.20(c) some of the different types of orbits used for communication satellites, the continuously circulating messenger missiles of the title of this section, are shown. For telephone communications, technologists use satellites in what are known as *geostationary* or *geosynchronous orbits*. These are located 35 784 kilometers above the surface of the Earth. Satellites in such an orbit take 24 hours to travel around the Earth. This is just the period in which the Earth rotates about its own axis. Therefore, this type of satellite moves at the same angular speed as the Earth and so to an observer on the Earth they do not appear to move. Hence the name geostationary which means that from the Earth ("geo"), the satellite appears to be stationary in the sky. The origin of the term *synchronized* is explained in Figure 7.21. In this wordweb the origin of various words derived from the name of the Greek god of time are shown. Things which are timed to operate together are said to be synchronized and thus the geosynchronous orbit is that in which a satellite rotates in a synchronous manner with the Earth. It can be shown that three geosynchronous satellites in complimentary orbits can cover the whole surface of the Earth with their "footprints." The term *"footprint"* is used in communications to describe the area covered by signals spreading out from the satellite as shown in Figure 7.20(b). When a message is sent by satellite from, say England to North America, the message travels up to the satellite and back a distance of approximately 71 600 kilometers. Even at the speed of light this takes 0.24 seconds. When speaking over the telephone via satellite, this delay in the signal can be very irritating. However, such a delay is actually a small price to pay for this new technique of sending many messages over long distances. Indeed, it is said that when the telephone companies explored the customer acceptance of the time delay involved in telephone transmission via satellites, the businessmen they asked were quite happy to accept the problem. When such telephone systems came into wider use, there were some problems with ordinary customers calling their loved ones. Thus when the businessman says, "Did you receive my cheque?", a 0.24 second hesitancy is not interpreted as a major problem. However, if the question is, "Do you still love me?", then the apparent hesitation by the other party could be misinterpreted as a change of emotions and not a technological delay due to the finite speed of light.

In the wordweb of Figure 7.21 the term of *anachronism* is illustrated. An anachronism is a description of a situation which is impossible because of the supposed time sequence involved. Thus if a student were to write in an essay that Queen Victoria of England did not approve of violence on television, this would be an anachronism since television had not been invented when Queen Victoria was alive. Makers of movies and television

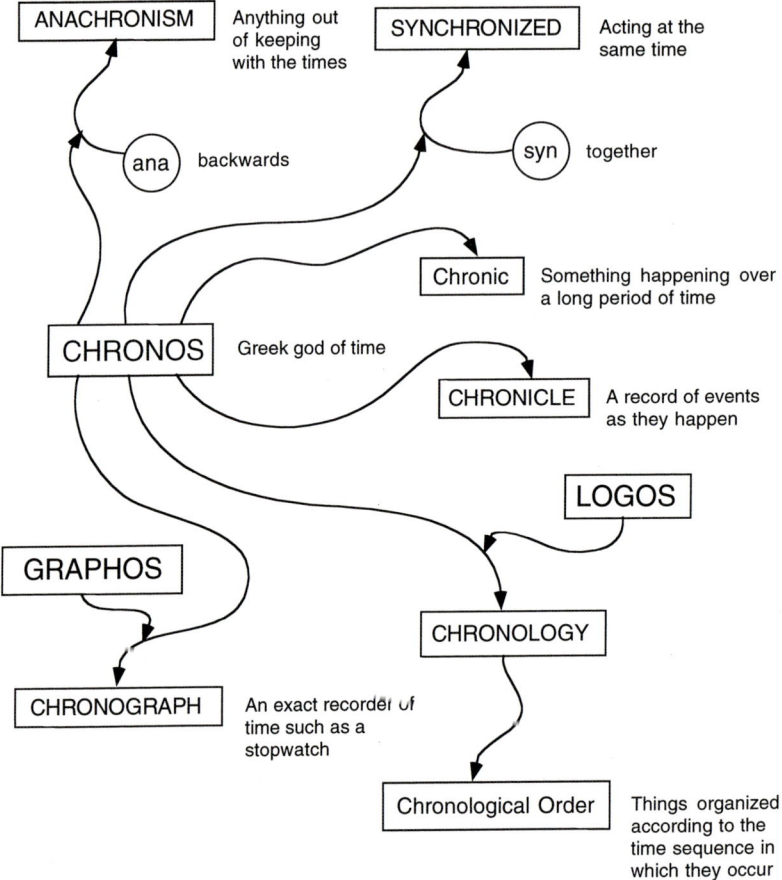

Figure 7.21 The story of the development of geosynchronous satellites is an interesting chronicle.

shows are usually very careful to avoid anachronism in the serious depiction of events. (The television show "The Flintstones" is full of anachronisms, but of course it was never intended to be an accurate portrayal of Stone Age man.) At the time of writing this book, I happened to watch a television film dealing with troops tracking an outlaw at the time of the American Revolution. The general in charge told the troops to use their "rifles." At that time, the troops would not have been equipped with "rifles" which, as discussed earlier, spin the bullets; they would have had smooth-bore muskets. The instruction given by the general was an anachronism that had slipped past the attention of the director of the movie.

Geosynchronous orbits are useful for many countries which happen to be located around the equator of the Earth. Countries like Russia which have large land masses in the northern part of the Earth's surface have to use satellites in the *polar orbits* or in elliptical orbits such as that shown in Figure 7.20(c).

The terms used to describe an elliptical orbit are illustrated in Figure 7.22. When the satellite is closest to the Earth, it is said to be at *perigee*. This word literally means "around the Earth" and in this case it means "in the vicinity" close to the Earth. The rootword "apo" in Greek means "away" and at the position of *apogee* the satellite is at its greatest distance from the Earth. (We will use these words later to describe the orbit of comets such as Halley's *comet*; see Section 8.3). The word "apo" is also to be found in words such as *apostle* and *apostrophe* as illustrated in the wordweb of Figure 7.23.

In an elliptical orbit the satellite does not move with uniform velocity around its orbit. As it moves toward the Earth, to its perigee, it speeds up; as it moves up to apogee, it slows down. This is useful from the Russian point of view since it means that the satellite spends more time over the skies of Russia, at apogee, than it does passing quickly underneath the Earth. (This is why Russian scientists choose an elliptical orbit rather than a circular orbit for their communication satellites.) For the western nations, one of the big advantages of the geosynchronous satellite is that the satellite is stationary in the sky and the receiving equipment does not have to move to follow the satellite; it can be pointed permanently at the satellite. The systems used to receive messages from communication satellites are usually dish-shaped and are familiar in the form of the satellite dishes used to receive television broadcasts. The size of the satellite dish used to capture messages is related to the wavelength of the electromagnetic waves carrying the messages or TV signals. A visitor to Britain will soon notice that the satellite dishes for television systems are much smaller than those used in North America. This is because the European TV transmissions are much shorter in wavelength and so a much smaller dish is sufficient to collect and focus the signals from the satellite. Similar small dishes are now beginning to appear in North America as new satellite systems are developed. The word *antenna* used to describe a system of wires to recover messages from the sky, is an extended meaning of the word used to describe the stick-like appendages used by insects to receive messages. In turn, these stick-like assemblies were called antennae from the Latin word for the main beam system or sailyard on which the sails of a ship were raised when it was to be moved along by the wind. (A ship without sails, from a distance, would look like a large water beetle sticking its antennae into the sky).

In the early development of communication satellites technologists used both *passive* and *active satellites*. A passive satellite only reflects signals sent to it from Earth. This type of satellite cannot be too high in the sky otherwise the reflected signals are too weak when they return to the ground. Active satellites receive the messages and amplify them before sending them back to the Earth. Sometimes active satellites have what are known as *transponder* systems. A transponder amplifier is not switched on all the time but is actually activated by the received message; hence its original name of *transmission responder*, a term which was quickly compressed to form the shorter word. Transponders save power when they are part of a communications system. Active satellites often also have solar-power cells to provide the energy to operate their amplifier system. Although passive satellites are no longer used because of advances in the devel-

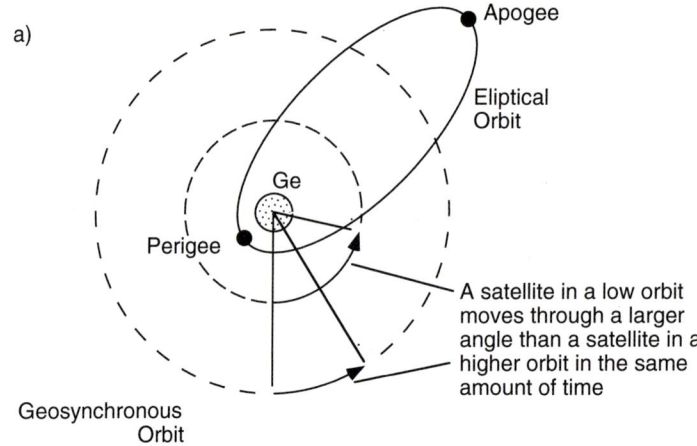

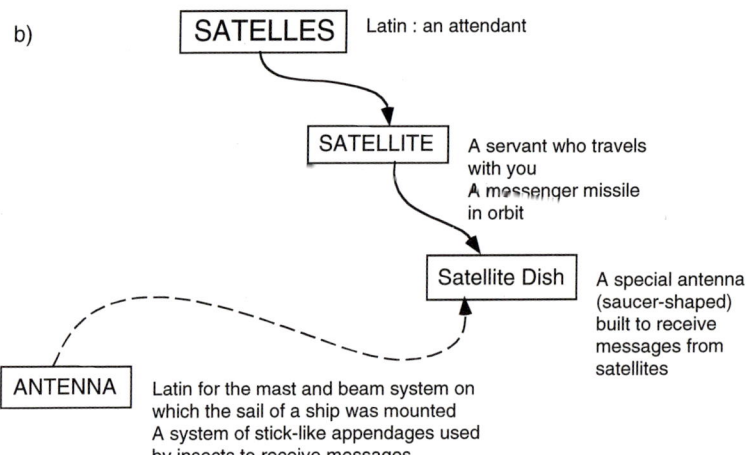

Figure 7.22 Satellite technology has given new meaning to old words. a) Satellites orbiting the Earth at different altitudes move at different speeds. Satellites in lower orbits circle the Earth in less time. A satellite in an elliptical orbit has a nearest point, perigee, and a farthest point, apogee, in its orbit. b) The origin of the word satellite.

opment of active satellite systems, it is useful to review the history of the use of such satellites because they were plagued with problems which were solved by an interesting application of physical principles.

During the period 1960–1964 two large passive satellites, Echo 1 and Echo II, were placed in orbit. These satellites were, in effect, very large balloons. Echo 1, when inflated, was about 30 meters (100 feet) in diameter and weighed 59 kilograms (130 pounds). Echo II was 41 meters (135 feet) in diameter and weighed 268 kilograms (570 pounds). When describing such systems as balloons, it is necessary to point out that

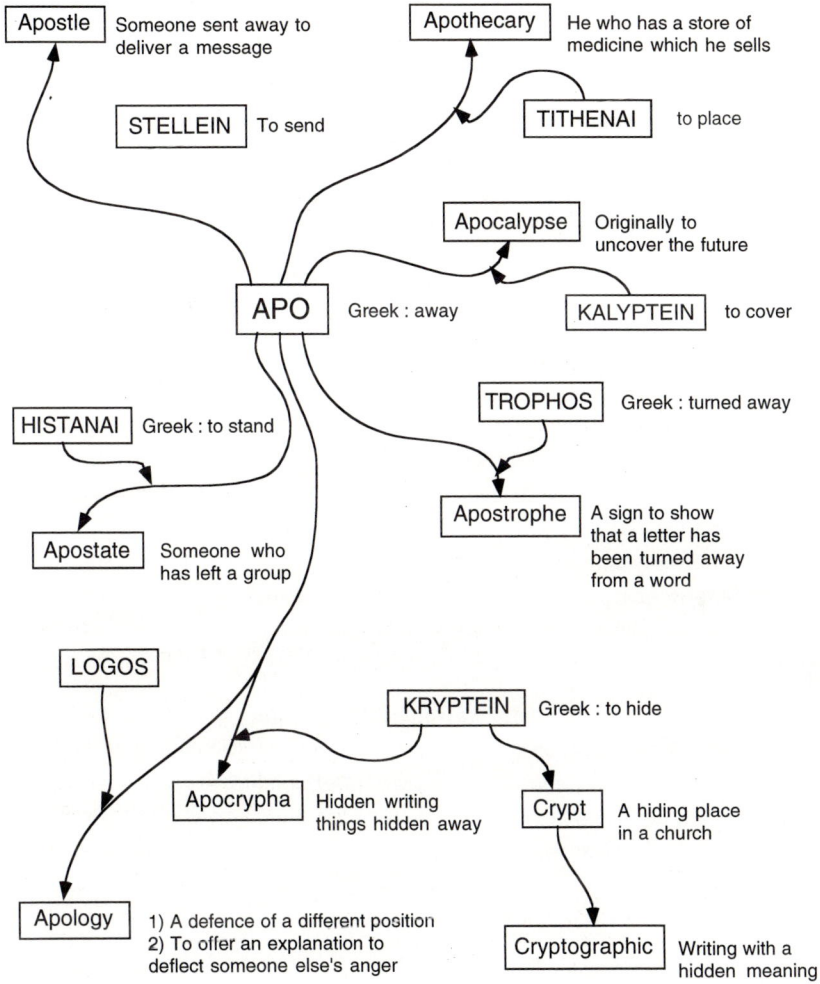

Figure 7.23 Apogee is one of the many scientific and literary terms created from the Greek prefix "apo" meaning "away".

they differed from the party balloons blown up on Earth. The skin of party balloons is expandable or elastic. As they are inflated, the tension built up in the rubber increases. When such a balloon has a hole made in it, it collapses because the pressurized gas in the balloon rushes out and the tension in the film of the rubber casing collapses the balloon. The balloons used in passive satellites are not very expandable and are made of a more rigid plastic; they are more like an inflated beach ball. Another difference between the satellite balloon and a party balloon is that at the height of 3200 kilometers (2000 miles), where satellites orbit the Earth, there is no gas pressure to resist the expansion of the balloon as it is inflated. The first time that an echo balloon was inflated, the inflation forces were too high and after had been expanded by pressurized gas, there was nothing to absorb the energy of inflation and the balloon ripped apart under the internal pressure. The technologists developing such systems then realized they needed to have a very slow inflation system when the folded plastic balloon was in orbit. The only satisfactory system that they could develop was to place a small amount of powder inside the folded balloon; simple vapor pressure from the surface of the powder very slowly expanded the balloon. Again, in the words of Jaffe [23],

> *Echo 1, when in orbit, was somewhat wrinkled after a short time because its plastic skin retained a memory of its folded condition in the canister placed on top of a rocket to place it in orbit.*

One of the questions often asked in a class discussion of the function of an echo balloon is; "What happens if it is punctured by micrometeorites hurtling toward the Earth?". In fact, once deployed, the satellite can be full of holes and still function since there is no tension in the surface of the balloon to tear it apart; neither is there an external gas pressure to collapse it.

In theory, when any satellite is placed in orbit around the Earth, it should stay up there forever because there is no atmosphere to slow it down. In practice, when a satellite is in very low orbit, there is some resistance from the very, very thin air of the upper atmosphere. In the words of Helen Gavaghan [9],

> *After a few years or even months this low-level resistance will slow a satellite so that it falls back to Earth. In higher orbits the irregular gravitational field of the Earth and the gravity from the Sun and Moon all act to push satellites out of their correct orbit. For this reason, active satellites have their own propulsion system. Computers continually monitor the satellites position and command the small rockets to fire so as to maintain the exact position of the communication satellite.*

Section 7.8 Interplanetary and Transgalactic Missiles with a Message

As mentioned earlier, Newton calculated that if a cannonball could be launched at an initial speed of 11.2 kilometers per second (approximately 7 miles per second), it would be able to escape from the gravitational pull of the Earth and go into orbit around the Sun. The place of that orbit around the Sun would depend upon the direction in which the cannonball was fired with respect to the Sun and also on how much the velocity of the ball exceeded 11.2 kilometers per second. To appreciate the vastness of the space around the Sun available for the orbiting cannonball to settle down into a trajectory, consider the size of the solar system as summarized in Figure 7.24. This summary does

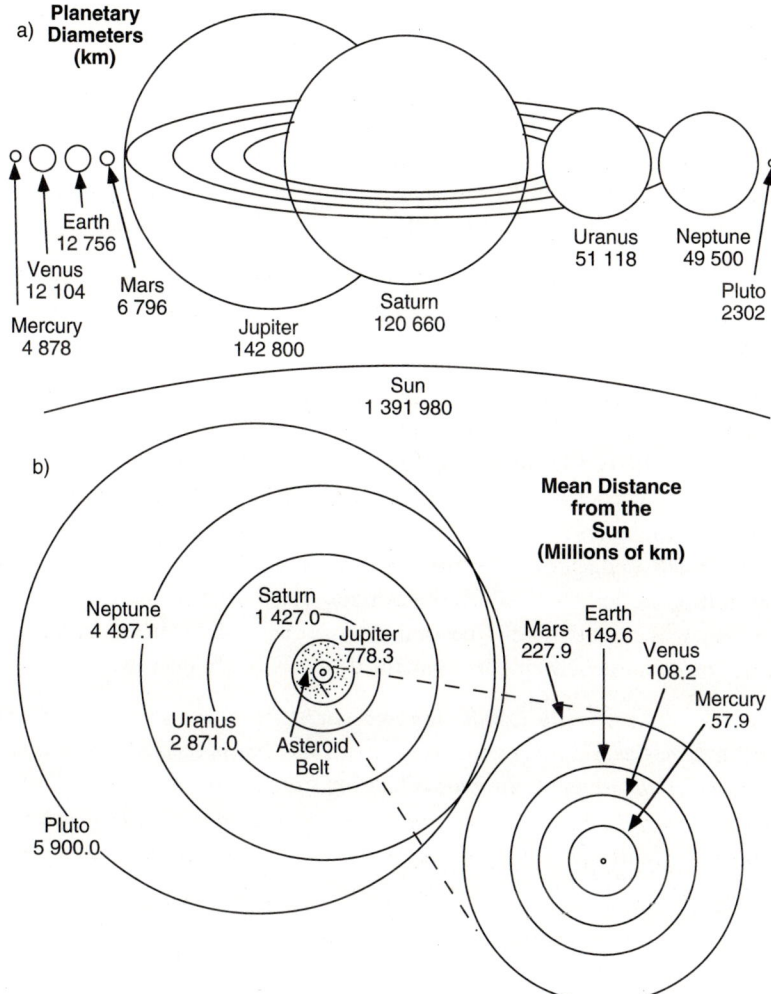

Figure 7.24 Our solar system consists of nine major planets, their moons, and innumerable minor planets (asteroids), comets, and small grains of material. a) Comparison of the sizes of the planets of our solar system and our Sun. Jupiter, Saturn, Uranus, and Neptune are known as "gas giants", while the other planets are solid and rocky. b) The orbits of the planets demonstrate the vastness of our solar system. Note that the planets are not shown in their orbits because, at this scale, they are vanishingly small.

not include the orbits of comets which we will discuss in Section 8.3. (At one time, scholars used to wonder about the possibility of intelligent life on some of the other planets of our solar system; however, space exploration carried out in the later part of the 20th century seems to have established that there are no other intelligent beings within our solar system.) Calculations show that a cannonball with an initial velocity of 12 kilometers per second would escape from the gravitational pull of the Earth and head out to regions beyond the solar system. In 1972 the United States *robot spaceship Pioneer 10* was sent off on a journey which would eventually carry it beyond the solar system. How far could such a robot travel? To answer that question we have to gain some appreciation of the size of the universe around us.

When the spacecraft Pioneer 10 was launched on the March 2, 1972 it was never envisaged that 20 years later it would still be able to send useful information back to Earth. However, more than 20 years after it was launched it was still sending information back, even though it had officially left the solar system in June 1983. In 1992 the signal strength from Pioneer 10 was 7.8 Watts as it left the spacecraft but by the time it reached the antennas used by NASA to communicate with the spacecraft the signal has diminished to less than a billionth of a trillionth of a watt (10^{-21}W). In 1992 Pioneer 10 was 8 billion kilometers away and it took 7.5 hours for a signal from Earth to reach the spacecraft and another 7.5 hours for NASA to receive confirmation that any new command had been received and implemented. Even at these great distances and with such weak signals the engineers at NASA are able to measure the Doppler shift in the signals from the spacecraft to diagnose the movement of the spacecraft in outer space. The signals from Pioneer 10 in 1992 enabled scientists to conclude that there was no evidence for the presence of a tenth planet beyond Pluto in the solar system. In the words of Booth who wrote an article on the continued travels of Pioneer 10

> *Pioneer Ten is destined to wander through interstellar space for ever. In 33 000 years time it will pass within 3.3 light-years of its first star, one of the Sun's nearest neighbors. After that it will continue its journey passing close to new stars every million years or so.*

Referring to such future travel, one of NASA's space scientists Fimmel says that someone may find the pioneer plaque and describe it as an interstellar cave painting [26].

As noted earlier, a star is defined in astronomy as a self-luminous body. Thus our Sun is described by astronomers as a star. Nonluminous bodies which orbit a star are known as planets and our Earth is a planet of the star named "the Sun." Venus and Mars which many people would call stars are actually dark planets orbiting the Sun. These are not self-luminous and are made visible only by the light that they reflect from the Sun. The Moon is a satellite of Earth which is also made visible to us by the light that it reflects from the Sun. Although Mercury and Venus are visible to us by their reflected light, the intensity of the light reflected is low in absolute terms relative to, say, the intensity of light emitted by the Sun. We are only able to see them because they are relatively close to Earth. Viewed from outer space the planets of the Sun will not be visible because they are not self-luminous. The Sun has nine planets, namely, Mercury, Venus, Earth, Mars, Jupiter, Saturn, Uranus, Neptune, and Pluto. The sizes of these planets and their distances from the Sun are shown in Figure 7.24. In the same figure, the diameter of the Sun is given. It can be seen that all the planets are small relative to the Sun. For example, Jupiter, the largest planet, is only one-tenth the diameter of the Sun. Earth has one-hundredth of the diameter of the Sun. The Sun and its planets are known as the Solar System.

From the investigations of astronomers we know that the solar system is part of a larger collection of stars known as the Milky Way. The Milky Way was the ancient name for the band of stars visible in the sky that we now know to be the plane of our own collection of stars; it is a disk-like assembly in space. When we look into the Milky Way we are looking across the disk of our galaxy. When astronomers discovered other massive groups of stars like the group that our solar system belongs to they needed a name for these groups of stars. They chose the term *galaxy* which comes from the Greek word for "milk." Thus galaxies are other "Milky Ways" spread out in the universe.

Although the sky appears to be full of stars we know that the universe is mainly empty space. In order to be able to appreciate the magnitude of the universe, we need a measurement system other than kilometers or miles because astronomical measurements quoted in kilometers or miles tend to have mind-boggling chains of zeros. A popular measurement unit used in astronomy is the light-year. Light travels at 3×10^5 kilometers per second. The light-year is the distance traveled if moving at the speed of light in a vacuum for one year. The light-year is approximately equal to 9.5×10^{12} kilometers.

To gain a physical appreciation of the magnitude of the light-year, we note that light traveling from the Moon takes 1.3 seconds to reach Earth. It takes approximately 5 hours for light to travel from the Sun to the edge of the solar system. Once we leave the solar system, the nearest star to us is four light-years away. For this reason it is extremely unlikely that space travel (which currently must take place at velocities far below the speed of light) will take place outside our Solar System. If we could keep traveling in a straight line at the speed of light we would only pass, on average, one star every five years (even though we would be traveling at 17.9 million kilometers a minute!). Our galaxy, the Milky Way, contains approximately 100 million stars. At the speed of light it would take 80 000 years to cross from one end of the Milky Way to the other. Once we left our own galaxy we would find that the nearest galaxy to ours, one which astronomers know as "*Andromeda*," is two million light-years away. Andromeda is actually one of several galaxies which form a cluster to which our Milky Way belongs. The name Andromeda comes from the name of a person in Greek mythology. Two million light-years away may not seem to be exactly a neighborly location; however, compared to the distances between groups of galaxies, this is a short distance. Astronomers have found out that the largest group of galaxies known as "the Hercules cluster" contains more than 10 000 galaxies and is three hundred million light-years away. It is estimated that there are at least 10 000 million galaxies in the known universe.

What is the probability that some of these stars in the universe contain intelligent life similar to that on Earth? It seems reasonable to assume that life as we know it could only flourish on planets orbiting stars. As pointed out earlier, we have no direct means of observing planets because they are not self-luminous; however, we can study the movement of stars in the sky and from slight wobbles in their paths deduce the probable

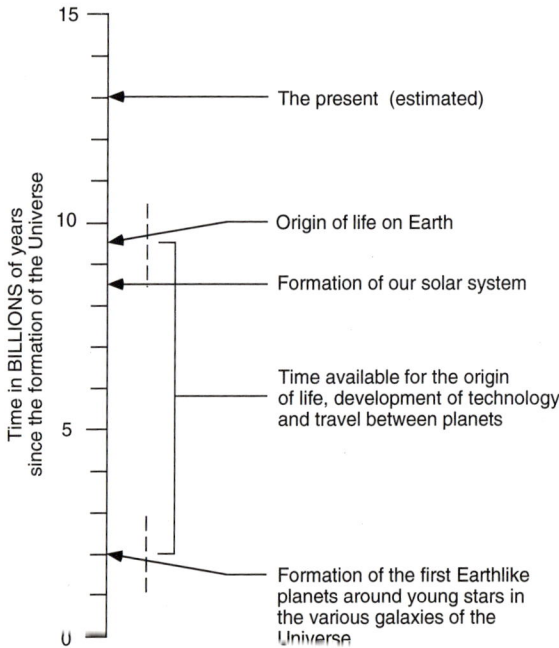

Figure 7.25 The time scale for the development of life on earth relative to the age of the universe would seem to indicate that there has been ample time for intelligent life forms to develop elsewhere in the universe.

presence of dark, cooler planets which could possibly sustain life. For example one of our stellar neighbors known as "*Barnard's Star*," which is approximately six light-years away, has been shown to have at least two dark companions, each of them about the size of Jupiter. Of the dozen or so stars nearest to the Sun, approximately half of them appear to have dark companions with a mass between one and ten times the mass of Jupiter. On the basis of measurements such as these, some astronomers believe that in the observable universe there may be many solar systems containing one or more planets similar to ours. An optimistic calculation, using the estimated number of stars per galaxy and the estimated number of galaxies in the universe, comes up with a figure of 100 000 000 000 000 000 000 (or 10^{20}) as the estimate of the number of planets capable of sustaining life similar to ours. Even if this estimate is grossly wrong, it does not seem unreasonable to assume that there are many millions of planets in the universe capable of sustaining life similar to that on Earth.

If these figures are correct, Stephen Dole of the Rand Corporation has estimated that there should be one habitable planet within 27 light-years of the Earth, five within 47 light-years, 10 within 59 light-years and 50 within 100 light-years. Thus even if we were able to converse by radio with the nearest probable location sustaining intelligent life, it would involve a two-way conversation that would have a 54 year gap between transmission and reception of messages.

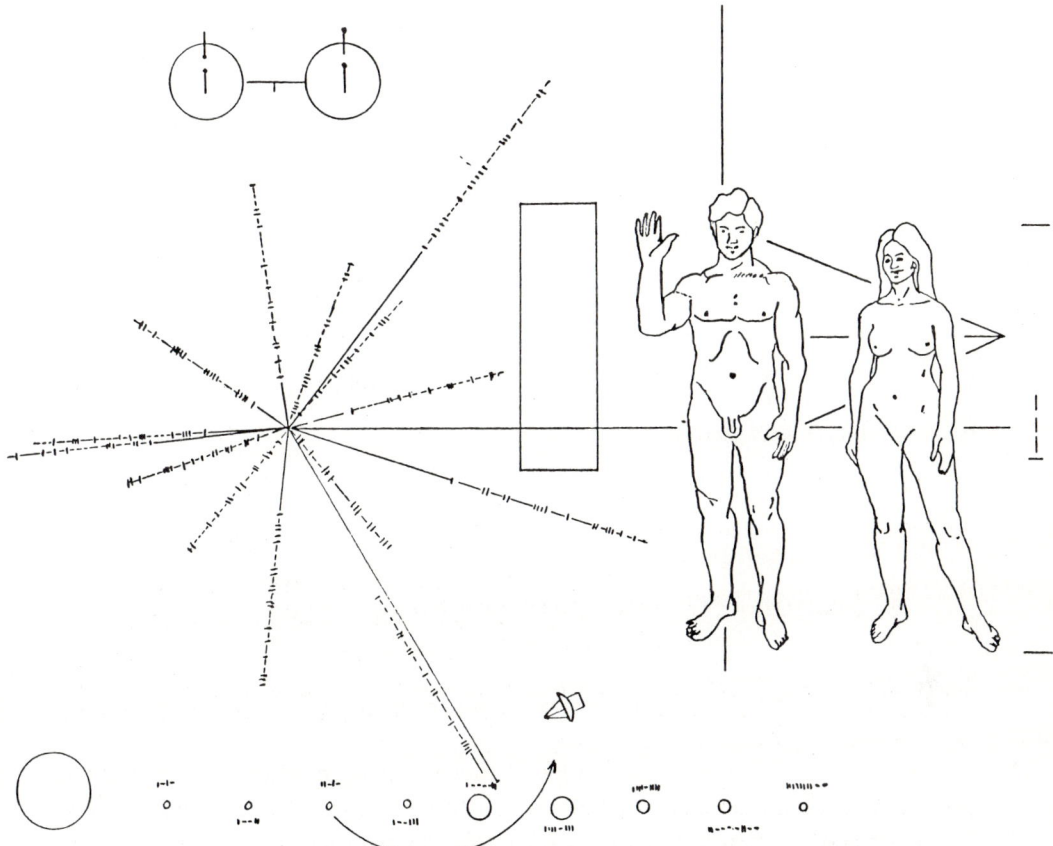

Figure 7.26 The first symbolic message dispatched outside the Earth's solar system aimed at intelligent life elsewhere in the universe was designed by Carl Sagan and is carried in the spaceship Pioneer 10.

If we accept that many planets could be capable of sustaining life, how likely is it that advanced intelligent beings have had time to evolve on these planets? In Figure 7.25 the time scale for events within our galaxy is shown. Notice that the time unit is a billion years. The first item on this graph indicates the probable time of the first formation of Earth-like planets in the oldest part of our galaxy. Our own solar system is farther out from the center of the galaxy and formed from interstellar dust relatively late in the history of the galaxy. Terrestrial life began about 1 billion years after the formation of our solar system. (Note that on this scale, the recorded history of the existence of mankind would be shorter than the thickness of the marker lines on the time scale.) This graph shows that the time available for the development of advanced civilizations of intelligent beings is greater than the total age of our solar system.

When all of these calculations have been completed, both with regard to the size, and the age of the universe, it seems reasonable to think that there are many millions of intelligent civilizations scattered about the universe — possibly interested in communicating with man [24, 25].

From our brief discussion of the size of the universe around us with its billions of galaxies and uncountable numbers of stars, we see that, in one sense, Pioneer 10 could travel forever. Just in case it was met or captured by intelligent extraterrestrial beings, the space scientists sending Pioneer on its way decided to make it a transgalactic missile with a message. A plaque bearing a symbolic message from Earth to other intelligent beings of the universe was placed on board the robot spacecraft. This plaque is shown in Figure 7.26. A detailed discussion of the symbols on this plaque is beyond the scope of this book but the interested reader can find a detailed discussion of the symbols and the reaction of the general public in North America to the plaque in Ref. [27]. The message sent on Pioneer 10 was part of a continuing effort being pursued by many scientists to communicate with extraterrestrial intelligence. The plaque of Figure 7.26 is probably only the first of many messages with a transgalactic destiny.

Section 7.9   The Future of Fireworks Displays

The public apparently has a insatiable appetite for firework displays. There are two basic types of display fireworks. One type has two stages with the top part generating golden sparks and the second stage is a large white flash with a bang. It generates a pattern looking like feathers spread out from the center of the explosion. In Asia a more popular type of firework is of spherical design with a fuse that ignites it in different stages. The typical rocket, like its cousin in space research, has propellant and its payload is the display part of the rocket. A typical display rocket has oxidizer, fuel binder, and special effects material. The oxidizer is usually something like potassium perchlorate with a powdered metal fuel. The ingredients are ground to a specific fineness and kept firmly packed with a binder such as dextrin. The secret of making a good firework is to ensure that the rocket reaches a good height before the rest of the display is triggered uniformly. Sodium compounds are added to create yellow flames. Strontium compounds give red flames, and a variety of compounds yield green flames. (As a student learning chemistry I used to use a wire dipped into various compounds which was then placed in the flame of a bunsen burner. The color of the flame generated when the wire was in the flame indicated whether the material to be analyzed contained sodium, barium, or strontium.) The sparks of a firework are made by either iron or aluminum filings. These metal filings are ejected as streaming molten particles and glow gold or silver during the explosion. Other ingredients produce the billows of smoke. Shock waves originating from the explosion create the firework's boom and the more fuel is packed into a shell the louder the noise. The recipes for making fireworks are usually kept a close secret within a company. Thus in an article written by Angier [28] it is said

The recipes for the formulas for making most fireworks have been handed down for generations.

In terms of their composition, the art of firework making has hardly changed over the years. Many formulations have been the same for hundreds of years. In the article by Angier a technique is discussed for making shapes such as hearts and elephants in the sky. In many countries the production of fireworks is closely monitored; nonetheless, a safety problem exists in many western countries because of the import of cheap fireworks from countries without the same standard of quality control.

Statistics on fireworks injuries are distressing. Thus even in Britain, where safety standards are tougher than elsewhere in 1991, 723 people were injured by fireworks, 247 of whom had eye injuries. 109 people had to miss work or school to recover and 26 spent more than one night in hospital. One third of the injuries were sustained by children under 13. Industrialized nations are drawing up common safety standards for fireworks and recently at least one major manufacturer of fireworks has moved to use the same type of technology as space rockets using a plastic binder to hold the ingredients together. Thus in another article on fireworks we are told that ordinary gunpowder is dusty and poses an explosive hazard and an inhalation problem. It is estimated that several hundred people die across the world through accidents in the fireworks industry. By adopting the type of technology used to make solid fuel rockets the dust is eliminated and the gunpowder propellant is no longer open to ignition through friction and impact during the manufacturing process. By using the resin binder to hold the ingredients together the manufacturer can extrude (force) the mixture through holes to make a spaghetti-like material which promises to improve the safety for both the manufacturer and the consumer [28].

References

[1] J. Orear, *Physics*, MacMillian Company Inc., New York, 1979.
[2] J.D. Cutnel and K.W. Johnson, *Physics*, 3rd Edition, John Wiley & Sons Inc., New York, 1995.
[3] C.B. Daish, *Learn Science Through Ball Games*, American Edition, Sterling Publishing Company, New York, 1972 (original edition published by the English University Press Limited, 1972).
[4] J.T. Shipley, *Dictionary of Word Origins*, Littlefield, Adams and Company, Trotowa, New Jersey, 1970.
[5] *Chambers Etymological English Dictionary*, ed. by A.M. MacDonald, W. and R. Chambers Ltd., II, Thistle Street, Edinburgh.
[6] See introduction in Goddard – The History of Rockets, in I. Asimov, *Biographical Encyclopedia of Science and Technology*, Doubleday, Garden City, New York, 1948.
[7] E. Stuhlinger, F.I. Ordway III, *Wernher von Braun: Crusader for Space*, Krieger, Florida, 1994.
[8] von Braun, Wernher in Encyclopedia Britannica Inc., Chicago, 1987.
[9] Gavaghan H., "Space Business", four-page supplement to *New Scientist*, December 3, 1987, one of a series known as "Inside Science".

[10] W. Cohen, *Solid Fuel Boosters*, Modern Science and Technology, edited by R. Colborne, G. Van Nostrand Company Inc., New York, 1965, pp. 526–535.
[11] S. Aftergood, "Poisoned Plumes," *New Scientist*, September 7, 1991, pp. 34–38.
[12] J.E. Allen, *Aerodynamics; the Science of Air in Motion*, Second Edition, Granada Publishing Limited – Technical Books Division, Saint Albans, Herts, England, 1982.
[13] Kaye, B. H., 1989, *A Random Walk Through Fractal Dimensions*, VCH, New York, p. 167. See the statement that "one can extrapolate the curve into the region of the probability of large particles to discover that there is a very small, but finite, chance that one could discover an elephant in a face powder container!"
[14] See entry in I. Asimov, *Biographical Encyclopedia of Science and Technology*, Doubleday, Garden City, New York, 1948.
[15] I. Asimov, Words of Science, Signet, Reference book published by The New American Library, New York, 1959.
[16] J. Ficke, "The Unbeatable Lightness of Aerogels," *New Scientist*, January 30, 1993, pp. 31–34.
[17] "Researchers Prepare New Aerogels for Expanding Uses," *R & D. Magazine*, January 1991, pp. 26–28.
[18] G.R. Noakes, *A Textbook of Light*, Second Edition, Macmillan, London, 1962.
[19] J. Gribbon, "The Greenhouse Effect," Inside Science No. 13, four-page supplement to *New Scientist*, October 22, 1988.
[20] A good reference for the concepts and ideas of astronomy is the book by K.F. Kuhn, *In Quest of the Universe*, 2nd Edition, West Publishing Company, St. Paul, Minnesota, 1994
[21] R. Fifield, "The Structure of the Earth," Inside Science, four-page supplement to *New Scientist*, February 25, 1988.
[22] I. Gruntfest and L. Schenker, *Ablation*, Modern Science and Technology, Ed. by Robert Coburn, Van Nostrand, 1965, pp. 435–542.
[23] L. Jaffe, *Communication by Satellite*, Modern Science and Technology, Ed. by Robert Coburn, Van Nostrand, 1965, pp. 543–550.
[24] For a summary of a co-ordinated international attempt to develop a search for communications coming from extraterrestrial beings, see the article by J. Gribbin, "Is There Anyone Out There", *New Scientist*, May 25, 1991.
[25] C. Sagan, F. Drake, "The Search For Extraterrestrial Intelligence", *Scientific American*, May, 1975.
[26] N. Booth, "Pioneer the Presistent Probe," *New Scientist*, February 29, 1992, pp. 48–50.
[27] For the public reaction to the Pioneer-10 plaque see "Message From Mankind", *Time*, March 6, 1972, p. 45.
[28] N. Angier, "The Hot New Science of Fireworks." *Discover*, July 1982, pp. 25–28.

Chapter 8

Cosmic Collisions

Chapter 8

Cosmic Collisions

Section 8.1   Target Earth

Laurentian University is situated on the shores of a beautiful lake. As one looks out of the university square across a landscape of trees and lakes it is hard to imagine what the Sudbury area looked like 1.8 billion years ago. Sudbury was originally a railway town formed at the confluence of a railway line from Toronto joining up with the trans-Canadian line being built from Halifax to Vancouver. Sudbury became an important mining town because surveyors for the railway discovered copper ore while mapping out the path of the railway through the area. Copper production is still a major activity in the Sudbury area. The ores here not only contain copper but they are also very rich in nickel. At one time, just after World War II, 63% of the world's nickel came from the Sudbury area. The ore of this area is unique in the world and contains valuable commodities such as gold, platinum, and iridium.

The nickel mines in Sudbury are located in an ellipse around an area known as the Sudbury basin. Although there is still some controversy concerning the way in which the Sudbury basin was formed, the general consensus of scientific opinion is it was formed 1.8 billion years ago when a large asteroid hit the Earth [1]. The force of the impact depressed the crust of the Earth to the point where nickel-rich magma flowed up from the inner parts of the Earth's crust to form the dark-edged rim of the Sudbury basin. Thus the quiet scene of today was a sea of boiling rock and shattered landscape 1.8 billion years ago.

The deformed ellipse linking the nickel mines of the Sudbury area does not appear to have too much in common with the sharp-edged type of crater produced by meteorites in recent times. A good example of a young crater is the *Barringer crater* in Arizona which is shown in Figure 8.1(a). The Barringer crater is young in relation to the billions of years in which the age of the Earth is measured. Its age is evidenced by the fact that it has a sharp rim, indicating that the forces of nature have had little time to erode and smoothen its features [2]. The Sudbury basin, because of erosion of the surface of the Earth and the movement of the Earth's crust, is a worn down version of its original structure.

One of the pieces of evidence that the Sudbury basin was formed by an enormous cosmic impact comes from the fact that geologists have discovered many examples of a rock formation called a *shatter cone* around the basin. A typical shatter cone is shown

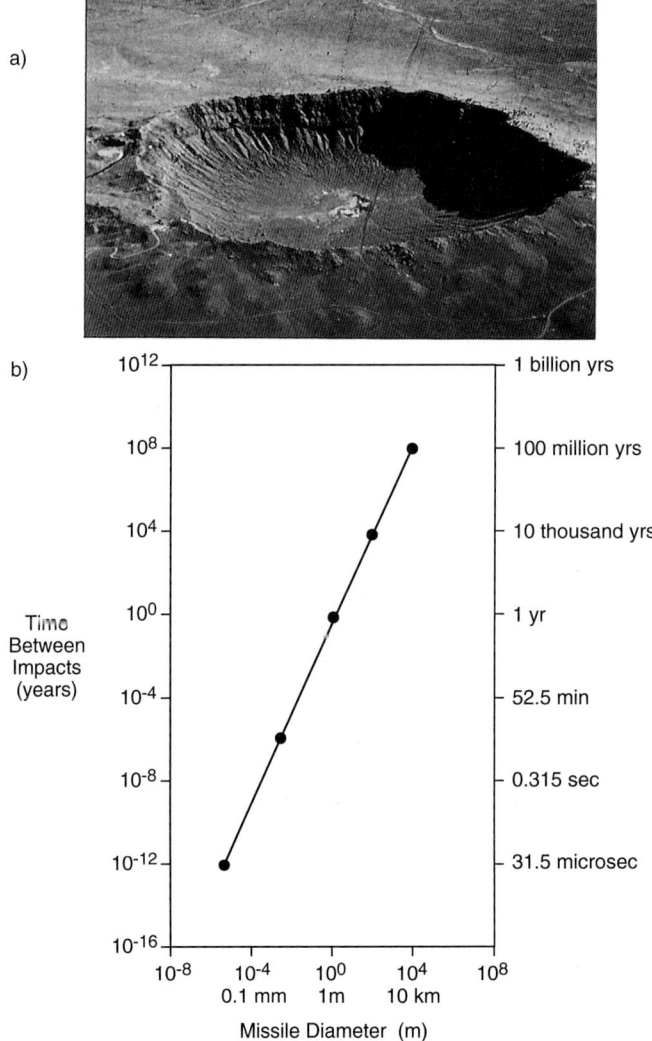

Figure 8.1 Cosmic collision of extraterrestrial bodies with the Earth has created some spectacular craters. a) Arizona crater formed only a few tens of thousands years ago; 1.2 km in diameter, it is as yet virtually uneroded. b) Graph of the frequency of collision of cosmic missiles with the Earth versus the diameter of the missile [4]. Reproduced with permission of the Minister of Supply and Services, Natural Resources Canada, Ottawa, 1996. From R. Grieve, "Impact Craters Shape Planet Surfaces", GEOS, Vol. 11, (4), fall 1982.

in Figure 8.2(a). The surface of the shatter cone has lines spreading out from the tip of the cone in a pattern which is known as a *striated surface*. The word striated comes from the Latin word "stria" meaning "a furrow" of the kind created in a field by a plough. Thus, from a distance, the striated markings on a shatter cone are not unlike the furrows in a ploughed field. Striated shatter cones are a unique feature of any area

Figure 8.2 The presence of shatter cones around the Sudbury Basin is strong evidence for the fact that the basin was formed by asteroid impact. The photograph shows a typical shatter cone found in the Sudbury region. Reproduced with permission of the Minister of Supply and Services, Natural Resources Canada, Ottawa, 1996. From R. Grieve, "Impact Craters Shape Planet Surfaces", GEOS, Vol. 11, (4), fall 1982.

which has been hit by a large meteorite. In the early 1970s, astronauts preparing for the Apollo 16 and 17 missions to the Moon traveled to Sudbury to learn how to recognize geological evidence of ancient meteorite impacts so that they could apply this knowledge when they were on the Moon.

In a review [3] of the probability of the Earth being hit by a cosmic missile Mallove, an astronautical engineer and science writer, makes the following statement:

> *Poised in the depths of space lurk deadly projectiles weighing millions of tons which will some day gouge the Earth. With roulette like inevitability Earth's number will come up, as it has many times in the past, and the planet will once again collide with an asteroid.*

Another science writer, Manfred Schroeder, discussing the frequency with which cosmic missiles hit the Earth makes the following statements [4]:

> *The mean frequency with which different kinds of interplanetary debris (shooting stars or meteors) slam into the Earth's atmosphere is inversely proportional to the squared diameter of the projectile and this is true over ten orders of magnitude. Whereas the space shuttle is hit at a rate of 1 particle every 30 microseconds with a diameter under 1 micron (10^{12} particles per year) the meteorites of the size that created the Arizona crater, with a diameter of 100 meters or more, are expected (thank heavens!) only once*

every 10^4 years and the next shooting star of the size that hit Sudbury, Ontario, with an astronomical 10 kilometers diameter, should not rock the Earth for another 10^8 years (100 million years).

The data referred to by Schroeder on the frequency with which various-sized cosmic projectiles collide with the Earth is summarized in Figure 8.1(b).

In a review [5] of the probable formation dynamics of the Sudbury basin, Cynthia Thompson describes the sequence of events leading up to the formation of the basin as follows:

> Shatter cones tell the story of a meteorite exploding against the Earth's crust and cracking it like an egg shell 1.8 billion years ago. The meteorite must have been from 1/2 to 10 kilometers wide. Shock waves radiating out from the impact site almost instantaneously melted and brecciated, crumpled and warped the surrounding country rock into mile high ridges. Split seconds later, the pressure of the impact subsided and the ridges rapidly flattened and slid back toward the initial excavation, then they rose in the center as masses of melted broken and fractured rock continued to be spewed into the stratosphere. A crater, that may have been up to 70 kilometers across, was excavated at the point of impact, rocks literally vaporized at temperatures that exceeded 2000°C. Deep below, nickel-rich magma began to well to the surface via cracks that had split the crust down to the mantle.

The word *brecciated* is a geological term for the creation of a rock composed of a mixture of broken rock which has been fused together by pressure over time, with individual fragments still recognizable in the mixture

Describing the present appearance of the Sudbury basin, Cynthia Thompson goes on to state [5]:

> A distorted 60 kilometer by 27 kilometer structure remains. Because of the concave bowl-like structure of the impact crater, extensive erosion makes the bowl more shallow while reducing the circumference. To understand this process visualize a melon cut in two with thin sections being sliced from the cut surface reducing both the depth of the center and the radius of the rind as slicing (similar to erosion) continues.

The word *crater* comes from the Greek word "krater" which was the name of a large bowl for mixing water and wine. In modern science the term crater is applied both to the saucer-like depressions left in the ground by meteorite impacts and to the deep throat of a volcano.

A unique science museum, known as Science North, is located in Sudbury. Opened in 1984, this museum has a design consisting of two snowflake-shaped buildings. The design is intended to represent the northern climate and the fact that the local terrain has been sculptured by the movement of glaciers. These two buildings are linked by an underground tunnel. As visitors walk along the tunnel they can see shatter cones in the 2.5 billion-year-old rock. Just before the main building of the museum is reached, there is a very large cavern in which three-dimensional films are shown, including one which portrays the meteor impact formation dynamics of the Sudbury basin. This movie recreates in a vivid manner the sequence of events leading to the formation of the ore body in the Sudbury region.

Running through the middle of Science North is a portion of a famous rock formation known as the *"Creighton Fault."* The shattered rock along this billion-year-old fault, long inactive but once a massive San Andreas-like fracture of the Earth's crust, lay in the path of the ice sheets during the ice age 20 000 years ago and features known as "glacial striae" can be seen on the rock surfaces. These striae are lines gouged into the rock surfaces by boulders being dragged along by the bottom surface of the glaciers as they moved over the area. The huge weight of the thick ice bearing down on the boulders caused them to scrape across the underlying rock surface. These various interesting geological formations can all be viewed in one visit to Science North.

It is interesting to note that currently some of the mines in the Sudbury area are deeper than 2000 meters. When mining at this depth, one of the major industrial problems is cooling the mine because of the heat coming from the interior of the Earth (at this depth the rocks have a temperature of 41 °C). Rock movements known as *rock bursts*, which are in fact minor earthquakes caused by mining action, become a more important problem as the miners go deeper into the Earth's crust because of the increasing stress caused by the weight of the rocks above the mine. However, as the miners go deeper, the ore discovered becomes richer. Currently plans are being made to go down to a depth of 3000 meters.

In recent years the International Nickel Company has explored the possibility of using the heat at the depths of the mines to grow cucumbers and other plants using artificial lights and *hydroponic culture*. In hydroponics the plants do not grow in soil but in water provided with the necessary nutrients in solution. In Chapter 10 we will also see how a deep mine is being used as a neutrino observatory.

As mentioned earlier, Schroeder, in a discussion of the bombardment of the Earth by cosmic missiles, stated that a missile of the size that created the Sudbury basin can be expected once every 10^8 years, that is, once every 100 million years [4]. That is not really the type of probability over which to lose any sleep, but it is natural to ask how often smaller objects hit the Earth. One of the ways of assessing the probability of future asteroid collisions with the Earth is to look at the pattern of past events. However, it must be noted that when the solar system was in its early stages, impacts were much more frequent as we will discuss when we look at the craters on the Moon. Past records of cosmic missile impacts must also be interpreted with caution because many of the earlier craters have been wiped out by erosion and surface movements. Figure 8.3 provides a summary of all known craters in the world with a diameter of more than one kilometer. Approximately 130 craters have been discovered so far. At first sight they appear to cluster in the northern parts of the world. However, it should be remembered that there is far more land in the northern hemisphere and also, that in the far northern climate, there are fewer mechanisms which are active in wiping out the record of the

Figure 8.3 Locations of known large meteorite/asteroid craters around the world [4].

craters. Grieve points out that one-third of the known craters of more than 1 kilometer in size are to be found in Canada, which covers only 7% of the Earth's surface [2]. However, half the area of Canada is covered by the Canadian Shield which is a surface composed of very ancient rocks that have been more or less geologically stable for the last 400–500 million years. It is one of the most favorable terrains on Earth for the preservation of craters from erosion, change, or burial. Dr. Grieve also points out that 70% of the Earth's surface is covered with water and that it is very difficult to discover old craters under the sea.

Section 8.2   The Dynamics of Asteroid Collisions on the Surface of the Earth

Before we look at the dynamics of an asteroid collision we must define what is meant by an asteroid. As astronomers started to map out the universe they were mystified by the fact that their calculations showed that a planet should exist between Mars and Jupiter, about 2.8 astronomical units from the Sun. One *astronomical unit* is defined as the Earth's average orbital distance from the Sun, about 150 million kilometers. When astronomers started to search for what they called the missing planet, they did indeed find a very small object. It was discovered in 1801 by Piazzi of Palermo in Italy who named the object Ceres. As several more lumps of rock were found in the same type of orbit, Sir William Herchel said that such objects could be called *asteroids*, a Greek word meaning star-like. The term is unfortunate because in astronomy a star is a self-luminous body and the asteroids are really planetoids. That is small planet-like objects which are only visible by reflected light. At first scientists thought that the asteroids were the fragments of a massive collision which destroyed a proto-planet. However, the modern theory of asteroid formation regards the asteroids as potential planet material that failed to coalesce and become part of a large planet. (The belt occupied by asteroids in the solar system is shown in Figure 8.4.) About a thousand asteroids are larger than 30 kilometers across, and of these, more than two hundred are larger than one-hundred kilometers across. Not all the asteroids smaller than 30 kilometers have been catalogued. In the words of Binzel and co-workers [6]

> *Such vast numbers of asteroids conjure images from popular films that show space craft weaving through fields of crashing boulders. The volume of space in the main belt is so large however that asteroids usually remain several million kilometers apart.*

Collisions are infrequent but do occur as we will discuss later in this chapter. Binzel and co-workers state that

> *Asteroids were once regarded as vermin of the skies because they interfered with the pictures that astronomers were taking of the stars.*

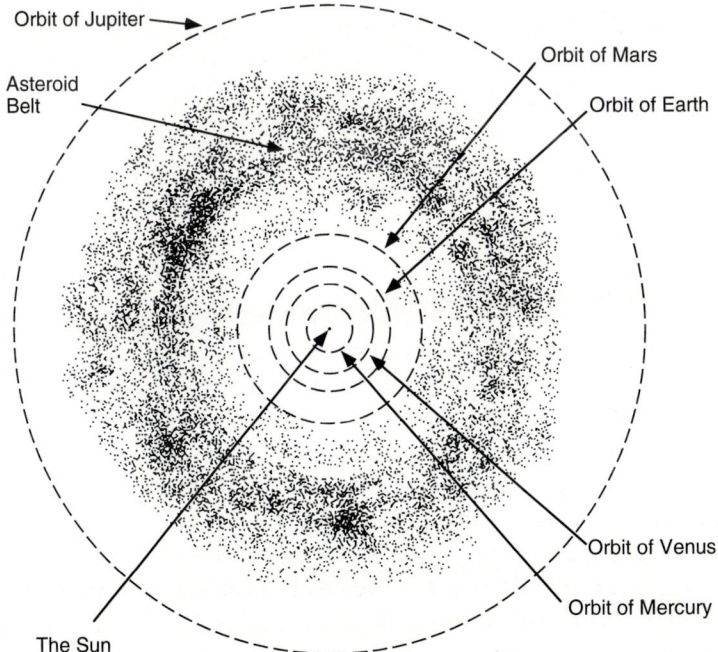

Figure 8.4 The "asteroid belt" lies within the solar system between the orbits of Mars and Jupiter.

Asteroids are now known to offer important clues to the birth of the solar system [7-9].

Long before the space program, astronomers used to refer to a certain group of asteroids as *Apollo objects*. Apollo was one of the names given by the Greeks to the god of the Sun, and was the name picked by NASA for the space shots that culminated in putting men on the Moon. An astronomer defines Apollo objects as a family of interplanetary bodies whose orbits cross that of the Earth. Their name comes from the fact that astronomers in the early 20th century had discovered a large asteroid which they called Apollo. It turned out that the orbit of this asteroid crossed the Earth's orbit, and the term Apollo bodies was extended to describe any asteroid with such an orbit. Current estimates place the number of Apollo objects in space with diameters greater than one kilometer, at about 1200. The average speed with which an Apollo body would hit the Earth is about 25 kilometers per second. Richard Grieve points out that speeds of this magnitude have little meaning in everyday life. To help visualize such enormous speeds, Grieve [2] gives the following illustration:

Suppose that you were on a flight from Ottawa to Montreal (160 kilometers) and an Apollo passed overhead just as the plane was beginning its takeoff; the plane would barely be off the ground by the time the Apollo body had passed over Montreal and hurtled beyond.

Grieve goes on to point out that, a body about 200 meters across hitting the Earth at these cosmic speeds releases an amount of energy equivalent to the explosion of about 100 million tons (100 Megatons) of TNT. A body of this size would be sufficient to create a crater of the same size as the Brent crater, an impact crater found in Algonquin Park about 200 miles from Sudbury. An aerial photograph of the Brent crater, which was formed 450 million years ago and originally 4 kilometers in diameter, is shown in Figure 8.5(a). Specialists in the study of impact craters formed by cosmic missiles, distinguish between simple impact craters and complex impact craters [2]. The Brent crater of Ontario is considered to be the largest known simple impact crater. Grieve has described the formation dynamics of a simple impact crater in the following way [2]:

When a body strikes the Earth at 25 kilometers per second, it penetrates into the ground to a depth of about the equivalent of the body's diameter. As the projectile penetrates and is being slowed down, most of the kinetic energy stored in the projectile is transferred to the Earth by means of an outward-moving compression shock wave. The shock wave compresses the rocks and accelerates some large fragments away from the point of impact. At the point of impact, the pressure can be millions of times the normal atmospheric pressure. Accelerations of rocks away from the crater are measured in kilometers per second squared. When this pressure wave dissipates, rocks close to the point of impact are vaporized and/or melted. These impact-melted rocks can have temperatures of several thousand degrees Celsius which is many times hotter than lava pouring out of volcanoes.

8.2 The Dynamics of Asteroid Collisions on the Surface of the Earth   295

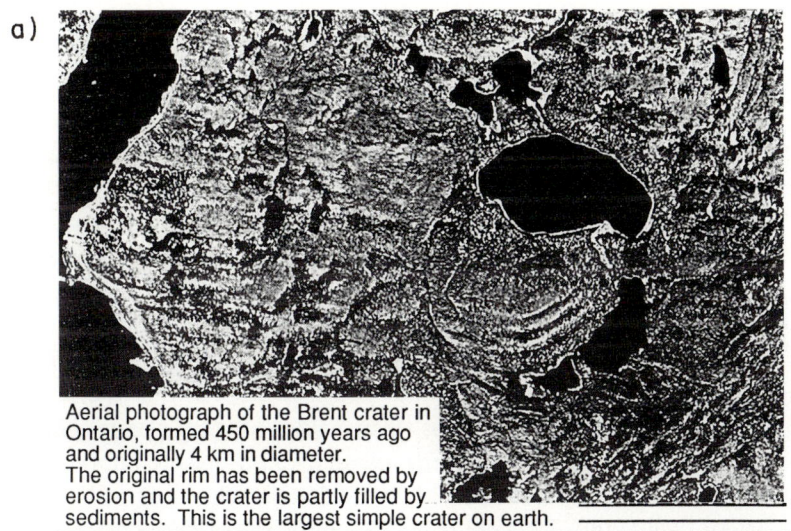

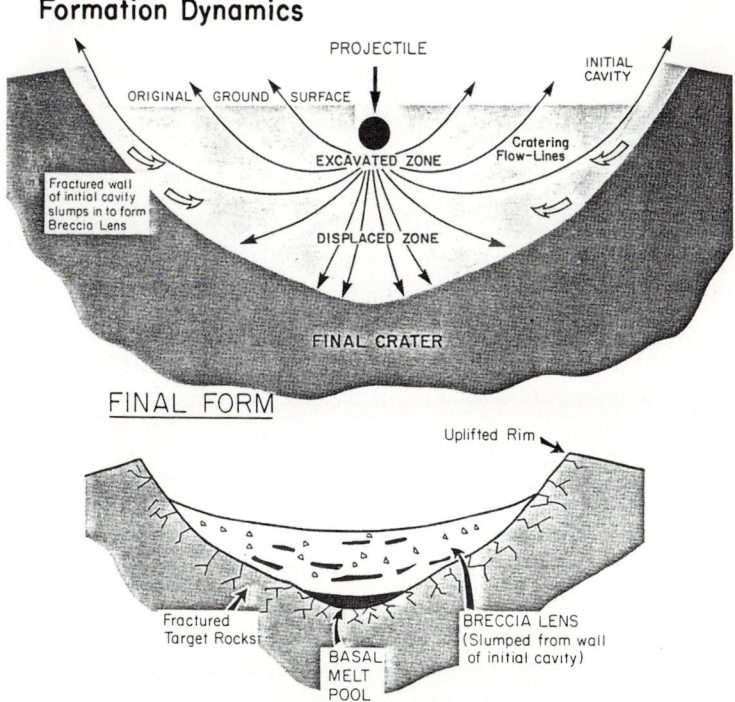

Figure 8.5 Craters formed by simple impact (relatively small meteorites) are known as simple craters [4]. a) Satellite photograph of the Brent crater. b) Diagram showing the formation of a simple impact crater. Reproduced with permission of the Minister of Supply and Services, Natural Resources Canada, Ottawa, 1996. From R. Grieve, "Impact Craters Shape Planet Surfaces", GEOS, Vol. 11, (4), fall 1982.

Lava is so named because that was the name the Italians gave to the streams of molten rock leaving Mount Etna in Sicily. The word "lavare" in Latin meant "to wash" (and hence the name lavatory, originally a place to wash oneself and attend to personal hygiene). Lava was the molten rock that was washing the sides of the volcano.

During the crater formation process some of the impact-melted rock remains at the bottom of the crater as shown by the item labeled *basal melt pool* in Figure 8.5(b) (basal meaning "at the base of"). Asimov tells us that the name *basalt* is a very ancient one which according to Pliny, a Roman writer, originated in Ethiopia where it was used to describe an especially dark variety of marble. The meaning has been extended in modern English to include different types of dark rock of igneous origin. We have also used the term "magma" several times; this term comes from a Greek word meaning "thick" like bread dough. The Greek word for bread dough is derived from the Greek verb "massein" meaning "to mold" (to knead) as when making bread. (The same Greek root-word has given us *massage* as the word for the kneading of the body to improve the muscle tone.) Magma is like a very hot moldable bread dough. When magma has solidified to make material such as the nickel ore around Sudbury, the rocks are described as *plutonic rocks* from the name of the Greek god of the underworld *Pluto*. Unfortunately, Walt Disney has made it hard for us to think of an underworld god when we use this word because in the minds of most people, it now conjures up the picture of a floppy eared dog.

Sometimes lava becomes full of bubbles and cools to form a stone known as *pumice* from the Latin words "pumex" meaning lava and "spuma" for "the foam on top of a wave". Pumice stone floats on water and can be used to rub away calluses on the skin. Pumice, basalt, plutonic rocks, and other solidified rocks, which at one time in their formation were melted by intense heat, are known collectively as *igneous rocks*. This word comes from the Latin word "ignis" meaning "fire." (This word came directly into the English language to give us the word *ignite* for the act of starting a fire.)

Considering the area farther from the point of impact when the simple crater is formed, it is found the rocks remain unmelted but are permanently changed, taking on peculiar features of deformation such as shatter cones and microscopic dislocations. To the geologist these changes are known as *shock metamorphic features*. Rocks which are thrown up into the air from the impact are known as *ejecta*. In a simple crater, the entire width of the crater will be of the order of 30 times the diameter of the missile. The crater is called the *footprint of the missile*. In a later section, we will find, when studying cosmic dust, that deciding just how big the missile is from the size of the footprint it leaves on the capture surface is an important branch of modern *astrophysics*.

The amount of material ejected at the point of strike or target in an asteroid impact is approximately 1000 times the mass of the projectile. This ejecta can end up many kilometers away from the original impact zone. The rim of the crater formed during the initial impact is unstable and the sharp fragments of broken rocks fall back into the cavity to create what the geologists call a *breccia lens*. At the interface between the breccia lens and the original rocks there are fractures and changed crystal structures produced

8.2 The Dynamics of Asteroid Collisions on the Surface of the Earth

by the energy of the shock wave. Again discussing the Brent crater, Grieve [2] describes this formation process by stating

> In the case of the Brent crater in Algonquin Park, the initial cavity is about 3 kilometers across and over 1 kilometer deep, as deep as part of the Grand Canyon. This initial cavity is extremely unstable and the weakened and fractured walls slump inwards partially filling the crater with broken rock and slightly enlarging the diameter of the final crater.

The sequence of events involved in the formation of the Brent crater as outlined above probably took place in one minute.

The crater formation process associated with the impact of much larger asteroids is more complex. The various stages for the larger impact are illustrated in Figure 8.6. The crater shown in Figure 8.7 is one of these complex craters. It is located in Manicouagan, Quebec and is 210 million years old. Grieve tells us that this event corresponded to 100 to 1000 times the energy of all the earthquakes occurring on the Earth in one year being concentrated at one spot on the Earth's surface. The original crater was between 75 and 100 kilometers in diameter. Rocks originally several kilometers below the surface were uplifted to form a central peak which today stands some 500 meters

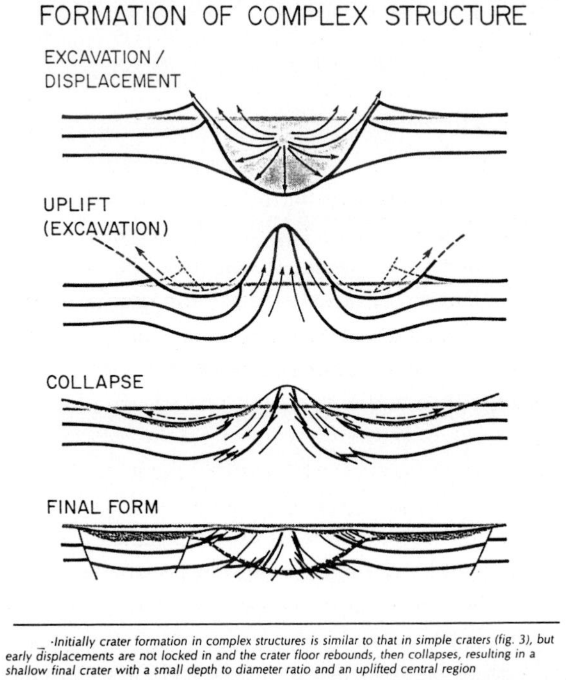

Figure 8.6 Initially crater formation in complex structures is similar to that in simple craters (Figure 8.5) but the initial displacement is not permanent and the crater floor rebounds, then collapses, resulting in a shallow final crater with a small depth to diameter ratio and an uplifted central region. Reproduced with permission of the Minister of Supply and Services, Natural Resources Canada, Ottawa, 1996. From R. Grieve, "Impact Craters Shape Planet Surfaces", GEOS, Vol. 11, (4), fall 1982.

Figure 8.7 LANDSAT image of the crater in Manicouagan, Québec. The crater was formed 210 million years ago and was originally between 75 and 100 kilometers in diameter. Reproduced with permission of the Minister of Supply and Services, Natural Resources Canada, Ottawa, 1996. From R. Grieve, "Impact Craters Shape Planet Surfaces", GEOS, Vol. 11, (4), fall 1982.

above the ground. The shape of the lake has been magnified in recent times by the fact that a hydro-electric dam has been built on the main river flowing out of the area. The flooding associated with this dam has magnified the circular structure of the impact zone. In complex crater formation, the crater is relatively shallow with a small depth/diameter ratio of 1 : 30 or less as compared to 1 : 10 for simple craters.

The central peak of the larger complex craters is formed by the processes outlined in Figure 8.6. After the initial impact, the floor of the cavity rebounds upward to form a relatively shallow structure with a central peak and/or rings. Geologists are not sure what really happens in these complex dynamic situations but they describe it as a hydrodynamic behavior because it is rather similar to the behavior of the surface of a pool of water when someone drops a large stone into the pool. When a stone is dropped into the water, ripples spread out from the impact but the water also springs back up at the point of impact.

Geologists assume that under the intense pressures of the very energetic impact, the rocks of the region at the point of impact, behave just like water. Some estimates place the depth of the original impact zone here in Sudbury at 10 kilometers. The whole process leading up to the third element of the sketch in Figure 8.6 takes place in two to three minutes. It has been calculated that for the case of the Manicouagan impact crater, 1000 cubic kilometers of the target rocks were melted. The effects of the impact are still visible in rocks over an area of 20 000 square kilometers, a region about half the size of the province of Nova Scotia.

Section 8.3 The Tunguska Attack: Cosmic Missile or Alien Spaceship?

The largest cosmic missile to blast the Earth this century did so in 1908 in an event known as the Tunguska disaster. This cosmic missile attack is described in a review article [10] by Ian Ridpath in the following manner:

> On the morning of the 30th of June, 1908, a blazing fireball descended to Earth in the valley of the stoney Tunguska River, Siberia, located 800 kilometers north-west of Lake Baikal. Its searing heat melted metal objects and burned reindeer to death in the target area. A farmer on his porch 60 kilometers away said that the heat from the fireball seemed to be burning his shirt and a neighbor clasped his hands over his ears to protect them from scorching. The blinding, bright blue bolide, trailing a column of dust, disintegrated explosively, producing a blast wave that knocked the first farmer off his porch. Sounds like thunder rumbled in the air. Farther north near the center of the fall, several of the Nomadic-Tungus people were thrown into the air by the blast, and their tents were carried away in a violent wind. Around them, the forest began to blaze. Locals cautiously inspected the site of the blast. They found scenes of terrifying devastation. Trees were felled like matchsticks for up to 30 kilometers around.

A *bolide* is defined in the dictionary as a bright, shooting meteor (fireball) especially one which explodes when it is near the end of its path in the atmosphere.

Since the area where this event occurred is very remote, it was some time before scientists were able to visit the site. The fireball phenomena associated with the Tunguska event was seen for thousands of miles around. In March, 1978 an article appeared in a scientific journal in which Akim Zaburunov, a retired member of the US Army Night Vision Research Laboratories, recalled seeing the Tunguska explosion when, as a boy, he was working in the fields in a Russian village 4200 kilometers away from the explosion site. He said [11]

> I was 10 years old and working with my parents and other villagers in the wheat fields located close to the Donetz River, 160 kilometers west of the junction of the Don and Donetz. Of necessity we worked in the fields until midnight or later to reap the wheat as quickly as possible when it ripened. Since we were harvesting the earliest of the fields, the time of the year would have been the end of June. About midnight local time our oxen were pulling wagons loaded with wheat when suddenly a luminous and purplish column appeared at the horizon in the easterly direction. Almost immediately a large purple and perfectly round luminous ball appeared on top of the column. The height was estimated to be 5–6° above the horizon. The ball had the apparent size of 1.5 times the diameter of the Moon. Then, without any perceivable motion, the ball and column disappeared after a short time. We casually took that as an unusual effect, then after about one minute, a second similar, but now red ball, appeared at the same height.

After a short period of time it disappeared. We now were concerned about the events. After about another minute a third ball appeared slightly to the right and again, after a short time, it also disappeared. After the third one we were scared but we continued to watch the horizon. Several more fireballs were observed. Each fireball was successively smaller. The last one appeared after about a two minute interval: it was about half the size of the Moon. It was somewhat lower and to the right of the previous balls. Curiously the last one had a small tail trailing downwards. I believe that altogether at least 5 fireballs appeared at regular time intervals. Each lasted for an estimated 3 second time period. After the series there was no more activity and we completed our work.

Explanations of this series of events in the Tunguska valley have ranged from the Earth being hit by a small comet or by a large meteorite to the idea that an alien spaceship had hit the Earth. Others speculate that a black hole hit Tunguska. Exactly what is meant by colliding with a black hole is very difficult to grasp and that explanation can probably be discounted.

The Russian scientist Leonid Kulik organized an expedition to the Tunguska region in 1927 hoping to solve the mystery of the blast of 1908. First of all Kulik and his assistant had to travel 3000 miles across Siberia from Leningrad. This took them to a place 400 miles south of their destination. Professor George Greenstein has described the difficulties faced by this expedition in an interesting article entitled "Heavenly Fire." In his account [12] Greenstein states

In mid-March, the little expedition left Tagset on horseback over snow. By the 19th, they had reached Kezhma where they paused a few days collecting supplies and information before setting off again. North of Kezhma, the road was little more than a beaten track. By the 25th, they had reached the tiny settlement of Vanavara on the Podkamennaja Tunguska River. After two weeks rest, the party set out with their equipment loaded onto reindeer. Two days later, even the narrow path they were traveling came to an end and they made their way through the Siberian forest. At times, the trees were so thick they were forced to clear their way with axes. On April 13th, they reached the edge of the devastated area. Here their guide was unwilling to go forward since the local people regarded this area as cursed. A place of fire. Underfoot were the weathered trunks of shattered trees each one pointed to due south. Two days later, they reached the edge of the burned area. The burns were unusual in that the bark of the fallen trees had been scorched, but the underlying wood was untouched by fire. It was clear that no forest fire had ever broken out; rather, the trees seemed to have been seared by a brief but brilliant flame.

After he had climbed a low ridge, Kulik records in his diary

From our observation point, no sign of forest can be seen for everything has been devastated and burned. Around the edge of this dead area, the young 20-year-old forest growth had moved furiously seeking sunshine and life. One has an uncanny feeling

when one sees 20 to 30 foot giant trees snapped across like twigs and their tops hurled across many meters away to the south.

Readers can pursue for themselves the amazing difficulties that the expedition faced in their 1927 expedition. (At one point they even contemplated eating their horses.) Kulik also went back to the area in 1928, 1929, and finally in 1939. The area is so remote that the scientists cannot even agree on whether or not there is a crater to be found near the impact center. Scientists have puzzled over the evidence for a long time, but many are of the opinion that the Tunguska disaster was caused by what the astronomers call a *dead comet*, approximately 100 meters across. Just exactly what we mean by a dead comet as distinct from an asteroid will be discussed in greater detail in the next section, but for now, it is sufficient to note that an asteroid is dense and rocky whereas a dead comet has relatively low density and is materially weak. Although the two types of bodies have different densities and origins, they can follow similar orbits; however, the denser asteroids are more likely to reach the Earth's surface than the dead comets. As a dead comet heats up in the atmosphere due to air friction, it will eventually explode in the air causing great damage from the blast wave that it creates but not necessarily leaving a direct crater. During World War II, aerial warfare used to employ mines dropped by parachutes above a large city. These mines were detonated a few hundred feet off the ground. The blast wave devastated a large area without creating a large crater. In the macabre arithmetic of warfare, exploding mines in the air is a more efficient use of explosive energy than wasting the explosives in the excavation of earth to form a crater. The difference in behavior between a dead comet and an asteroid can be appreciated from the fact that the asteroid which created the Barringer Crater in Arizona was probably 150 meters across. It created a crater of 1.2 kilometers as compared to the 100 kilometers circle of devastation created by the Tunguska body.

In an attempt to gain more knowledge about the cosmic missiles hitting the Earth, scientists have begun a watch for such impacts. Specialists studying earthquakes use an instrument known as a *seismograph*. The name comes from the greek word "seismos" meaning "an earthquake". The basic principles employed in such an instrument are illustrated in Figure 8.8. In a simple seismograph, a massive object, such as a cylinder of steel, is hung from two strong, thin wires inside a frame. If an earthquake happens near the seismograph, the outside of the frame moves but there is nothing to exert a force on the suspended mass except the two thin wires. As a consequence, the frame moves whereas the suspended mass stays virtually in its original position. A long thin pen from the suspended mass touches a rotating cylinder and its position is marked on a rotating chart. When an earthquake moves the outside of the seismograph, the pen appears to move in the opposite direction because of the relative motion of the Earth with respect to the cylinder. The record of the earthquake drawn on the chart is known as a *seismogram*. Obviously the more violent the shaking of the outside of the container, the larger the movement of the pen. Geologists use what is known as the *Richter scale* to measure the amount of energy involved in an earthquake. This scale

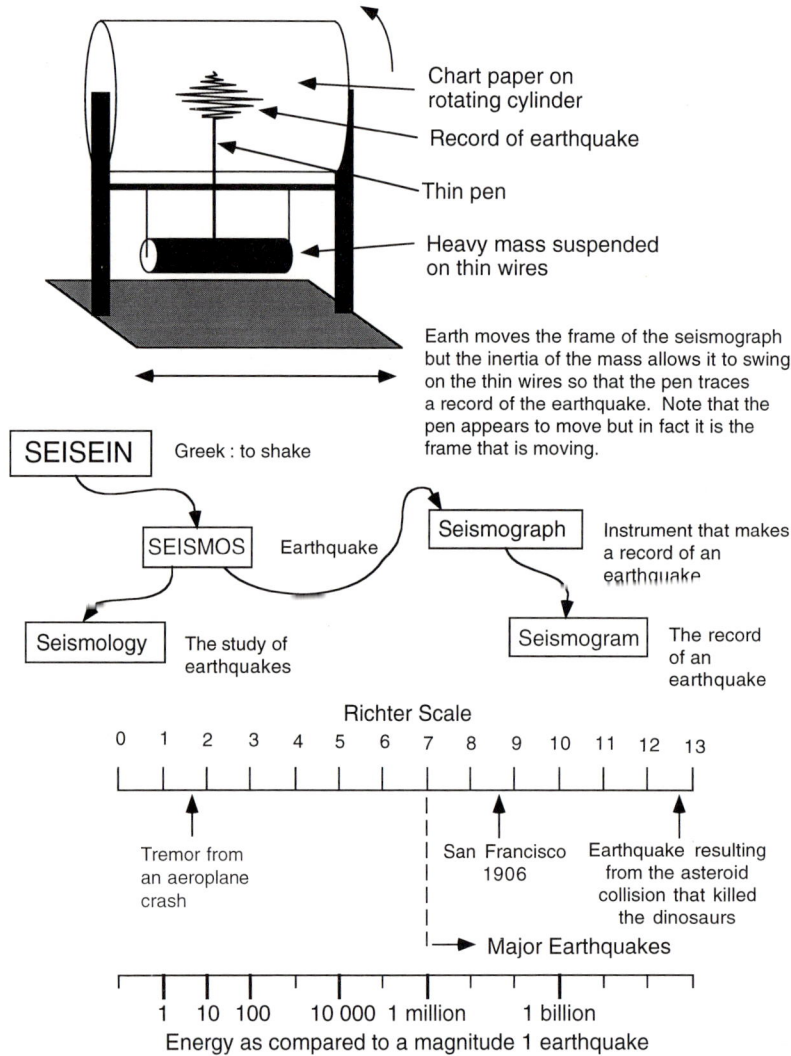

Figure 8.8 The scientific study of earthquakes known as seismology employs an instrument called a seismograph. This makes use of the inertia of a heavy mass on thin wires to create a graph of the shaking of the earth.

is named after its inventor Charles F. *Richter* and is a logarithmic scale as shown in Figure 8.8. On this scale, an increase of one unit represents a tenfold increase in energy. Thus an earthquake measuring 5 on the Richter scale has ten times more energy than an earthquake that registers 4. An earthquake of magnitude 6 has one hundred times more energy than one measuring 4, and so on. Geologists who study earthquakes consider anything above 7 on the Richter scale to be a major earthquake, although considerable damage can be caused by those of the order of 6. So it was on October 10, 1986, when San Salvador was struck by an earthquake that measured only 5.4 on the Richter scale, but mud slides from the shaken hills, saturated by violent rain storms, killed 1500 people.

Seismographs can register relatively small events; for instance, when terrorists blew up the Pan Am flight over Lockerby, Scotland in 1988, local seismographs registered the impact of the falling aircraft on the Earth as 1.6 on the Richter scale. By looking at the records from several seismographs, scientists can pinpoint the center of an event. British scientists would be alerted to the impact of a meteorite from seismographs in various universities. The shape of the wiggly line on the seismogram can reveal whether an event is due to the impact of a cosmic missile or to a small earthquake. The records can also indicate where to look for the cosmic missile. The scale shown in Figure 8.8 allows some physical appreciation of what is meant be an earthquake of a given magnitude.

Seismographs around the world recorded the Tunguska event. Geophysicist Ari Ben-Menahem of the Weizmann Institute in Israel has compared old seismograms of the Tunguska event with records of modern nuclear explosions in the air. From this comparison, he has deduced that the Tunguska body exploded with a force of 12.5 Megatons at an altitude of 8.5 kilometers. (The Megaton is a measure of explosive energy corresponding to the energy released by one million tons of tnt.) The energy of the atomic bomb that was exploded at Hiroshima in 1945 was the equivalent of 13 000 tons of tnt. The Tunguska explosion was therefore almost 1000 times more powerful than the Hiroshima blast. If a similar event to the Tunguska explosion were to happen today, John Pike of the Federation of American Scientists has calculated that such an explosion over a rural area of the United States would kill 68 000 people and cause $4.5 billion worth of damage.

Brian Marsden of the Harvard Smithsonian Center for Astrophysics has calculated that Earth/cosmic missile events as large as the Tunguska collision should occur at a frequency of between once a century and once in a thousand years. (This range is hard to narrow down further because of our lack of knowledge of the near-Earth objects in space.)

If events on the scale of the Tunguska collision have happened once a century, then records of such events may be part of the folklore of people around the world. Duncan Steel of the Anglo-Australian Observatory in Coonabarabran, New South Wales, Australia believes that there is evidence of a Tunguska-like explosion on New Zealand's South Island about 800 years ago. At the time widespread fires destroyed forests across the southern part of the island contributing to the extinction of the giant flightless bird, the Moa. According to Steel, the Maoris, the aboriginal inhabitants of New Zealand, have a myth describing the falling of the skies, raging of the winds, upheaval of the earth and mysterious devastating fires from space. Anthropologist Atholl Anderson has challenged this idea saying there is not reliable evidence for such an event. Steel suggests that the problem can be resolved by searching for carbon deposits from ancient fires which can be dated using modern methods to see when there were ancient fires in New Zealand. Steel continues to search the world's ancient records for evidence of cosmic collisions and he believes a similar event may have occurred above the Amazon jungle as recently as 1930 and also above New Guinea in the 13th century. In support of the latter event, Steel has discovered an Australian aboriginal story which tells of a huge explosion of a falling star near Wilcannia in the far west of New South Wales. Dave Rodey of the US Geological Survey in Flagstaff, Arizona, says that part of the problem is that, since the world is 70% covered by water, most of the events would not be recorded in local folklore. This is obviously an area of research where anthropologists and astronomers can cooperate in their search for records of past cosmic collisions [13–16].

Section 8.4 Hairy Stars and Telltale Tails

A reader dipping into this book might suspect from the title of this section that we are now going to discuss the lifestyle of a modern rock star dressed in animal skin clothes which include a fascinating tail. Our discussion, however, will be less sensational than the study of modern rock stars. It will concern itself with heavenly bodies that streak across the sky leaving behind trails as illustrated in Figure 8.9. The name for such bodies, comets, comes from the Greek word "kometes" meaning "long haired." A well-veloped comet close to the Sun has two tails as illustrated in Figure 8.9(c). The long thin tail labeled plasma tail consists of gas molecules which have been excited by energetic fragments of atoms moving from the Sun so that they give out light. The stream of energetic atomic fragments from the Sun which makes the gas molecules in the tail of the comet glow is known as the *solar wind*. The solar wind is not a flow of air like an earthly wind and so the name is slightly misleading. The actual center of the comet is usually a relatively small body known as the *nucleus of the comet*. The word nucleus in Latin means "a little nut." The solid body at the head of the comet is like a little nut that streaks across the skies buried in the trail of light and dust. The comet only becomes visible because of the light reflected off the dust-like debris which it leaves behind and the light given out by the gas molecules as they are bombarded by the solar wind. The comet is brightest when it is closest to the Sun, a position known as *perihelion*. When it is at the opposite end of its elliptical orbit, it is invisible because so little energy reaches it from the Sun that no material is released from the nucleus and there is no tail to the comet which would reflect light and make it visible. As the comet swings round and makes its way back to the Sun, there is a point at which the solar energy reaching the nucleus of the comet starts to generate gas and emit dust from the nucleus which then form the visible tails. The part of the tail which emits light due to the interaction of ist gas molecules with the solar wind is described by astronomers as the *plasma tail* or the *gas tail*. Although the glowing tails of comets look substantial,

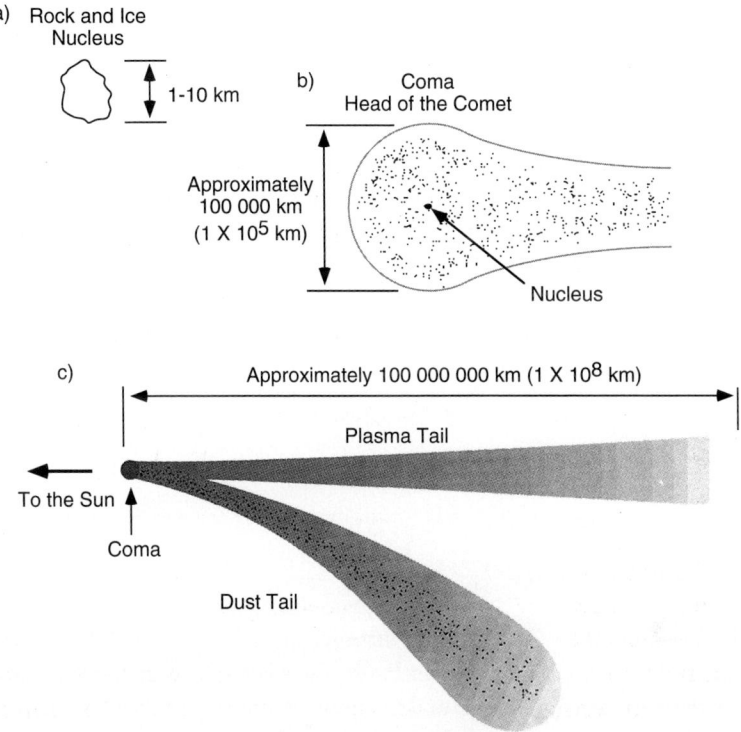

Figure 8.9 The trails streaming out from a comet give rise to its name since "kometes" in Greek meant, "long haired." The appearance of a comet in the sky was often interpreted by primitive people as a sign from angry gods that a disaster was imminent. a) The nucleus of a comet is a small body consisting of dust, rocks, and ice. b) The "head" of the comet, known as the coma, consists of gas and dust that has been melted from the surface of the nucleus by solar heating. c) Often a comet will display two distinct tails

they are still mainly empty space. They are only clearly visible because they are hundreds of thousands of kilometers long. The gas tail always streaks out into space in a direction opposite to the Sun as illustrated in Figure 8.9(c). The other part of the tail consisting of dust thrown off from the surface of the comet is made visible by reflected light from the Sun. Since the dust fine-particles have a measurable mass, they continue to travel around the orbit with the nucleus of the comet but lag behind the plasma tail as shown in Figures 8.9 and 8.10. The diffuse area that forms the head of a comet is technically known as the *coma*. This is easily remembered because of the resemblance between this word and the word comma, and the coincidence that the dusty tail of the nucleus of a comet looks somewhat like a comma.

Human beings have observed comets in the sky since the beginning of recorded history. Primitive people have often believed that the appearance of a comet indicated that the gods were about to devastate the Earth with plagues and other disasters. Thus there is an engraving representing the famous comet which we now know as Halley's comet

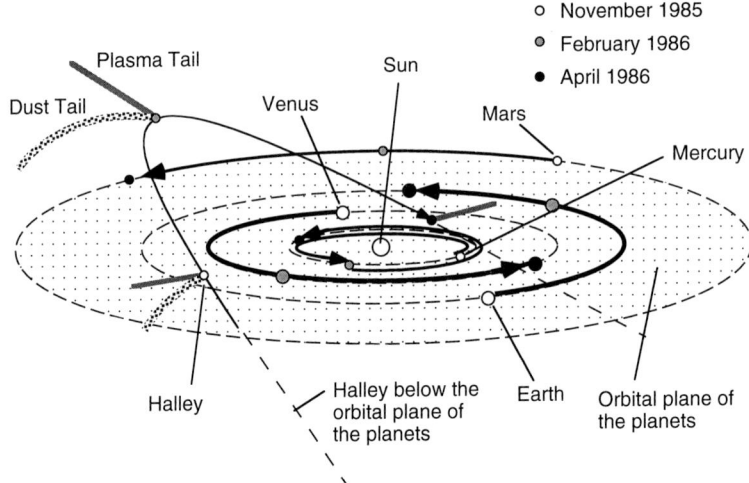

Figure 8.10 As a comet travels around the sun, its tail changes in size and direction. The plasma tail always points away from the Sun.

in a book known as the "Nuremburg Chronicle". This book, published in 1493, recounts how in AD 684, a year in which Halley's comet appeared, there had been disasters such as three months of rain, thunder, and lightning, during which people and animals died, grain withered in the fields, and the eclipse of the Sun and Moon was followed by a plague. Milton, in the poem, "Paradise Lost," published in 1667, had this to say about comets:

> *Incensed with indignations Satan stood unterrified, and like a comet burned the fires the length of Ophiucus, huge in the Arctic sky, and from his horrid hair shakes pestilence and war.*

Some artists have imagined that Halley's comet was the Star of Bethlehem heralding the birth of the Christ Child [17]. The Bayeux tapestry which was woven to commemorate the invasion of England in 1066 by William the Conqueror includes a stylized representation of Halley's comet. Apparently, King Harold of England, when he saw the comet in the sky, believed that it was an omen of his defeat. The word *omen* means "a sign of some future event" and comes from the Latin word for "a sign" [18, 19].

Astronomers over the centuries have recorded the appearance of approximately 600 comets, but only a few of them have been as spectacular as Halley's comet. Until 1577, the movement of comets across the sky was considered to be an event in the atmosphere of the Earth, but in that year the great Danish astronomer Tycho *Brahe* proved that the comet was beyond the Earth's atmosphere and out in interplanetary space. His discovery of the orbital motion of comets was a further blow to the medieval idea that planets were embedded in crystal spheres that moved around each other to create the observed motion of the planets.

Medieval engineers were aware that man-made machinery creaked and groaned because of the lack of smooth motion and lubrication. Poets, however, refused to believe that the *crystal spheres* carrying the stars would screech and groan as they were moved over each other by angelic propulsion. Poets therefore assumed that the crystal spheres carrying the planets made music instead of the squeaky noises given out by earthly machinery. Thus the poet John Milton has the line in his poetry

Ring out ye crystal spheres.

Poets are not always quick to adopt the new views of science and a well-known hymn written in the middle of the last century, that starts off with "This is my father's world," contains the line,

All around me is the music of the spheres.

Halley's comet is named after the English astronomer Edmond *Halley* (1656-1742). Halley was the first astronomer to travel into the southern hemisphere to look at the stars visible in the southern sky. He established an observatory on St. Helena. (This island in the south Atlantic later became famous as the last home of Napoleon Bonaparte.) Later in life, Halley was appointed professor of geometry at Oxford (1703). He worked with his close friend Isaac Newton on the mathematics of the paths of the various comets that had been studied for several hundreds of years. One of the comets he studied appeared in 1682. In 1705, after he had listed the movements of some two dozen comets, he suddenly realized that the comets of 1456, 1531 and 1607 had paths very similar to each other and to the 1682 comet that he had viewed. He realized that all of these comets were in fact the same comet, returning every 76 years and that this comet followed a very elongated orbit around the Sun. He realized that it was visible only when close to the Earth and that between appearances it must travel away beyond Saturn, the most distant planet known to the astronomers of the time, before turning back to reappear 76 years later. Halley wrote a book in 1705 in which he predicted that this comet would return about 1758. (Halley recognized that the gravitational interference from the planets of the solar system would change the orbit slightly and the time of appearance of the comet would vary slightly from trip to trip.) Halley died before the predicted return of the comet, but other astronomers excitedly awaited its reappearance which was essentially on time. It returned again in 1835 and in 1910 and 1986.

The appearance of Halley's comet in 1910 was perhaps the most brilliant display of the comet throughout recorded history because this time the Earth actually passed through the tail of Halley's comet as it sped around the Sun. When astronomers announced to the public that in 1910 the Earth would pass through the 50 million mile long tail, some people panicked even though the astronomers told them that the tail consisted, as we have already noted, mostly of empty space. Expecting that there would be deadly gases in the tail, some people sealed their windows in fear of the comet's effect. Needless to say, the only real effect of passing through the tail was that on some nights at the time, there were spectacular displays of meteorites in the sky as the dust from the comet's tail became shooting stars in the Earth's atmosphere.

Beginning in the early 1980s, scientists anticipating the return of Halley's comet in 1986, planned a series of experiments in which they sent probes to explore the dusty tail of Halley's comet [20]. Somewhat earlier than 1980, scientists had been very actively trying to assess how much dust there was in the space between the Earth and the Moon. They recognized that one of the major dangers facing astronauts in space was the high-speed sandblasting of their space capsules by dust traveling at high speed. In a very minor way, I was involved in planning possible experiments to be placed upon capsules traveling to the Moon to measure the density and energy of cosmic dust bombarding the space module. The group I was working with at the Illinois Institute of Technology put forward a proposal to NASA for looking at the dust captured on the surface of the spacecraft using television cameras, but the experiments were not very feasible at the time. The early studies of cosmic dust were made by attaching microphones to certain parts of the spacecraft to listen to the impact of the dust fine-particles. By measuring the energy of the audio signal generated by such impacts, the size and/or energy of the impinging dust fine-particles could be deduced.

One of the major probes sent to Halley's comet, a spacecraft called Giotto, used a system originally developed to protect spacecraft from high-speed cosmic dust to study the energy and size of the dust in the comet's tail. This robotic explorer of Halley's tail was named after the Italian artist *Giotto de Bondone* (1267–1337) who painted what we know as Halley's comet into a picture showing the adoration of the Christ Child [17].

To be able to understand the design of the dust monitor mounted on the Giotto probe, it is helpful to discuss the work of Oort and Whipple, two astronomers who made suggestions about the structure of comets. In 1950, the Dutch astronomer *Jan Oort* analyzed the paths of several comets and deduced that they had all originated far beyond the orbit of Pluto, the outermost planet. He calculated that there could be as many as 100 million comets in a spherical shell surrounding, but trillions of miles away from, the Sun (approximately 1 light year away).

> *Out there, 1000 times further from the Sun than Pluto, the comets wait in a cosmic deepfreeze until the gravitational nudge of an occasional passing star sends them on a 2 million year journey in toward the Sun.*

The above quotation is from an article written by Dennis Overbye [20], who goes on to say most of these comets never make it back to the spherical shell of distant comets, now known as the *Oort Cloud*. Many of them become trapped within the planetary zone of the solar system and some eventually collide with planets or the Sun.

Thus on July 17 1951, A year after Oort's discovery, *Fred Whipple*, a well-known Harvard astronomer, realized that if comets did indeed come from the Oort Cloud, then they would be essentially large dirty snowballs made up of clumps of gas and dust left over from the formation of the solar system. It is believed that the solar system was

8.4 Hairy Stars and Telltale Tails

formed out of a cloud of interstellar gas and dust which, under the influence of the mutual gravitational attraction of the myriad of grains, curdled into the planets of the solar system. Within the solar system, the original dust present at the formation of the solar system was swept clean by the solar wind, but the dust spilled by the orbiting comets continues to populate interplanetary space with a sparse distribution of comet debris. Whipple's idea that comets were dirty snowballs explained one of their puzzling habits. They often arrive slightly ahead of or behind schedule. Whipple suggested that this was because gases, ejected from holes in the comet's dusty crust, act as propelling jets accelerating or slowing the comet depending on their direction.

Astronomers have calculated that a comet loses about 1% of its mass each time it travels around the Sun. Eventually a stage is reached where there is no more easily evaporated material on the comet's nucleus. In its further travels around the Sun it is no longer visible and behaves like a lump of dead material which the astronomers describe as a *dead comet*. However, as we will discuss later in this section, there is some suggestion that the dead comet nucleus can split up [21]. Some astronomers believe that a fragment of a dead comet that had been previously observed was responsible for the Tunguska event.

As we have already mentioned, NASA was well aware of the fact that the spacecraft they would launch toward the Moon and other parts of the planetary system would be bombarded by cosmic dust traveling at very high speeds. In their planning, the scientists estimated the dust that would be encountered by spacecraft would be traveling at speeds of up to 20 kilometers per second. We have already discussed that when pounding a nail with a hammer, some of the kinetic energy of the hammer is transformed into heat, which warms up the nail and the hammerhead. If a piece of dust traveling at 20 kilometers per second were to be abruptly stopped, the transformation of the kinetic energy into thermal energy would be sufficient to vaporize the dust fine-particle [22]. Early in the development of space technology, F.L. Whipple suggested that spacecraft could be protected from high-speed dust by a foil bumper put in front of the spacecraft. Whipple's idea was that an impacting micrometeorite would vaporize as it hit the outer foil of the bumper and that the melted debris from the micrometeorite would condense in the cold space between the outer foil and another protective foil directly above the spacecraft shell. This system became known as the *Whipple Meteor Bumper*.

A group of workers at the Space Sciences Laboratory of the University of Kent, Great Britain, J.A.M. McDonnell, W.C. Carey, and D.J. Dixon took the basic idea of the Whipple Meteor Bumper and turned it into a device for studying cosmic dust in interplanetary space. The basic concept employed in this device is illustrated in Figure 8.11(a). The first version of the instrument placed on the Columbia space shuttle had a surface area of approximately 1 square meter of 5 micrometers thick aluminum foil bonded to a gold-coated brass support mesh which in turn was bonded to a sheet of Kapton (a plastic with a metallic appearance chosen for its good thermal properties) as illustrated in Figure 8.11(a). This device for studying cosmic dust was described as the *microabrasion foil experiment* and referred to as *MFE* equipment. The sequence of events

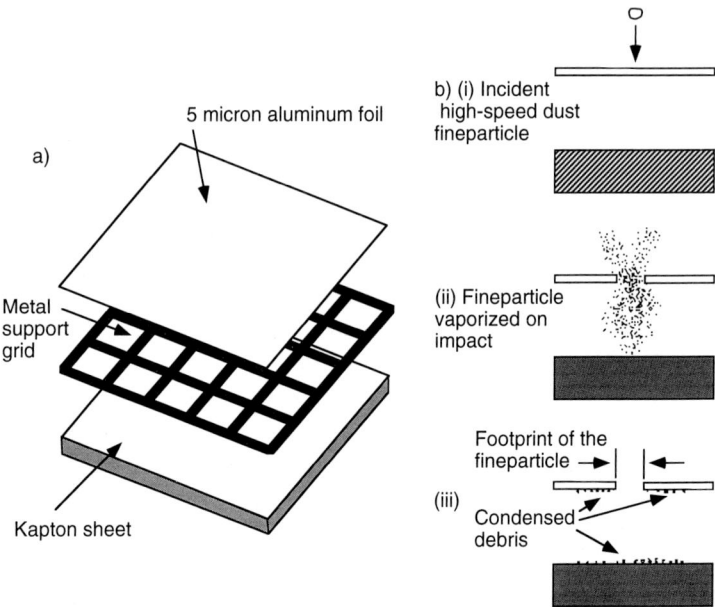

Figure 8.11 Scientists at the University of Kent, Canterbury, England, designed what is known as the "microabrasion foil experiment" for studying cosmic dust in interplanetary space. a) Sandwich construction of the microabrasion foil experiment. b) Impact sequence of an encounter of a grain of cosmic dust with the foil.

which happens when a cosmic dust fine-particle hits the foil is illustrated in Figure 8.11(b). (It must be remembered that although the conversion of kinetic energy to thermal energy results in the vaporization of the dust fine-particle, the encounter and conversion takes place in the chill of outer space. Thus the vaporized material is immediately condensed and frozen onto the surrounding surfaces after it has penetrated the upper foil.) The engineers designing this system faced an interesting dilemma. If they made the foil too thick, the majority of the smaller dust fine-particles would not be able to penetrate the upper surface of the foil. It is not possible, however, to examine one square meter for every microscopic crater which did not go through the foil. But, in a thinner foil, the cosmic dust would go through the material and then the holes in the foil could be found using an optical method to generate pinpoints of light to guide the scientist to the locations of the events that occurred on the foil. The 5 micrometer foil was selected on the basis of calculations indicating the likely range of energetic cosmic dust that would be encountered.

The first MFE system for studying cosmic dust was sent into space on the space shuttle Columbia in March, 1982. It was exposed to cosmic dust for 8 days before returning to the Earth. The hole punched in the foil is described as the *footprint of a dust fine-particle*. Scientists were aware that the footprint of the dust fine-particle would be much bigger than the actual dust fine-particle. The only way that they could relate the hole to the impinging fine-particle was to carry out experiments on Earth using specially developed guns to fire tiny missiles at typical sheets of aluminum foil. In experiments at the University of Kent in Canterbury, investigators built machines capable of firing iron dust, consisting of fine-particles 1 micrometer in size, at velocities of over 20 kilometers per second. Other scientists in the United States have used compressed gas in special guns to carry out similar experiments for calibrating footprints in the foil of cosmic dust collectors. The scientists who developed the equipment for studying dust in Halley's comet realized that the same dust would probably destroy the robotic probe as it moved through the comet's tail. Using the known information on the speed of Halley's comet and the speed that the robotic probe would have as it entered the comet tail, scientists calculated that the Giotto probe would meet the dust at 77 kilometers per second (50 times faster than the fastest bullets used in modern guns). The Giotto probe was planned to travel within 1000 kilometers of the comet's nucleus. Even though it was protected by a dust shield, the scientists expected Giotto to be able to send back data for only about 4 hours until, in the words of Overbye [20],

It was sandblasted into silence.

By early 1991, the preliminary results of the 1986 Giotto probe and other probes were beginning to become available in the scientific literature. Using the information from the various space probes, scientists have deduced that the nucleus of Halley's comet is roughly 16 kilometers long and 8 kilometers wide, and that it has a low density. Since the deduced density of the agglomerated ice and dust is approximately one-fifth the density of water, it must be porous. The measurements also indicate that Halley's comet is losing 100 million tons of material in each orbit. Since, however, it has also been calculated to have a mass of 100 billion tons, Halley's comet will appear another several hundred times as a visible comet before it degenerates into a dark "dead comet" moving around the Sun. Giotto actually traveled to within 600 kilometers of Halley's nucleus before it was hit by an energetic dust fine-particle. The "wipeout" collision came two seconds before the probe reached its closest point to the comet nucleus. Scientists were surprised and pleased when, twenty-one seconds after the probe was hit, it began to receive and send signals to Earth again. From these signals, scientists were able to discover that 600 grams of the spacecraft were missing and that half of the instruments were still functioning. When Giotto was launched, this was done so accurately that it still had energy in its maneuvering jets. In August, 1991, it was announced that Giotto would be given a final task of inspecting another comet known as Grigg-Skjellerup. After this last task, Giotto will be steered into a final resting place — a stable solar orbit [19–23].

Halley's comet is now on its journey into outer space and will only turn back toward Earth in the year 2024. By early 1991, scientists expected that Halley's comet would be only a faint speck at the limit of visibility of their telescopes. However on the 12th of February, 1991, two scientists checking up on Halley's comet were surprised to find it in their pictures as a large bright patch of light. From their calculations of the brightness observed over the next five nights, they deduced that Halley's comet had suddenly been surrounded by a cloud of dust stretching more than 300 000 kilometers in space. An astronomer of the University of Sheffield, David Hughes, has calculated that a cloud of this size would have been created in outer space if Halley's comet had been hit by a small meteorite weighing 1 gram. Therefore, as it moves into outer space, the possibility exists that Halley's comet has itself been a victim of a cosmic missile attack and that when it returns in the year 2062, its appearance may surprise the astronomers of that time [24].

Section 8.5 Apprehending Cosmic Drifters

Donald *Brownlee*, in an article on cosmic dust [25], gives us the following information:

> Small cosmic particles, as fine as dust, are actually very common although invisible to the unaided eye. Each year more than ten thousand tons of extra-terrestrial particles smaller than one millimeter in diameter enter the atmosphere and fall to the surface of the Earth.

Later in the same article he tells us,

> Most of the particles larger than one hundred microns do not slow down until they reach high enough air densities to cause melting by friction. (A typical shooting star is only the size of a grain of sand as it begins its brilliant path to the Earth.)

Brownlee points out that if you spend an appreciable amount of time outdoors you will be hit by several miniature cosmic particles a week. However, with speeds of descent of one centimeter per second and masses of a billionth of a gram, impacts are less than noticeable! These particles are in the air we breathe, the food we eat, and the water we drink.

Some of the dust fine-particles are fragments of larger objects that broke up while entering the Earth's atmosphere, but the majority of the cosmic dust falling to the Earth is thought to be brought into the inner part of the solar system from the Oort Cloud by comets. Some of the first studies of cosmic dust hitting the Earth were carried out by British scientists traveling on the survey ship Challenger in 1870. These scientist dredged up from the ocean floor some microscopic spheres which they collected by dragging a magnet through the sediment. Recently scientists at the University of Washington carried out a similar experiment. They employed a piece of equipment they called a "cosmic muck rake" which was a sled with magnetic collectors. When this rake was dragged through water at depths of about five kilometers, it picked up tens of thousands of magnetic spheres per day.

8.5 Apprehending Cosmic Drifters

It is believed that many of these magnetic spheres found in the ocean sediments are actually the fragments created by colliding asteroids in the asteroid belt which are then pulled down to Earth by gravity. In earlier studies of cosmic dust, the astronomer Whipple predicted that some cosmic dust smaller than one hundred micrometers in size should, because of their very small size, reach the Earth's surface without melting. In fact, they enter the atmosphere at about 11 km per second but, because of their small size and relatively large surface area, they radiate away heat without reaching their melting point. Dr. Brownlee mounted a major experimental investigation into the capture of these cosmic drifters as they flutter down through the Earth's atmosphere. In his article on cosmic dust [25], Dr. Brownlee says

> *On average a square meter of New York will collect one cosmic dust fine-particle every day, but it will also collect up to ten billion other particles of similar size from air pollution. In an optical microscope the cosmic particles are often indistinguishable from terrestrial particles.*

Rather than attempt to differentiate between the cosmic dust and the other more earthly dust, Dr. Brownlee attempted to intercept these drifters high in the upper atmosphere before they mingled with their more common cousins. In his first experiment, carried out in the early 1960s, Brownlee hung a ten-foot-long rotating arm with a collector on the end beneath a high altitude balloon. This rotating rod failed to capture any cosmic dust. Brownlee then devised a piece of equipment he called the *Vacuum Monster*. This instrument was a large pump slung underneath a balloon and powered by 150 pounds of rocket fuel. It sucked air at half the speed of sound through a five inch opening crisscrossed by cigarette-sized capture rods. In its first flight, the Vacuum Monster collected one fine-particle that was different from all the others and had the structure anticipated for cosmic fine-particles. On its second mission, the Vacuum Monster collected nine cosmic fine-particles. The third mission had to be canceled because the balloon leaked. The fourth time out, the Monster's parachute failed, causing it to self-destruct at the end of its fall [25].

In 1971 a more extensive program to collect cosmic dust was begun using U2 aircraft [26]. U2 aircraft were designed to fly at very high altitudes to spy on military installations in what was then Soviet Russia. These aircraft became less vital with the development of observation satellites. At about the time, NASA began to worry heed public concern over the number of fine-particles being left in the atmosphere by rocket fuels and disintegrating components of space shot equipment. Therefore, NASA undertook a project to study stratospheric aerosols using a U2 nicknamed *Dragon Lady* in which the U2 aircraft was flown high in the air carrying sampling equipment to look for cosmic dust. This sampling equipment consisted of small glass plates coated with a sticky substance mounted on pylons bolted to the tips of the wings. During the ascent through the lower atmosphere these remained folded into airtight compartments to prevent them from being contaminated by atmospheric fine-particles, such as the smoke associated with atmospheric pollution. It is estimated that between 1974 and

1983, the high flying aircraft captured three thousand cosmic dust fine-particles. A typical cosmic drifter captured by the Dragon Lady experiment is shown in Figure 8.12. As mentioned earlier, Brownlee and many other astronomers believe that such cosmic drifters represent virgin dust from the cloud from which the solar system condensed. Such dust particles are representative of the cosmic dust filling interstellar space and the raw materials from which sprung the stars spread out in the galaxies [27]. It will be noted that the cosmic drifter of Figure 8.12 is itself an aggregate of many small fragments, each possibly an individual grain of interstellar dust.

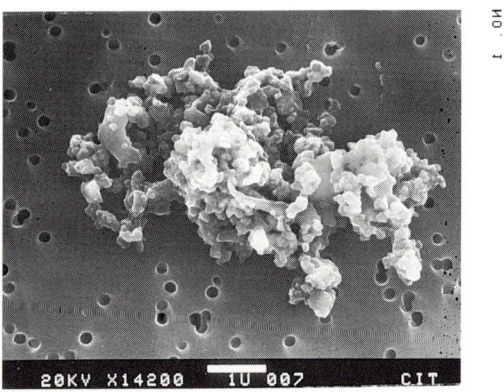

Figure 8.12 Scientists are studying cosmic dust hitting the Earth's surface to gain information about the birth of the solar system. A cosmic drifter captured by Dr. Brownlee in the stratosphere of the Earth. (Photos courtesy of Dr. Donald Brownlee, Department of Astronomy, University of Washington.)

Some scientists believe that life itself began on the Earth when Brownlee-like drifters brought organic chemicals from outer space down through the Earth's atmosphere. It has been estimated that in the early days of the Earth's evolution, before the atmosphere built up, 60 000 tons of organic material fell to the Earth from space each year. Some scientists have described such organic material as a *molecular drizzle* of organic molecules bringing life to the Earth.

Some types of cosmic missiles are found to contain what are obviously condensed droplets of molten rock. This indicates that at some time in their history, before they became incorporated into an asteroid as it is built up from the colliding dust particles of outer space, they existed at temperatures above 1200 °C. This type of cosmic missile is called a *chondrite*. This name comes from the tiny droplets of rock in the chondrite which look like seeds in a fruit. "Chondros" in Greek means "seed." The tiny droplets themselves are called *chondrules* [26]. The scientist H.C. Sorby called the chondrules "drops of fiery rain." The scientific study of chondrules is still a very active area of research for geologists and astronomers. Parkin has shown that fine-particles resembling the iron cosmic spherules can be reproduced by quenching the sparks of material generated by eroding iron with a grinding wheel if the sparks are quenched close to the wheel. Some very beautiful photographs of the different types of cosmic spherules found in ocean sediment are presented in a scientific paper by Parkin and colleagues [28].

Section 8.6   Cosmic Missiles and the Disappearing Dinosaurs

One of the largest animals that has ever lived was the dinosaur known as Tyrannosaurus rex. In the words of Asimov [29],

> This creature grew to as much as 45 feet in length and stood as high as a giraffe on tremendous hind legs with a skull 4 feet long and cavernous jaw equipped with foot long teeth. It was the largest meat eater ever to inhabit dry land; its name is very descriptive.

We can all probably heave a sigh of relief that such creatures no longer roam the jungles of the world. Indeed we only know of the former existence of such creatures from the fact that geologists have discovered the bones and footprints of such creatures in ancient rocks. As the geologists began to unravel the story of how the Earth used to be populated by various creatures — now extinct — they discovered a mystery. Dinosaurs suddenly disappeared from the history of the various layers of rock making up the crust of the Earth. This sudden demise of the dinosaur and its relatives remained a mystery until the early 1980s when it was suggested that these animals were wiped out by catastrophic events connected with the collision of a large asteroid with the Earth.

To be able to understand the details of the dinosaur mystery and other similar catastrophes we need to understand the basic history of the rocks of the Earth's crust. The specialist who studies the structure of the Earth's crust is the *geologist*, literally a person who writes about the Earth. As the early geologists dug down into the Earth they discovered rock structures that looked like strange animal bones. These were called *fossils*. For a long time people thought these fossils were rocks that accidentally resembled living things or perhaps were the remnants of animals drowned in Noah's flood. In 1791 the English land surveyor William Smith demonstrated that different rock layers contained different types of fossils and that a given layer could be traced across broken ground by following the location of its characteristic fossils. In 1796 the French scientist George *Cuvier* studied the structure of the fossils and pointed out that they could have functioned as living animals but that many of the fossils represented animals completely different from any existing at that time. Gradually, over the last two centuries, geologists have constructed a time map of the rocks present in the Earth's crust as shown in Figure 8.13. Scientists estimate that the Earth is 4.5 billion years old. They think that for the first billion years of its existence the surface of the Earth was a cauldron of chemicals and erupting lava and that living creatures did not begin to exist until approximately 3.5 billion years ago. The names adopted for various periods of time in the Earth's his-

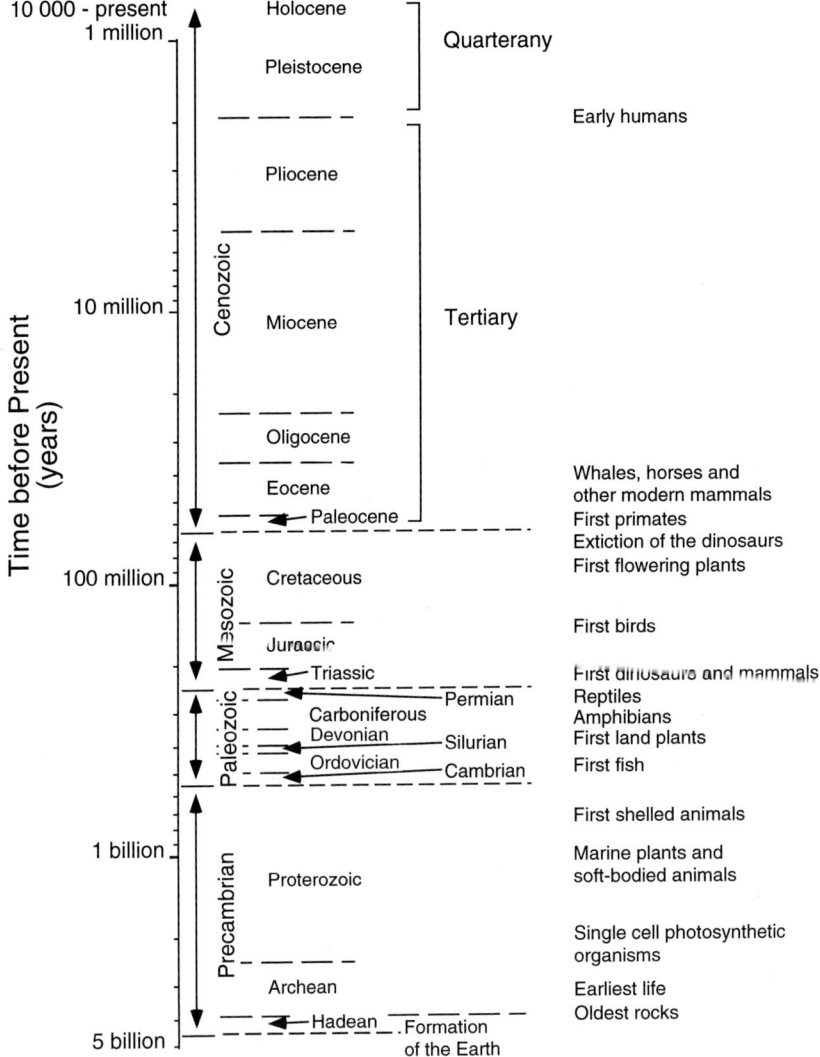

Figure 8.13 The chronology of the rock history of the Earth's crust is divided into several distinct time periods.

tory are illustrated in Figure 8.13. Moving backwards from recent times, the first three major time groups were called *Cenozoic*, *Mesozoic* and *Paleozoic*. The ending, zoic, comes from the Greek word "zoon" for "animal,", and the front parts of the three time zones come from the Greek: "kainos" meaning "new," "mesos" meaning "middle," and "palaios" meaning "old." Each time zone is split into what is known as an era and the Cenozoic has two eras, the Quarternary and the Tertiary which mean fourth and third. Each of these eras is split into epochs. For example, the Quaternary era of the Cenozoic

8.6 Cosmic Missiles and the Disappearing Dinosaurs

period is split into the *Halocene* and *Pleistocene* epochs. The length of the epochs is shown in Figure 8.13. The ending, cene, also comes from the Greek word "kainos". The various epoch names are created from the Greek words: "eos" meaning "dawn," "oligos" meaning "few," "meion" meaning "less," "pleion" meaning "more," "pleistos" meaning "most," and "holos" meaning "whole." Thus the epoch name referred to the number of animals that were found in the respective fossil records.

When we come to look at the names for the periods in the Mesozoic era, the term *Cretaceous* takes its name from the Latin word "creta" for "chalk," indicating that this was the time in the history of the Earth when the great chalk beds were created from the skeletons of tiny sea animals. The Jurassic period takes its name from the fact that rocks of this age were first studied in layers of rock exposed in the Jura mountains in Switzerland and Triassic simply means the third period in the Mesozoic rocks. When we move into the Paleozoic era, the term Permian is named after the region of Perm in the Ural Mountains. The term Devonian is named after rocks found in the county of Devon in England. The terms Silurian, and Ordovician come from the names of ancient tribes that used to live in Wales and the third name, Cambrian, comes from the older name for Wales itself.

It is difficult to gain a physical appreciation of such enormous time periods. It is helpful if one attempts to comprehend the very short period of time in the history of the Earth that is occupied by the last ten thousand years. If we represent the entire existence of the planet Earth on a timescale covering one calendar year, we would find that on January 2nd the crust of the Earth would form from the molten protomass. On March 6th the first living creatures would appear on the surface of the Earth. During October, plants and animals of various kinds would develop. In November, the coal supply of the Earth would form. On December 1st, dinosaurs were abundant but disappeared on December 20th. On December 31st, the human species would have appeared. At 11:59 and 30 seconds on the last day of the year, the Roman Empire would be at its peak. North America was discovered by Columbus with 8 seconds left on the clock and the United States of America became independent from England with less than a second to go.

As the geologists increased their knowledge of the different types of fossils that occurred in the various levels of rocks, it became apparent that what had previously been assumed to be a gradual evolution of the various forms of life from tiny sea creatures to human beings had been punctuated by "murderous" episodes in which many different types of life were wiped out off the surface of the Earth, never to reappear again. An episodal map of the intermittent disasters interrupting evolution is shown in Figure 8.14. The most famous of these episodes is known to geologists as the *K-T disaster*. (This disaster occurred at the boundary between the Cretaceous and the Tertiary periods of the geological time map.) The source of such interruptions to the evolution of life remained a mystery until a startling discovery was made by a group of scientists led by Louis Alvarez (1911–1986) who was awarded the Nobel Prize for physics in 1968. Louis Alvarez was the scientist who designed a method for sending X-rays through the Great

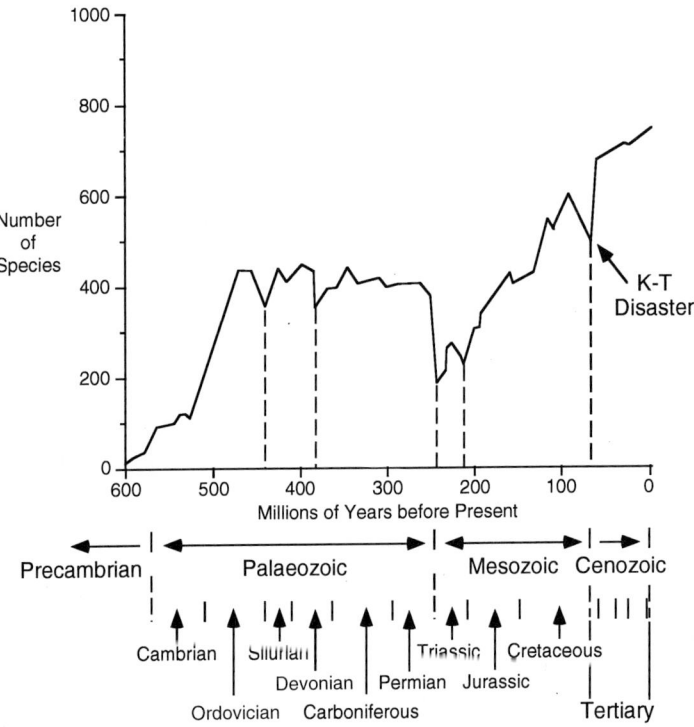

Figure 8.14 As our knowledge of the previous inhabitants of the earth grew, scientists began to suspect that murderous episodes occurred in the evolution of life. In one of them, called the K-T disaster, the dinosaurs disappeared.

Pyramid in Egypt to see if any undiscovered chambers existed within it. After he retired from his post as Professor of Physics at the University of California he worked with his son, Walter Alvarez, who was a Professor in the Department of Geology and Geophysics at the University of California. Louis and Walter Alvarez, along with some other colleagues, made a study of some of the layers of rock exposed in a deep gorge in Gubbio, a small city in central Italy. In particular, the Alvarez team was interested in a layer of red clay that separated two deposits of chalk at the K-T boundary of the rock strata. When they measured how much iridium (a rare precious metal) was in the clay, they were surprised to find thirty times the anticipated levels. The Alvarez team knew that much more iridium was present in the rocks of meteorites than in the Earth's surface. Therefore the discovery of this mysterious layer of iridium-rich clay led them to the idea that what they had discovered was the result of a catastrophe on Earth caused by the impact of an asteroid, ten kilometers in diameter. They estimated that the asteroid must have hit the Earth at 72000 kph. They hypothesized that such a collision would send tons of rock into the atmosphere, blocking out the Sun for months or years and inhibiting photosynthesis. Although this original hypothesis is considered to be an oversimplification, many scientists believe that this catastrophe theory for the extinction of the dinosaurs is essentially correct. Some scientists point out that the collision would create tremendous fires all over the surface of the Earth and that the smoke from such fires would be swept into the stratosphere by the high winds associated with the burning of the forests. This super soot cloud would be the actual trigger for the devastation of the Earth's climate. Scientists have found soot layers associated with the iridium-rich layer of clay at various sites around the world.

When scientists started to contemplate what the effects of a massive missile colliding with the Earth would be on the long-term climate, they also began to worry about what would happen if nuclear warfare broke out on Earth. Up until the 1980s it had always been assumed that the major effect of nuclear war would be the actual devastation from the explosion of the nuclear weapons. But now scientists realize that the soot from the massive fires triggered by nuclear warfare would have a long-term effect by cooling the climate of the Earth for two to three years. This would cause famine and devastation to many parts of the world. Scientists studying such an aftermath of nuclear war described the long-term effects of such devastation as a *nuclear winter*. It should be noted that forest fires can create fierce convection currents that send dust up into the stratosphere which, when it drifts across the sky, can cause the Moon to look blue because of the way in which the dust scatters light. This has led to the saying in everyday English that rare events happen "once in a blue Moon."

As the idea that the extinction of the dinosaurs was due to a cosmic collision began to stimulate scientific thought, scientists began to look for craters on the Earth's surface of the right age and magnitude to have triggered the death of the dinosaurs. Some scientists have suggested that Iceland, which is a volcanic island in the North Sea, was produced by a cosmic collision with what was then the ocean. The collision, they speculate, would open up a pathway for volcanic lava to form the island. However, the general consensus of opinion by early 1992 was that the dinosaur catastrophe was caused by one or two blows from asteroids or comets, with the largest hitting an area in what is now the *Yucatan* in Mexico, as shown in Figure 8.15, and a subsidiary collision Iowa, USA. Scientists of the University of Arizona have concluded that the comet or asteroid hit the edge of an ocean basin and that the impact on the water produced monstrous waves as high as five kilometers that raced across the sea tearing up the bottom sediments and sweeping rocky debris inland. These scientists searched through the scientific literature and uncovered reports of chaotic mixes of large rocks at the 65 million year boundary level in Texas, Mexico, Cuba and northern South America [30-32].

Other scientists reject the comet/asteroid theory for the death of the dinosaurs altogether and suggest that the whole climatic shift that killed the dinosaurs was a result of volcanic action that threw large quantities of dust into the atmosphere. It is well known that volcanic dust can modify the Earth's climate, but many geologists feel that this is not an adequate explanation for the disaster that wiped out the dinosaurs. W. Glen, in an excellent review article [32] of the various theories about the death of the dinosaurs points out that the impact of a ten kilometer asteroid with the Earth would produce an earthquake of about magnitude 13 on the Richter scale (see discussion of this scale in Section 8.3 [33]). Many scientists think that such an earthquake would have caused the Earth to vibrate like a large bell setting off every sleeping volcano on the globe. Glen also points out that it is unlikely that the discussion about the extinction of the dinosaurs will be settled in the near future. Indeed, the stimulus given to science by the Alvarez hypothesis has produced extremely fruitful research in many different areas of science. David Raup and John Sepkoski of the University of Chicago suggest that the extinction of life by catastrophe is a periodic event that occurs once every 26 million years [34]. Some scientists suggest that the crater in Manicouagan, Québec, shown in Figure 8.7, is the result of the impact that caused the most severe extinction in history; others dispute this claim. In case there is indeed a catastrophe every 26 million years in which the Earth is nearly destroyed, scientists are now looking for possible causes for such periodic disasters. One hypothesis that has been put forward is that the Sun has a small companion star which is known as Nemesis. (In Greek mythology the name *Nemesis* was given to the goddess of retribution. Nemesis was originally thought to be the goddess who watched out for anyone who became too prosperous. When anyone became proud and insolent she would arrange for bad fortune to even things out. In our language, Nemesis has come to mean an unavoidable turn of fate.) Some astronomers suggest that if the star Nemesis exists, then every 26 million years, as it passes through the Oort cloud, it dislodges many comets and sends them on a collision course with the Earth. If such a star exists, it will be very difficult to detect; at its closest to the Earth, it would still be one hundred times too dim to be seen by the naked eye [33].

Other astronomers do not believe in the existence of Nemesis. They suggest that as the Sun rotates about the center of the galaxy, it bobs in and out of clouds of dust and that every 26 million years, it encounters dense clouds of dust. This, they claim, would produce the same effect as the passage of Nemesis through the Oort cloud. Periodic disasters could prove to be a feature of the Earth's history, but it has been calculated that the next massive killer comet attack, of the type that wiped out the dinosaurs, is not expected to occur for another 13 million years. These long periods of time make disaster tracking difficult at best. Recently, however, a comet 10 kilometers in diameter called Swift-Tuttle has been estimated to have a chance of between 1-in-400 and 1-in-10 000 of hitting the Earth in August of the year 2126. Scientists suddenly have a potential disaster to contemplate [35].

Although a major attack of the type that wiped out the dinosaurs is not an immediate threat to the Earth, scientists have calculated that we could be attacked by comets of the size of the Tunguska event once every century. Some scientists have therefore urged that the Earth should be ready to deflect any asteroid that appears to be on a collision course with the Earth by sending an atomic warhead to explode near the approaching asteroid and deflect it away from the impending impact [35]. Others have suggested schemes for altering the Earth's climate and/or exploiting the rich mineral wealth of asteroids. Thus in 1971, Samuel Herrick, an authority on celestial mechanics, proposed that a portion of *Geographos*, a large asteroid, be explosively cleaved and, with rocket

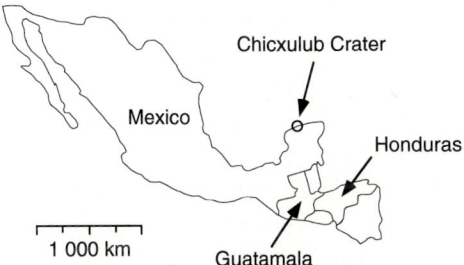

Figure 8.15 A comet or asteroid that could have caused the death of the dinosaurs may have hit what is modern day Mexico.

propulsion, gently set on a course to strike Earth on August 25th, 1994. It was to be aimed at the North Western Columbia River to form an interocean crater canal. It was estimated by Herrick that in this collision, the asteroid would bring to Earth more than 900 billion dollars worth of nickel and the rare elements osmium, iridium, platinum, and gold. Herrick suggested that we should do this because, in any case, Geographos if left to itself might strike the Earth by the year 2000. Other scientists feel that this type of suggestion has no merit. Such action is unlikely to be taken because the resources of the Earth can be better employed in other projects.

Section 8.7 Moon Struck

The picture of the surface of the Moon shown in Figure 8.16 shows that the Moon has been the victim of many cosmic collisions. Scientists who have made a study of such craters estimate that the Moon must have been bombarded much more frequently in the past than in current times. Several of these craters display the central blobs that result from the molten rock conditions created by the energy of impact. As explained earlier, the molten rock behaved like water does when a stone is dropped into it and forms ripples [36].

One of the things that the Apollo astronauts left on the surface of the Moon was a set of reflector panels. From the MacDonald Observatory near Fort Davis, Texas, technicians could then fire laser beams at the Moon using a telescope. Using this reflector equipment, technicians have been able to measure the Moon's distance at any moment to within four inches. By analyzing the measurements from a series of reflected signals, the French astronomer Odel Calame and her collaborator J.D. Mulholland at the University of Texas found that the surface of the Moon was moving back and forth with an amplitude of 80 feet. In other words, the surface of the Moon was vibrating like a bell. By checking historic records they found a report of an event on the Moon that

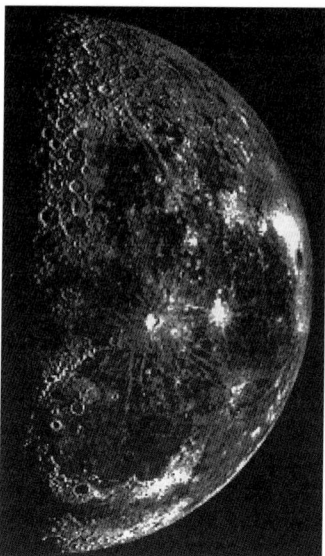

Figure 8.16 The surface of the Moon shows that it has been the victim of cosmic collisions of various magnitudes over billions of years. Since there is no air or water on the Moon, there is less erosion to wipe out the record of cosmic collisions.

they think was an asteroid collision that set the Moon in vibration. It is reported that, shortly after sunset on June 18 in 1178, near Canterbury, England, a group of men stood admiring the new Moon, a bright sliver hanging low in the west. As they watched the Moon's upper horn, it suddenly "split in two as a flaming torch sprang out of the division spewing fire, hot coals, and sparks." In the words of the observers, "the Moon writhed and throbbed like a wounded snake" and finally took on a blackish appearance [36].

Jack Harton, upon reading this historic record, thought the men might have seen an asteroid smack into the Moon. He found enough clues in the ancient records to predict an impact site on the back side of the Moon just beyond the northwest horizon. He checked a photograph of the far side of the Moon taken by Russian and American astronauts and there, in the right place, he found a remarkably fresh crater, twelve miles across, twice as deep as the Grand Canyon, and radiating white splatter marks for a hundred miles. Harton guessed that to gouge a hole of that size, a boulder the size of the Houston Astrodome would have had to slam into the Moon at 40 000 mph. An article about Harton's work appeared in the New York Times where it was spotted by Mulholland [37].

Scientists are now using the information on the vibration of the Moon and the probable date of collision to work out the physical properties of the inside of the Moon. Such a collision may also have sprinkled Moon dust on the Earth. This dust may show up in the polar ice cores used to track the records of climatic history. When astronomers brought back Moon dust, they found that the small grains of dust were themselves cratered. These impact craters were created by high-speed bombardment of the dust by micrometeorites. A typical crater of this kind is shown in Figure 8.17(a) [25].

Scarlett and Buxton believe that such high-speed cosmic dust impacts are the source of tiny glass spheres found in samples of Moon dust brought back to the Earth by astronauts [38]. The size distribution of the spheres they found in the dust is shown in Figure 8.17(b). Scarlett and Buxton believe that when the micrometeorites hit the Moon dust,

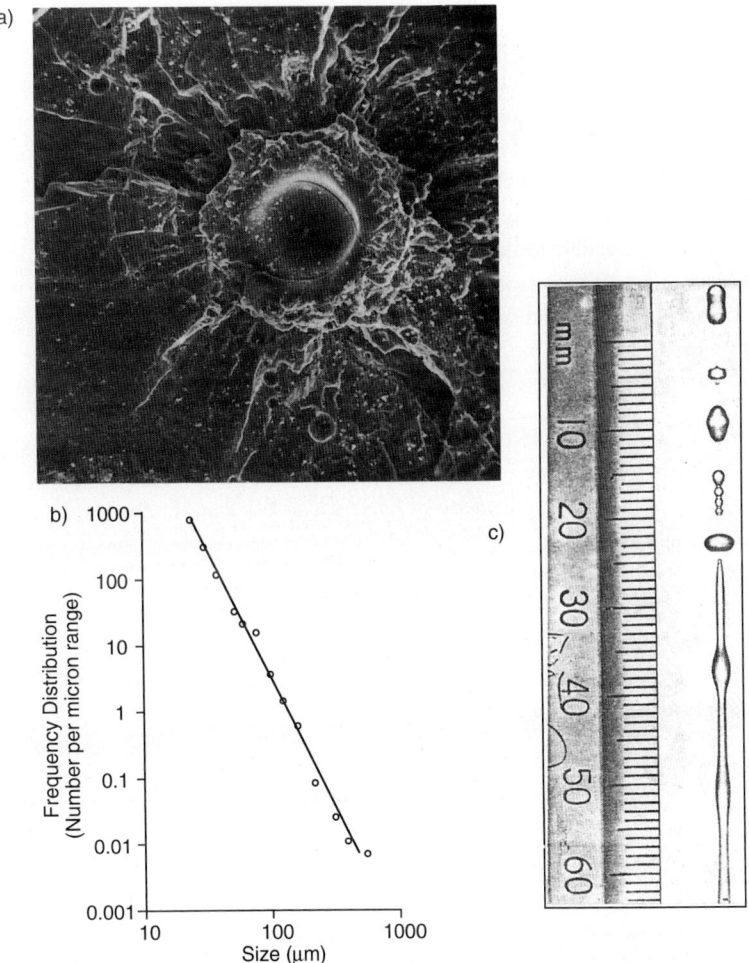

Figure 8.17 High speed cosmic dust collisions with lunar dust produce glass spherules as the kinetic energy is converted into heat during the collision. a) A crater discovered in a lunar rock sample, when viewed with an electron microscope, was created by a high speed dust particle striking the rock. b) The size distribution of spherical dust from the Moon. c) The process by which the glass spheres found in moon dust are formed can be modelled on Earth using water jets. (Courtesy of B. Scarlett and R. E. Buxton [38].)

the kinetic energy of the bombarding particles is turned into sufficient heat to create a liquid jet of molten rock, which is thrown up from the point of impact. They modeled the formation of such spheres by studying the break-up of a water jet as illustrated in Figure 8.17(c).

Although there are many theories, scientists have never been sure just how the Moon was created. One early theory was that the Earth and the Moon were once a single planet which spun so quickly that part of the surface was thrown off to become a separate element of the solar system. Another theory was that the Moon was a separate, small planet that had been captured by the Earth in the early days of the solar system. When scientists were able to examine the nature of rocks brought back from the Moon by the Apollo astronauts, they found their composition to be too different from those of the Earth to have formed either by the same process or by a later spinoff. In a discussion of the nature of Moon rocks, Henbest makes the statement that the Moon's composition is so strange that astronomers could not think where in the solar system it could have formed [39]. One thing that puzzled the scientists was the fact that the Moon contained little iron. Secondly, it lacked most volatile substances such as chlorine, potassium, and water, indicating that it must once have been heated to the point of incandescence. In 1975, astronomers Hartmann and Davis of the Planetary Science Institute in Arizona suggested what has become known as the *Big Splash Theory* to explain what we know about the Moon. The Big Splash Theory is described by Henbest [39]

> *A world as large as the planet Mars, fully half the diameter of Earth, smashed into the Earth at ten times the speed of a rifle bullet. The impact smashed whatever crust the Earth had, melting rock to a depth of hundreds of kilometers. The missile planet was completely destroyed. The iron that had formed at its core sank through the ocean of molten rock into the Earth's own core. At the same time, the surface rocks exploded into space in an incandescent plume to form a fiery ring around the Earth. Within a day these drops of molten rock came together to form a new world in orbit around the Earth. This body had very little iron because the iron of the impacting world had ended up in the Earth's core. It was also bone dry with little left of the water or volatile elements that had boiled away from the fiery plume. The Earth had acquired its Moon.*

In retrospect it appears that the astronomers had failed to come up with some of the possible explanations of specific features of the solar system because they implicitly wanted things to evolve smoothly. The concepts of catastrophe introduced into solar system history by the Alvarez hypothesis concerning the death of the dinosaurs stimulated the search for other possible catastrophic events to understand features of the solar system which were difficult to explain by evolutionary concepts. Henbest [39] also states that the Earth and its fellow planets may be survivors from "a time when planets ricocheted around the Sun like ball bearings on a pinball table."

There are other aspects of the solar system which Henbest claims could be the outcomes of cosmic collisions. He says that most planets rotate in the same sense in which

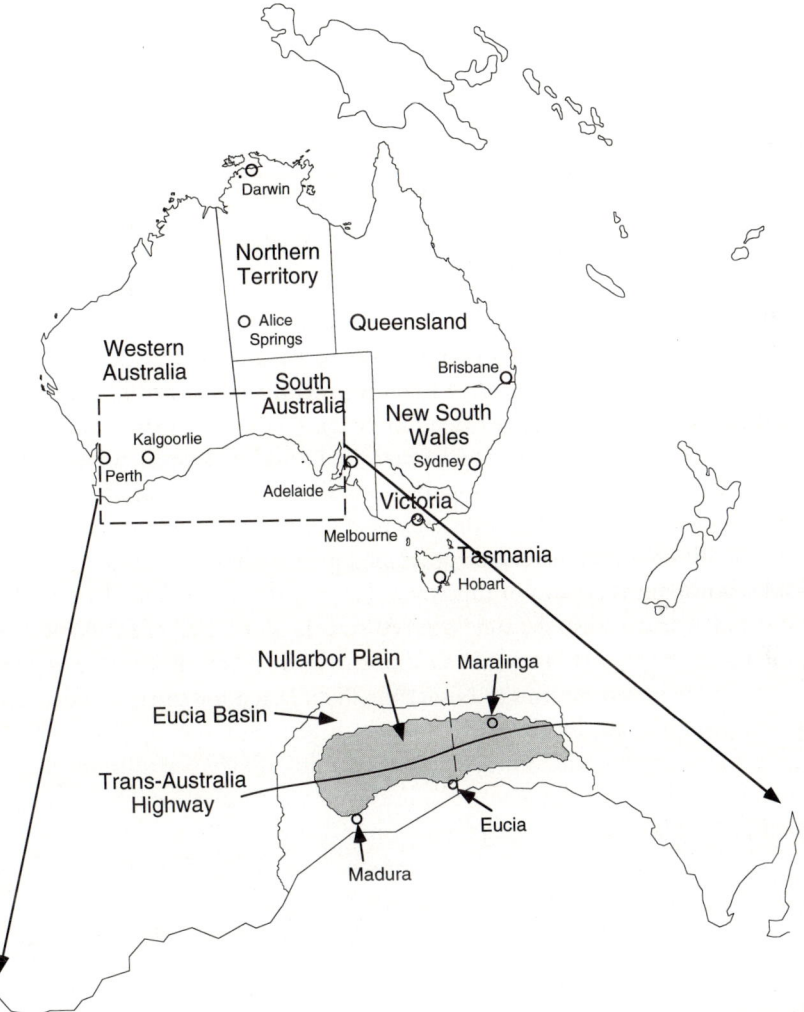

Figure 8.18 Scientists have recently discovered a treasure trove of meteorites in a desert area of Australia known as the Nullarbor Plain.

they orbit the Sun. The exceptions are Venus, Uranus, and Pluto. Henbest states that recent studies aimed at modeling the growth of the universe have come up with startling results [39]

> *The final stages in the growth of the Earth and the other inner planets was not the gentle sweeping up of small meteorites; instead, the scene was dominated by high-speed collisions between fully grown planets.*

Thus Cameron and colleagues at the Harvard Smithsonian Center for Astrophysics think that Mercury was originally a planet twice as big as it is today. They believe a smaller world ran into it at 20 kilometers per second, vaporizing much of the planet's rocky outer layers without disrupting its interior, to leave a planet with a disproportionately large metallic core. The reason why Venus is spinning backwards relative to ist orbital motion can again be explained by a giant collision [39]

> *If a world the size of Mars plowed into Venus in a direction opposite to the planet's original spin, it would have delivered a sufficiently large kick to make the planet rotate the other way. A similar collision may have tipped up the planet Uranus which orbits the Sun while spinning on its side.*

The Earth is known to nutate in the way that a spinning top nods if someone attempts to push it over from the top. The nutation of the Earth, which influences the ice ages, may well have been caused by a collision with a large asteroid which disturbed the uniform spinning of the Earth (see discussion of the spinning top in Chapter 4).

In their search for more meteorites to help them understand the bombardment history of the Earth, scientists are finding that one of the best places to look is the Nullarbor Plain in South Australia. It is unusually easy to find meteorites on the plain because the pale smooth limestone pavement provides a perfect background for spotting them. The Nullarbor Plain is an area of limestone that stretches for 600 kilometers along the south coast of Western Australia and South Australia. Alex Bevan of the Western Australian Museum in Perth has been leading the search for meteorites in this region.

One search for meteorites in the Nullarbor Plain lasted for three months. This area is also the approximate location of the British weapons testing ground known as *Maralinga*. (This name is an Aborigine word meaning the "voice of thunder.") I spent three months working in the desert conditions of the Nullarbor Plain (a Latin word which means literally no trees) and can confirm that the area is very dry with no shifting sands to cover up the meteorites which, unattacked by rain or chemical reactions, lie where they fall for very long periods. Bevan states that some of the meteorites found on the Nullarbor Plain may have been in that position for 16 to 18 thousand years.

Section 8.8   Space Junk

Not all of the cosmic collisions of concern to the modern scientist are created by fragments of the solar system bombarding Earth. Over the past several decades human space sciences have managed to leave a lot of celestial litter orbiting the Earth and posing a threat to tomorrow's spacecraft. In November 1991 the space shuttle had to make an avoidance maneuver to avoid colliding with the remains of a Soviet booster rocket named Cosmos 851, believed to have been launched into space in 1976. The incident was reported by the press [40]

At about 4:10 pm Eastern Standard time on Thursday November 25th the US space command in Colorado Springs, Colorado, an organization which operates the space surveillance network, flashed a warning to the Johnson Space Center in Houston that a near collision of the rocket and the space shuttle would occur within 11 hours. The warning said that the rocket would be about 2.5 kilometers above the Atlantis at its closest approach but NASA officials said such predictions have a built-in error margin of about a kilometer. A spokesman for NASA said that given the uncertainty of tracking abilities, 'we took the conservative approach and Atlantis swerved out of the way.'

Considering the speed with which the two objects were moving through space, a collision would have been catastrophic. Because of the kinetic energy of the orbiting debris from previous space shots, physically small objects can have devastating effects on a spacecraft. It has been estimated that a half-inch fragment could kill an astronaut working outside the space shuttle. In October 1990 it was estimated that there was a 1% chance that a fragment of debris ten centimeters or larger could destroy the Hubble space telescope. Scientists also estimate that trying to protect objects such as the Hubble space telescope from space junk larger than 1 centimeter would make them too heavy to launch. The space debris poses a second type of hazard to instruments such as the Hubble space telescope. The Hubble telescope is equipped with computers and sensors to help it use stars to navigate its path around the Earth and to help to orient the telescope. However, some of the pieces of space debris reflect light such that they appear similar to distant stars and can fool the guidance system of the space telescope. Scientists are beginning to work on the design of possible clean-up systems to sweep near-Earth space clear of the junk left by previous generations as they attempted to launch satellites and climb up into outer space [41–43].

Section 8.9   What are the Chances of a Person being hit by a Cosmic Missile?

One of the most recent incidences of a meteorite nearly hitting someone occurred in a suburb of New York in October 1992. A ten-kilogram stone plunged straight through the trunk of a Chevrolet Malibu parked in a driveway, turning the driveway underneath the car into a crater. The owner of the car, an 18-year-old high-school student Michelle Napp, is rumored to have sold the one-hundred-dollar second-hand car, and the meteorite, for $56,000 to a consortium led by Marlin Cilz of the Montana Metereorite Laboratory. [44].

In Canada there are only five recorded cases of people being literally within a stone's throw of a rock arriving from outer space. In the rest of the world, so Dickinson [44] tells us, there are about a dozen cases where incoming meteorites

whizzed within a meter or two of somebody's head.

He states

Despite legends to the contrary there is no authenticated case of anyone being killed by a meterorite although several domestic animals have been killed and at least two people in modern times have been hit by these extraterrestial missiles.

See Ref. [44] for a detailed discussion of your chances of being hit by a meteorite.

References

[1] D.H. Rousell, "Sudbury and the Meteorite Theory," *Geoscience Canada*, 8 (4) December 1981, pp. 167-169
[2] R. Grieve, "Impact Craters Shape Planet Surfaces," *GEOS* Vol. 11 (4), Fall 1982. See also, Grieve R., "Impact Craters Shape Planet Surfaces," *New Scientist*, November 17, 1983, pp. 516-519.
[3] E.F. Mallove, "The Bombarded Earth," A special report in *Technology Review*, Vol. 88 No. 5, July 1985, pp. 64-69.
[4] M. Schroeder, *Fractals, Chaos, and Power Laws; Minutes from an Infinite Paradise*, W.H. Freeman and Company, New York, 1991
[5] C. Thompson, "Shatter Cones: Epilogue to a Dynamic Story," *GEOS*, Vol. 13 (2), Spring 1984, pp. 1-4.
[6] R.P. Binzel, M. Barucci, M. Fulchignoni, "The Origin of the Asteroid," *Scientific American*, Vol. 265 No. 4. October 1991, pp. 88-94.
[7] K. Kroswell, "Vermin of the Skies," *New Scientist*, August 27, 1994, pp. 26-29.
[8] C.J. Cunningham, *Introduction to Asteroids*, Willmann-Bell, Richmond, VA, 1988.
[9] For a spectacular picture of an asteroid with its tiny moon see news item "Asteroid Ida is First with a Moon," *New Scientist*, April 2, 1994, p. 5.
[10] I. Ridpath, "Tunguska, The Final Answer," *New Scientist*, August 11, 1977, pp. 346-347.
[11] A.S. Zaburunov, "Eye Witness To A Mystery," *Industrial Research and Development*, March 1978, pp. 103-105.
[12] G. Greenstein, "Heavenly Fire," *Science 85*, Vol. 6, No.6, July-August, 1985, pp. 70-77.
[13] J.D. Fernie, "The Tungusta Event," *American Scientist*, Vol. 81, No. 5, Sept./Oct. 1993, pp. 412-415.
[14] "Fireball Over Siberia: 1908," News story in *Time*, March 3, 1968. p. 43.
[15] J. Gribbin, "Cosmic Disaster Shock," *New Scientist*, March 6, 1980, pp. 750-752.
[16] See letter to the *New Scientist* from Jim Thomas of Australia, April 18th, 1992, p. 52.
[17] R.J.M. Olson, "Giotto's Portrait of Halley's Comet," *Scientific American*, Vol. 240 No. 5, May, 1979, pp. 160-170.
[18] R. Stephenson, K. Yau, "Oriental Tales of Halley's Comet," *New Scientist*, September 27, 1984, pp. 30-32.
[19] H. Balsiger, H. Fechtig, J. Geiss, "A Close Look At Halley's Comet," *Scientific American*, Vol. 259 No. 3, September, 1988, pp. 96-103.
[20] D. Overbye, "Great Balls of Fire," *Discover*, Vol. 2, No. 12, December, 1981, pp. 20-26.
[21] J.D. Fernie, "Comets Again," *American Scientist*, Vol. 82, No. 2, January/February, 1994, pp. 104-106.
[22] L.E. Murr, W.H. Cunard, "Effects of Low Earth Orbit on Spacecraft," *American Scientist*, Vol. 81, No. 2, March/April, 1993, pp. 152-165; D. Dixon, T. McDonnell, B. Carey, "The Dust that Lights up the Zodiac," *New Scientist*, January 10, 1985, pp. 26-29.
[23] "Europe's Battered Space Probe Sent On Final Mission," News story in *New Scientist*, August 10, 1991, p. 13.

[24] N. Henbest, "Has Halley's Comet Split In Two?" *New Scientist*, March 9, 1991, p. 22.
[25] D.E. Brownlee, "Cosmic Dust," *Natural History*, 4, 1981, pp. 73–77.
[26] See E. Ashpole, "Capturing Cosmic Dust" in *Science Spectrum Notes*, Number 181, 1982, published by the British Embassy in Ottawa, Canada.
[27] G. Taubes, "U-2 Mission: To Catch the Dust of Comets," *Discover*, Vol. 14 No. 10, October, 1983, pp. 74–77.
[28] D.W. Parkin, R.A.L. Sullivan, J.N. Andrews, "Further Studies on Cosmic Spherules from Deep-Sea Sediments". *Philosophical Transactions of the Royal Society of London, A: Mathematical and Physical Sciences*, Vol. 297, No. 1432, 1980, pp. 495–518
[29] I. Asimov, *Biographical Encyclopaedia of Science and Technology*. Doubleday, New York, 1948.
[30] C. Zimmer, "The Smoking Crater," *Discover*, Vol. 13, No. 1, January 1992, p. 48.
[31] L. Jaroff, "At Last, the Smoking Gun?," *Time*, July 1, 1991, pp. 42,43.
[32] Newsitem, "Were the Dinosaurs Dealt a Killer Blow at Yucatan?" *New Scientist*, November 17, 1990, p. 14.
[33] W. Glen, "What Killed the Dinosaurs?" *American Scientist*, Vol. 78, No. 4, 1990, July/August, pp. 354–370.
[34] D. Raup, "Extinction: Bad Genes or Bad Luck?". *New Scientist*, September 14, 1991, pp. 46–49.
[35] L. Jaroff, "Look Out!" *Time*, February 15, 1993, pp. 46–48.
[36] A.G.W. Cameron, "The Moon," in *Modern Science and Technology*, (Ed. R. Colborne) Van Nostrand, New York, 1965, pp. 492–502.
[37] T. Dunkle, "Why is the Moon Ringing?" *Science*, Vol. 3, No. 2, March 1982, p. 104.
[38] B. Scarlett, R.E. Buxton, "Particle Size Distribution of Spherical Particles in Apollo 12 Samples", *Earth Planet. Sci. Lett.* 22, 1974, pp. 177–187.
[39] N. Henbest, "Birth of the Planets," *New Scientist*, August 24, 1991, pp. 30–35.
[40] "Close Encounter," *New Scientist*, November 29, 1991, p. 48.
[41] B. Wood-Koczmar, "The Junkyard in the Sky," *New Scientist*, October 13, 1990, pp. 37–40.
[42] "Sweeping up Space Junk," *Discover*, Vol. 13, No. 1, January 1992, p. 11.
[43] "Laser Blasts a Path through Orbiting Litter," *New Scientist*, October 20, 1990, p. 27.
[44] T. Dickinson, "It Came From Outer Space," *Canadian Geographic*, May/June 1995, pp. 31–40.

Chapter 9

Some Down to Earth Missiles

Chapter 9

Some Down to Earth Missiles

Section 9.1   Ice Bullets and Hailstones

In the 1940s an adventure comic known as *The Wizard* published an exciting serial story about the mysterious deaths of people found to have holes in their bodies but no bullets. At the end of the story it was revealed that an embittered astronomer had been firing ice bullets from his observatory tower using one of his telescopes to view victims from a distance. I have never carried out detailed calculations to decide whether or not such a scheme could work, but the theme of ice bullets as mysterious weapons of death does surface from time to time in thriller stories. Of course we do know icicles become dangerous missiles when they detach themselves from roofs and fall on innocent passersby, but whether or not ice bullets from a gun could be used to achieve mysterious deaths and baffle detectives we will have to leave for others to discuss. What is certain, however, is that ice bullets from the clouds, otherwise known as hailstones, present a very real hazard for our earthly existence. In the days when people used to believe in a multiplicity of gods, the Norse god Thor (after whom Thursday is named) was the god of thunder. Lightning was said to have been caused by his throwing a hammer at his victims. If lightning was Thor's hammer, then hailstones were ice missiles from the clouds [1].

A few years ago I visited a research station in Denver shortly after a massive hail storm. It was as if every car in the parking lot had been fabricated from dimpled steel. I was told that, about a week earlier, hailstones the size of small oranges had rained down from the sky damaging all of the cars. Hailstones damage millions of dollars worth of crops in many parts of the world every year. One area where hailstones are a problem is Alberta. The hardest hit areas in Alberta are the Lethbridge-High River Corridor, the Ponoka-Penn Corridor, and the area bounded by Drumheller Hills and Calgary. In fact, the latter area is known as "hailstone alley." Other areas of the world where heavy crop damage occurs are Switzerland, Argentina, the North Caucasus area of the former Soviet Union, and Northern Italy.

The Alberta government has a hailstone abatement program in which an emergency team tracks thunderstorms capable of turning into hailstorms. When such storms are detected, the scientists rush to the site to pick up hailstones for study to gain information on how the hailstones grew. Later in this chapter we will look at how farmers attempt to fight back by firing rockets into the region of the cloud where hailstones are produced. A possible reason for the effectiveness of such rocket attacks in combatting Thor's ice bullets will be explored.

Section 9.2   Latent Heat and Nucleated Ice

To be able to understand the possible methods for hailstone abatement we have to gain an appreciation of how rain falls to earth and of the physical changes involved when water turns into ice.

Sometimes the vocabulary of science helps us remember that what is considered elementary today was considered difficult and obscure by those who first investigated the branch of physics. The vocabulary can also demonstrate how scientists sometimes hold incorrect opinions about reality for long periods of time. For instance, when scientists began to look into heat theory, they completely misunderstood the nature of heat. They thought that the hotness of a body was really a liquid called caloric that could be squeezed out of the material (see word web in Figure 2.4). As already discussed in Chapter 2, the instrument used by scientists to measure quantities of heat is called a calorimeter and the classical unit for heat energy was the calorie.

I have a very clear memory of the first experiment with a calorimeter that I carried out as a schoolboy. The equipment used is illustrated in Figure 9.1(a). A calorimeter partially filled with water was placed inside a container of crushed ice and salt which had a temperature well below 0 °C. (In Canada we put salt on our roads in winter because a mixture of salt and water has a much lower freezing point than zero degrees Celcius.) The water inside the calorimeter could be stirred. The calorimeter was equipped with an insulated lid to reduce heat exchange with the outside environment. We timed the fall in temperature of the water inside the calorimeter as ice began to form. The purpose of the experiment was to show that the temperature of the water remained steady while the water turned to ice as illustrated by the experimental graph of Figure 9.1(b). We were then challenged to explain why the temperature remained the same until all of the water had changed into ice. It was then explained to us that the pioneer scientists who first investigated this effect were baffled by the fact that the temperature did not change as the ice formed. They thought that somewhere in the ice/water mixture there was some hidden caloric which was squeezed out of the water as it changed into ice. Because they could not discover the source of this caloric, they used the term *latent heat* from the Latin word "latere" meaning "to lie hidden" to describe the situation. (When a camera is used to take a picture, the pattern created by the photons of light in the photographic film is called a *latent image* until the development process makes the image visible to the human eye. In the same way, a fingerprint left at the scene of a crime is described as a *latent fingerprint* until dusting powder makes it visible.) We now know that the mol-

9.2 Latent Heat and Nucleated Ice

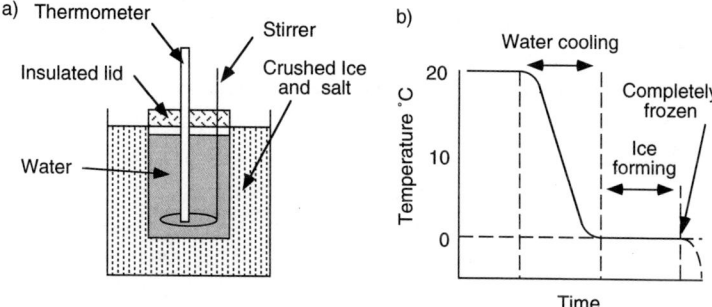

1 gram of water, starting at 20°C, must lose 1 calorie to cool 1°C.
1 gram of water at 0°C must lose 80 calories to become ice at 0°C
1 gram of water at 0°C must absorb 100 calories to reach 100°C
1 gram of water at 100°C requires 550 calories to become vapour at 100°C

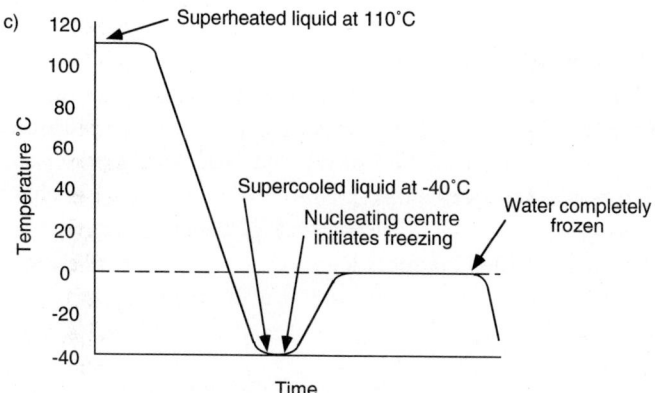

Figure 9.1 Stating that water freezes at 0°C and boils at 100°C is a simplification of reality. a) A calorimeter can be used to determine the amount of heat that must be removed from a quantity of water in order for it to freeze. b) The temperature of cooling water does not change until it is completely frozen. c) Very pure, clean water can be raised to temperatures above 100°C or cooled to temperatures below 0°C.

ecules in the liquid water are rushing around with lots of kinetic energy and that as they change into the ice state they have to give up some of this kinetic energy before they can become co-operative members of an ice crystal. In round numbers, we know that one gram of water at 0°C loses 80 calories before it becomes one gram of ice. Because this same amount of energy must be absorbed to make ice change into water, ice cubes can effectively cool a drink. Scientists use the terms *latent heat of freezing* (or fusion), and *latent heat of melting* to describe the amount of heat required to bring about the changes of state from water to ice and vice versa. One gram of water at 100°C requires 550 calories to become steam at a 100°C. This is referred to as the *latent heat of vaporization* or, for the reverse transformation, as the *latent heat of condensation*. All substances require latent heat to bring about a change of state from liquid to solid or from liquid to vapor and vice versa [1].

Before the microwave oven was invented, the fact that steam turning into water releases tremendous amounts of heat was widely used in restaurants to heat drinks and food. So it was that when purchasing a cup of coffee in a British railway station 20 years ago, the customer watched the operator pour the ingredients into a cup and then place a steam pipe into the liquid in the cup to let steam condense inside the cup to warm up the drink. For the same reason, if a person inadvertently puts their hand in a steam jet, the condensation of the liquid releases tremendous heat causing serious burns and scalds. By the way, you cannot see water vapor: steam from a kettle only becomes visible when the vapor condenses to produce water droplets in the jet of steam.

Because ice requires so much latent heat to change it into water, icebergs dissolve very slowly. Since the density of ice is only nine-tenths that of water, an iceberg floats; however, most of the iceberg is under water with only the tip showing above the surface. It has been seriously conjectured that icebergs might solve the water problems of desert countries such as Saudi Arabia. When an iceberg breaks away from the antarctic ice cover, it has been suggested its top be sprayed with insulation. Then, by placing a small propulsion unit on the edge of the iceberg, it could be driven all the way to Saudi Arabia. It would then be melted to form irrigation water for the desert areas. By the way, the ice of an iceberg is not a mixture of salt and water but is pure water formed upon the surface of the sea as the water changes into ice or falls as snow. It has been calculated that moving icebergs would be a cheaper way of bringing water to Los Angeles than improving the dams on the Columbia River. However, scientists are worried about the cooling effect that the iceberg would have as it changed to water. The cooling could change the microclimate (that is the local climate) of a city such as Los Angeles and cause widespread fog.

As often happens in our growing knowledge of physics, we now know that the results summarized in Figure 9.1 are a simplification of the complex transformation of water into ice and vice versa. If we carried out the experiment with water which had been filtered to take out all of the small invisible fine-particles and placed in a vacuum for a short time to let all gas bubbles leap out of the water, and if we carried out the experiment very carefully, we would find that the temperature could drop to as low as $-40\,°C$ before ice formed. In the range of 0 to $-40\,°C$, the water is referred to as a *supercooled liquid*. If we very carefully cooled a quantity of pure water to approximately $-10\,°C$, we would find that we could turn it into ice crystals very quickly by sprinkling very fine sand into the water. We now know that it is very difficult for ice crystals to form in pure water, but if we provide small nucleating centers, such as dust fine-particles in the liquid, we can encourage ice crystals to form. Even so-called pure water used in the laboratory contains many tiny invisible fine-particles which act as nucleating centers for the formation of ice crystals.

Pure water can also be superheated to temperatures as high as $110\,°C$ (see Figure 9.1 (c)). This fact explains, for example, the geyser action of hot springs in Yellowstone

Park. A *geyser* is a periodic eruption of hot water and steam from a rock crevice. The word comes from an Icelandic word "geysa" meaning "to gush." The water seeping into the narrow crevices of the hot rocks of the region is very pure. In the crevice it heats up to a temperature of about 107 °C before suddenly erupting into a mixture of steam and water driving the spectacular fountain out of the rocks to create the visible geyser. The fact that water can become heated above its boiling point is the reason why scientists add small pieces of broken pottery to any liquid that is being boiled so that bubbles can form easily in the liquid, and potentially explosive superheating is avoided. The process of providing a tiny center around which an ice crystal or vapor bubble can form is known as *nucleating technology* from the Latin word "nucleus" meaning "a tiny nut." [2, 3]

As already noted in our brief discussion of the nature of an iceberg, ice occupies more space than the corresponding amount of water so that, when ice crystals form in a substance, their larger volume can rupture the structure of the material. Thus if we freeze a piece of meat, the expansion of the ice crystals breaks down the walls of the cells of the meat. Because the tenderized meat cells produced by the expanding ice encourage bacterial growth and degradation of the meat, a piece of meat that has been thawed should not be refrozen. It is also the reason why, when shipping human organs for transplants, the organs are sometimes flushed out with an antifreeze solution to allow shipment at low temperatures without damaging the cell structure. There is a story that, in Chicago, a person rescued from the street on a cold winter night was found to have a very low body temperature. But the individual was so drunk that the alcohol in the body had acted as antifreeze and prevented frostbite damage.

In a similar way, when a plant freezes, the formation of ice crystals can rupture its cell walls and this is why a plant exposed to frost loses its structure and flops onto the ground. It is well known that some plants are better able to resist frost than others. It is thought that in frost-resistant plants the internal liquids are very pure and can be cooled down without significant crystal formation. One of the crops which traditionally has been damaged by light frosts is strawberries. When investigating why light frost damages such plants, it was found that the leaves are eaten by bacteria which excrete materials that encourage frost formation by nucleation. The bacteria excrete this material to promote the production of frost-damaged leaves which are more easily digestible than the fresh, undamaged leaves. Scientists have developed a variant of this bacteria which does not produce the nucleating chemical and experiments had been planned to test the effectiveness of the altered bacteria in preventing frost damage to strawberries by spraying some fields with the new bacteria. These would compete with the natural strain of bacteria responsible for the early frost damage. However, concerns were raised about possible unforeseen effects of releasing non-native species into the environment.

The type of bacteria that produces the frost damage to strawberries is now being used in freeze-dried form to help make snow for ski resorts by promoting nucleation of the water sprays used to make snow.

Section 9.3 Rainmaking and Hailstone Abatement

In the early 1930s the Norwegian meteorologist Tor *Bergeron* found that most clouds do not produce rain because the droplets of water in them are too small to fall to earth. In many clouds, the droplets are of the order of 20 micrometers in diameter. Such droplets cannot fall to the earth but are continually held in circulation by air currents. For a raindrop to fall to the ground it must have a diameter of several hundred micrometers. Bergeron also noted that the droplets in common clouds grow very slowly and that they can be supercooled to as low as $-40\,°C$. He observed that if some nucleating centers could be provided in the cloud, the droplets would turn into ice crystals and grow very quickly. In 1946 *Schaefer* dropped some small pellets of *dry ice* into a thin cloud of water droplets. (Dry ice is the popular name for *solid carbon dioxide* which has a freezing point of $-78.5\,°C$.) In a matter of minutes vast numbers of ice crystals formed and grew rapidly. Some of them became big enough to fall down to earth. This is regarded as being the first ever artificial rainmaking experiment. We now know that the powdered dry ice produced ice crystals by a process known as thermal shock. Since those early experiments carried out by Schaefer, the preferred material for making rain has become a smoke of silver iodide. This chemical has the advantage of having a crystal structure very similar to that of ice and this fact further encourages the formation of ice when the smoke is injected into a cloud [4].

Canadians were quick to try to reproduce the success of Schaefer's experiment in seeding clouds to create rain. The first cloud-seeding experiment in Canada took place over Sudbury. Thus the Sudbury Star of Friday June 11, 1948, carried the headline

> *Dry Ice Pellets Bring Downpour — Rainmaking Trial Checks Orfino Blaze.*

The story goes on to detail how

> *An aircraft flying out of Sudbury dumped 75 pounds of powdered dry ice (solid carbon dioxide) into a cloud over a forest fire burning in Northern Ontario at a location 40 miles east of Sudbury. It is reported that at 6 o'clock the dry ice was spread on the clouds and that within a few minutes a drenching downpour had started. For fifteen minutes the rain poured down and for the next three hours a drizzle fell. The plane from which the ice particles were laid on the clouds was at 12 000 feet during the operation and came down below the cloud when its task was completed. It met the heavy rain.*

The story tells us that the plane used in the cloud-seeding experiments was piloted by Rusty Blakey, a well-known pilot in the Sudbury district and a veteran bush pilot. Rusty Blakey continued to fly bush aircraft for many years after his participation in the rain making experiment. I had the pleasure of knowing him as he played an active part in the life of Sudbury and Northern Ontario. I am indebted to his son Richard Blakey, a local high school teacher, for a copy of the original news story detailing the experiment described above.

9.3 Rainmaking and Hailstone Abatement

Now that we have a basic grasp of the physics of rain clouds and of human techniques of rainmaking, we can look at how the same technology is being applied to modification of hailstone growth. Figure 9.2 illustrates the various stages in the growth of a hailstone-generating thundercloud [5, 6]. The formation of the thunderstorm begins when a large mass of cold air, called a *cold front*, collides with a mass of warm air. The cold air is denser than the warm air and pushes underneath it forcing it upwards as shown in Figure 9.2(b). As the warm air rises, it cools until it can no longer contain the moisture which it carried upwards with it from ground level. Precipitation of the vapor as supercooled liquid releases latent heat of condensation which provides energy for the upward growth of the cloud. This upward growth, which produces the towering clouds seen just before a thunderstorm strikes, is known as *convective growth*. The rapid convective growth of a thundercloud draws further supplies of warm air up into the colder regions

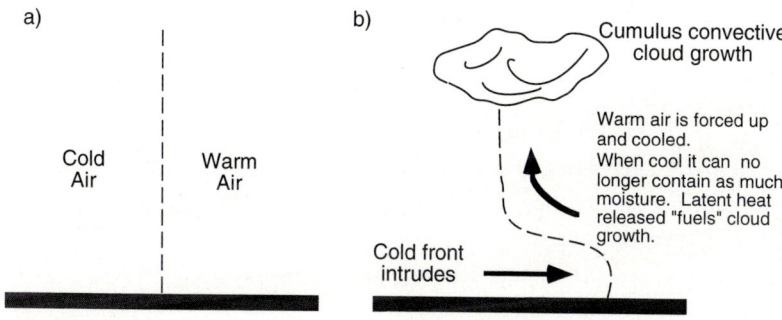

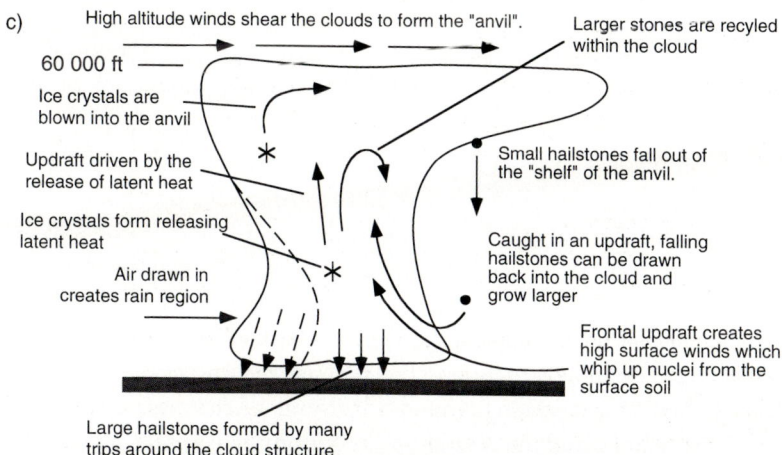

Figure 9.2 Hailstone-bearing clouds form when a "cold front" attempts to displace a mass of warm moist air. a) The cold front is where the cold and warm air meet. b) Cumulus clouds grow as cold air intrudes under warm air. c) Pattern of "fallout" from the storm cloud.

of the cloud causing further formation of supercooled droplets and more upward growth of the cloud. The arrival of the cold front is usually accompanied by high altitude winds which tend to shear off the top of the growing thunderclouds as shown in Figure 9.2(c). This blown-over, high-level part of a thundercloud formation is called the *anvil* of the cloud. In part, the rising updraft of air accompanied by a small amount of air sucked in at the rear of the cloud causes rain to form as shown in Figure 9.2(c). Dust fine-particles in the rising air initiate the nucleation of ice crystals in the rising convection cloud. When the ice crystals reach the top of the column of the cloud these are blown over into the shelf of the anvil cloud by the high altitude wind. In this part of the cloud there is no longer a turbulent updraft and the crystals start to fall toward the ground. If the high altitude wind is of sufficient velocity, these small hailstones are carried well forward of the turbulent region and fall at the front of the storm as relatively small hailstones or even as rain. However, under certain conditions the small hailstones are trapped in the updraft at the front of the storm and are recycled as illustrated in Figure 9.2(c). As they sweep upward through the storm they scavenge more supercooled droplets, releasing more latent heat of evaporation which increases the turbulence of the cloud. Hailstones can travel back up through the growing cloud and out on the anvil shelf many times, growing very large in the process. As the hailstones grow, a stage is reached at which they can no longer be supported by the strength of the updraft and they then fall out of the cloud [5–7].

It should be realized that hail-generating thunderstorms are primarily a problem of the temperate climates of the world. They are rarely found in polar regions because the air here is seldom unstable enough to generate strong vertical wind gradients and the surface of the earth is too cold. Hail is also rare in the tropics because the height at which water freezes is too far above the sea level for clouds to be able to produce chunks of ice.

Studies of the structure of individual hailstones have established that two basic mechanisms can be involved in their growth. First, as the hailstone moves in the cloud, it scavenges small drops of supercooled water. As these droplets hit the surface of the hailstone they change to ice. In changing to ice however, they must give up their latent heat of fusion. If the rate of arrival of droplets on the surface of a hailstone is low, the hailstone has sufficient thermal capacity to absorb the latent heat of fusion almost instantaneously and the drop freezes very rapidly. Ice cannot hold dissolved gases in the way that water can and the rapidly freezing droplets contain many air bubbles which form as the gas is released from solution. The scattered arrival of the various droplets also results in a relatively loose structure of ice containing gaps between the frozen drops. Therefore, a shell of ice growing around a hailstone under these conditions contains many small air bubbles and the crystals of ice within the layer are very small. For this reason, ice growing under these conditions looks very milky or, when examined under the microscope, the bubbles of gas absorb the light and these appear as dark rings. If a growing hailstone moves into a part of the cloud rich in droplets, the droplets cannot

9.3 Rainmaking and Hailstone Abatement

all freeze immediately. Therefore before they are turned to ice, they spread out over the surface of the hailstone forming uniform layers of ice which have a relatively coarse crystalline structure. Sometimes hailstones reaching the ground have recently traveled through such a region of the cloud and appear to be slushy on impact.

Scientists have developed techniques for sectioning hailstones to study the formation conditions for each layer of ice and they are able to count the number of layers of ice in the ball; these layers are related to the number of round trips that a hailstone has made in the turbulent region of the thunderstorm [6].

Hailstones can develop various shapes. For example, those that are smaller than 2 centimeters tend to have a conical shape. This is because, in the final exit from the cloud, a falling stone grows by sweeping up water droplets on its lower face. Larger stones tend to be flattened spheres and although there is some doubt as to the mechanism which produces this shape, the majority opinion tends to support the theory that they tumble in the cloud and grow almost equally all around their periphery. Occasionally, well-formed lobes develop on the sides of a hailstone. The shape and size of the hailstone can tell the scientist a great deal about the internal dynamics of a thundercloud [5, 6].

Various techniques have been proposed for reducing the size and structure of hailstones in thunderclouds. These techniques are illustrated in Figure 9.3. One possibility is "cloud seeding" the thunderstorm from above to increase precipitation. This strips the cloud of its supply of supercooled liquid droplets available for feeding hailstone growth as they circulate around the storm system. However, it is difficult to seed a

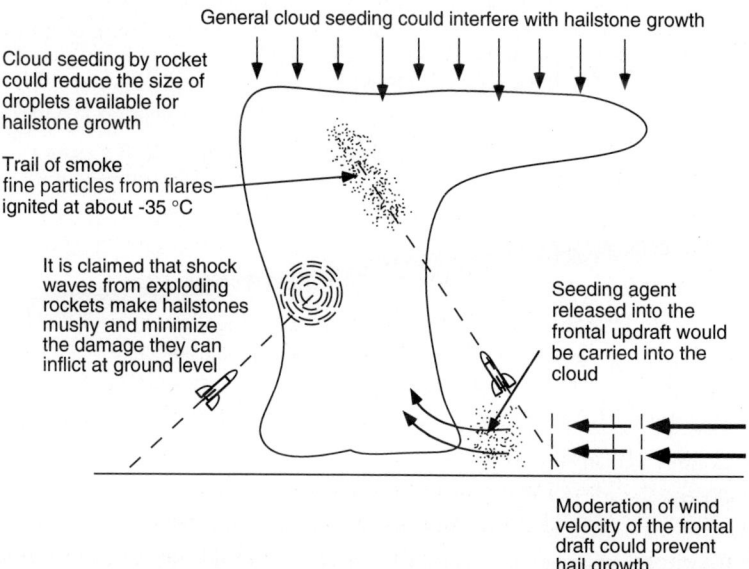

Figure 9.3 Several techniques have been proposed for interfering with the structure of hailstone-generating thunderclouds.

storm's cloud system from above. A more efficient and safer way of interfering with a hailstone cloud is to deliver the cloud-seeding nuclei by way of a rocket system to the center of the turbulent region where the hailstones are known to grow to maximum size. The use of cloud-seeding rockets to disturb the central region of a hailstone-generating cloud has been developed in Russia by Georgii Sulakvelidze. In the Russian technique, radar is used to locate clouds where hailstones are forming. A rocket is then fired into the cloud. The rocket carries 100 grams of lead iodide which is released as a smoke cloud in the heart of the thunderstorm. Russian scientists claim that this seeding technique reduces the size of hailstones and often prevents precipitation altogether. The whole operation takes about 2 minutes with one radar installation covering an area of 1500 square kilometers. It is claimed that in 1968 and 1969, cloud-seeding methods using rockets successfully prevented any really damaging hailstorms in the North Caucasus. Some western scientists have expressed doubts concerning the efficacy of the Russian techniques. However, further experiments using the Russian methods are underway in Argentina, Lebanon, Hungary, and Yugoslavia [7].

Studies seem to establish that the successful seeding of a hailstone cloud depends on discharging the nucleating smoke fine-particles in the region of the cloud where the temperature is $-35\,°C$ to $-40\,°C$. Attempts were made to improve the efficiency of the Russian cloud-seeding rockets by using delayed ignition rockets hoping that firing would commence in the critical area of the cloud. Austrian scientists have patented the design of a rocket equipped with a temperature-sensitive device which is able to sense the temperature of the cloud as it moves into the thunderstorm. When this device detects that the external temperature is that of the critical seeding zone, it ignites the smoke-releasing system [7].

Italian farmers claim that they have been successful in reducing damage from hailstones by using many small rockets which explode in the cloud. These rockets are usually 3 inches in diameter, 5 feet long, and constructed from cardboard. They reach an altitude of between 1000 and 1500 meters before they explode. At the peak of their trajectory, a charge of 800 grams of gunpowder located in the head of the rocket is exploded. In the 1959 growing season Italian farmers used 100000 of these rockets. The efficiency of these devices has been investigated by Octavio Vittori, a well-known atmospheric chemist. During his investigation he was told by many eyewitnesses that within minutes of a rocket launching the hailstones became mushy. To explain this claimed success, Vittori has proposed that the explosion of the gunpowder rocket causes shock waves to pass through the water and air pockets inside rapidly growing hailstones resulting in the development of many tiny cracks in the ice and the crumbling of the overall structure of large hailstones.

It appears that the idea of interfering with hailstone production by firing rockets into a cloud originated in China. Thus Needham reports in his extensive study of Chinese science and technology that the firing of rockets into thunderstorm clouds appeared to have developed many centuries ago [8]. Perhaps the practice developed from a chance observation during a festival for a thunder god when modified hailstones fell from a cloud after rockets had been fired into it. Fighting back against the thunder gods with rockets became a widespread custom even if its effectiveness was never entirely proven. Perhaps this is another technological innovation brought back from China to Italy by the famous explorer Marco Polo [8].

Another possibility which could explain the claimed success of the rocket attack on hailstone-generating clouds is that the shock waves from the exploding rocket cause the freezing of small droplets by mechanical shocking of the supercooled liquid droplets. This in turn could provide many more nuclei for the formation of hailstones, thus reducing their average size. Furthermore, in rapidly creating many small hailstones rather than one large one over a long period of time, there is less time for the droplets joining a growing hailstone to give up their latent heat of fusion and the many small hailstones are more likely to be slushy than is the single larger hailstone formed over a longer period in the cooler regions of the cloud.

It has been suggested that cloud seeding could be effectively carried out by exploiting the very strong updraft which feeds the growth of the hailstone-generating cloud by releasing the cloud-seeding agent in the frontal draft as shown in Figure 9.3. There also have been suggestions that perhaps the most efficient way of preventing a thunderstorm cloud from developing its hailstone capacity is to interfere mechanically, through explosions or other systems, with the updraft current feeding the base of the thunderstorm. No feasible systems for interfering with the updraft air current have yet been developed. Scientists are beginning to explore the use of composite rockets which not only explode in the cloud but also deliver silver iodide for cloud-seeding purposes. As yet, little information is available on the efficiency of these devices [4–6].

A hail reduction program in Alberta is typical of many western attempts to suppress hailstones. The program in Alberta started as a private venture sponsored by a group of farmers. The initial event triggering this activity was the visit by a farmer from Manitoba to the United States Midwest in 1951. He noticed that the fields around a certain small town were greener than elsewhere. At a gas station the farmer was told by the attendant that these fields fell within an area where experimental cloud-seeding had been carried out under the direction of Dr. Irving P. Krick. In the period between 1953 and 1955, farmers from Deloraine, Manitoba, the area where the farmer lived, hired Dr. Krick to carry out experiments in their region. After a major hail disaster in Alberta in 1956, a committee of farmers hired Dr. Krick to seed hailstone-bearing clouds with silver iodide. The experiment was funded by a levy of $15 per quarter section from farmers in the direct area of hailstones and $10 for the farmers in surrounding areas (a quarter section is 160 acres of land). Experiments in which aircraft delivered silver iodide seeding seemed to be successful in 1958; however, a severe hailstorm in 1959 in the hail-suppression area caused many people to become skeptical of the efficiency of the program. Dr. Krick and his team discontinued their efforts in Canada and returned to the United States [9].

At the same time, as a result of public pressure, the Alberta Hail Studies group was formed. This group included scientists from McGill University in Montreal and from the University of Alberta, with the Federal and Alberta governments providing funds for the experiments. Preliminary studies carried out by this group, which is now a Crown corporation, indicated that present technology might be able to reduce crop losses from hail by 10–25%. It is estimated that if the $1 million program could reduce hail damage by 10%, the crop savings could be worth $4 million. The program, started in 1969, operates out of the Canadian Forces Base Penhold just south of Red Deer. Storm clouds are monitored with radar as they form over the foothills and the clouds' life histories are recorded as they develop and move east. The radar information is used to direct mobile meteorology stations and to position them in locations to obtain measurements of temperatures, humidity, and wind in the vicinity of the storm. They also collect hailstones and plunge them into deep-freeze conditions for later analysis. Because of the damage done by hailstones, we can expect efforts to control hailstone-forming clouds to continue until it can be fully established whether or not the technologies being used are effective [9].

Section 9.4   In Search of Nothing?

The type of down to earth missile to be discussed in this section is very different from the hailstones. It is a cosmic missile, the *neutrino*. Neutrinos are amongst the most elusive and enigmatic entities in the bizarre world of fundamental particles. They are electrically neutral, massless, or virtually so, and are therefore almost undetectable.

Fundamental particles to a physicist are fragments of atoms and constituents of the atomic structure of the universe whereas items such as dust, or paint pigments should be described as "fine-particles" rather than as "particles". However actual usage of the term particle is sloppy and the reader of any particular article must determine which type of particle is being discussed in the article. Like other sub-atomic particles, the neutrino's existence was predicted before it was actually discovered. Scientists are interested in neutrinos because they are known to be produced in large quantities in the energy-generating reactions taking place in the Sun. They are also produced by other spectacular events in the universe. Scientists hope that, by studying the rain of neutrinos that showers down upon the Earth from energetic stars and fantastic events in the universe, they will gain a better appreciation of the structure of the universe and of the Sun's energy-generating reactions. The neutrinos are so energetic that they can travel through the Earth without being captured. Therefore it is necessary to use very special equipment to detect their presence. As mentioned in Chapter 8, the Earth is also being continually bombarded by different types of cosmic rays. This means that ways must be devised to filter out all of the other cosmic rays reaching the Earth in order to be able to detect the neutrinos.

The main features of the Sudbury Neutrino Observatory (SNO), construction of which started in the early 1990s, are shown in Figure 9.4. To protect the detection sys-

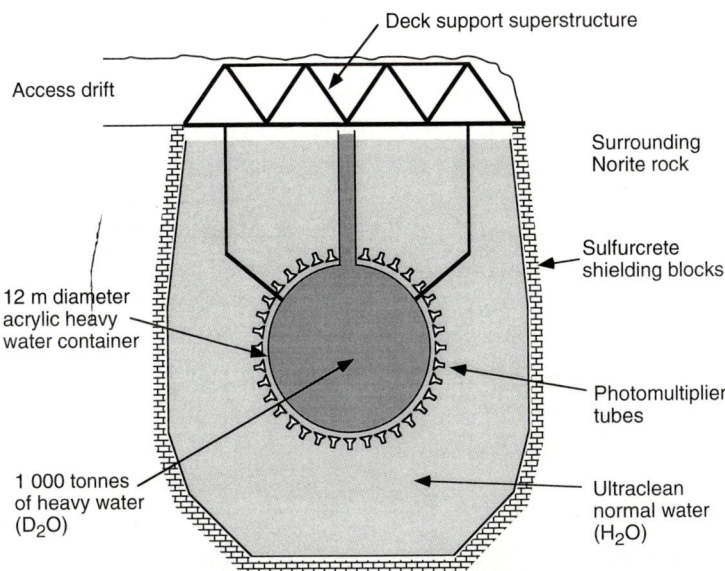

Figure 9.4 Structure of the detector core of the Sudbury Neutrino Observatory (SNO) built two kilometers below the surface of the earth in Sudbury, Ontario.

tem from ordinary cosmic rays, it is being built at a depth of 2 kilometers under the surface of the Earth in an abandoned nickel mine of the International Nickel Company, INCO. The availability of a suitable underground cavity in a worked-out part of the nickel mines in Sudbury was one of the main reasons for developing the *SNO Project* in Sudbury. The heart of the new neutrino detector, a facility equivalent to a ten-story-high building, is seen in Figure 9.4. At the center of the structure is a tank holding 1000 tons of *heavy water*. The availability of heavy water worth $300 million from Atomic Energy of Canada was another important factor in bringing this scientific research project to Sudbury, Ontario. Funding is being provided by a group of agencies from Great Britain, The United States, and Canada [10–13].

To understand what we mean by heavy water, we have to briefly review the structure of different types of atom. For the purposes of this discussion we can regard elements such as hydrogen, helium, carbon, and oxygen as being made up of three basic components. These are known as neutrons, protons, and electrons. An *electron* is much smaller than a neutron or a proton and orbits around the center of an atom which is composed of a mixture of neutrons and protons. The center of the atom is known as the *nucleus*. This comes from the fact that when scientists started to probe the structure of the atom, the first physicists were surprised to find how the mass of the various types of atoms was concentrated in a very tiny volume at the center of the atom — like the kernel of a nut. (Physicists who study the structure of the center of the atom are known as *nuclear physicists*.). The *proton* differs from the neutron in that the proton has a small positive electric charge equal and opposite to the negative charge of an electron. The electrons gather in certain groups of orbits as they travel around the center of the atom. The *neutron* has no electric charge and hence its name which means neutral (see Figure 9.5). As knowledge of the structure of the atoms of various elements increased, it was found that they can be arranged in a table, which became known as the *periodic table of the elements*, as shown in Figure 9.6.

The number of outer electrons determines the chemical activity of an atom whereas the structure of the nucleus determines its physical properties. Ordinary water is formed when two atoms of hydrogen combine with one atom of oxygen to share their electron clouds as shown in Figure 9.5(c). The incredibly small size of atoms can be appreciated by the fact that one of the largest atoms that can exist, uranium, has a size of several ten billionths of a meter and that atoms such as hydrogen are much smaller than the

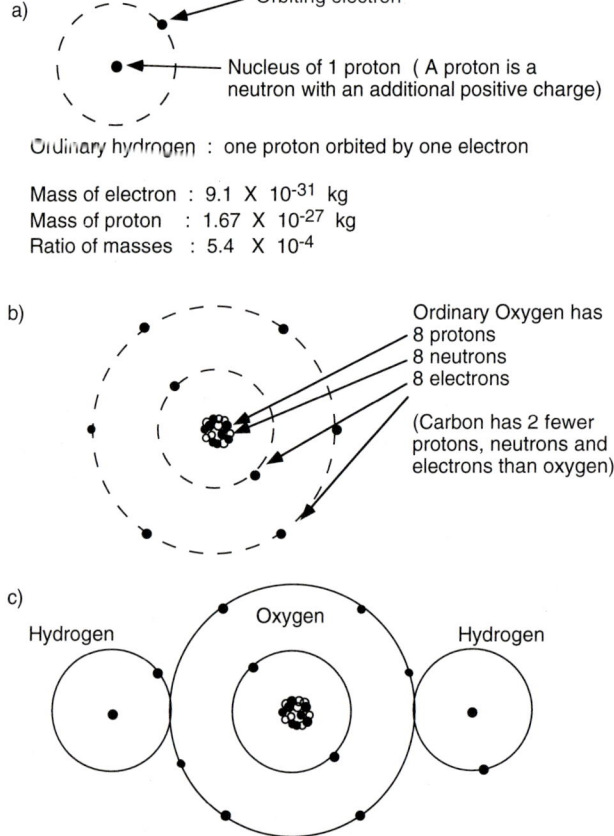

Figure 9.5 To understand what is meant by heavy water one must first understand the classic concepts of the atomic structures of the atoms of an element. a) Atomic structure of a hydrogen atom. b) Atomic structure of an oxygen atom. c) Schematic representation of a molecule of water (H_2O).

	2A 2 IIA		3A 3 IIIB	4A 4 IVB	5A 5 VB	6A 6 VIB	7A 7 VIIB	8 8 VIII	8 9 VIII	8 10 VIII	1B 11 IB	2B 12 IIB	3B 13 IIIA	4B 14 IVA	5B 15 VA	6B 16 VIA	7B 17 VIIA	0 18 VIIIA
1.0079 $_1$H																		4.0026 $_2$He
6.941 $_3$Li	9.0122 $_4$Be												10.811 $_5$B	12.011 $_6$C	14.007 $_7$N	15.9994 $_8$O	18.998 $_9$F	20.180 $_{10}$Ne
22.990 $_{11}$Na	24.305 $_{12}$Mg												26.982 $_{13}$Al	28.086 $_{14}$Si	30.974 $_{15}$P	32.066 $_{16}$S	35.453 $_{17}$Cl	39.948 $_{18}$Ar
39.098 $_{19}$K	40.078 $_{20}$Ca		44.956 $_{21}$Sc	47.88 $_{22}$Ti	50.942 $_{23}$V	51.996 $_{24}$Cr	54.938 $_{25}$Mn	55.847 $_{26}$Fe	58.933 $_{27}$Co	58.69 $_{28}$Ni	63.546 $_{29}$Cu	65.39 $_{30}$Zn	69.723 $_{31}$Ga	72.61 $_{32}$Ge	74.922 $_{33}$As	78.96 $_{34}$Se	79.904 $_{35}$Br	83.80 $_{36}$Kr
85.468 $_{37}$Rb	87.62 $_{38}$Sr		88.906 $_{39}$Y	91.224 $_{40}$Zr	92.906 $_{41}$Nb	95.94 $_{42}$Mo	98.906 $_{43}$Tc*	101.07 $_{44}$Ru	102.91 $_{45}$Rh	106.42 $_{46}$Pd	107.87 $_{47}$Ag	112.41 $_{48}$Cd	114.82 $_{49}$In	118.71 $_{50}$Sn	121.75 $_{51}$Sb	127.60 $_{52}$Te	126.90 $_{53}$I	131.29 $_{54}$Xe
132.91 $_{55}$Cs	137.33 $_{56}$Ba			178.49 $_{72}$Hf	180.95 $_{73}$Ta	183.85 $_{74}$W	186.21 $_{75}$Re	190.2 $_{76}$Os	192.22 $_{77}$Ir	195.08 $_{78}$Pt	196.97 $_{79}$Au	200.59 $_{80}$Hg	204.38 $_{81}$Tl	207.2 $_{82}$Pb	208.98 $_{83}$Bi	208.98 $_{84}$Po*	209.99 $_{85}$At*	222.02 $_{86}$Rn*
223.02 $_{87}$Fr*	226.03 $_{88}$Ra*																	

138.91 $_{57}$La	140.12 $_{58}$Ce	140.91 $_{59}$Pr	144.24 $_{60}$Nd	146.92 $_{61}$Pm*	150.36 $_{62}$Sm	151.97 $_{63}$Eu	157.25 $_{64}$Gd	158.93 $_{65}$Tb	162.50 $_{66}$Dy	164.93 $_{67}$Ho	167.26 $_{68}$Er	168.93 $_{69}$Tm	173.04 $_{70}$Yb	174.97 $_{71}$Lu
227.03 $_{89}$Ac*	232.04 $_{90}$Th*	231.04 $_{91}$Pa*	238.03 $_{92}$U*	237.05 $_{93}$Np*	244.06 $_{94}$Pu*	243.06 $_{95}$Am*	247.07 $_{96}$Cm*	247.07 $_{97}$Bk*	251.08 $_{98}$Cf*	252.08 $_{99}$Es*	257.10 $_{100}$Fm*	258.10 $_{101}$Md*	259.10 $_{102}$No*	260.11 $_{103}$Lr*

Figure 9.6 The elements of the universe can be arranged in a display known as the periodic table. This table was originally developed by the Russian scientist Dmitri Mendeleyev (1834–1907). Atoms of elements which differ from each other only in the number of neutrons in their nucleus occupy the same place in the periodic table and are described as different isotopes of that element.

uranium atom. Different elements correspond to atoms with different numbers of protons in their nuclei. Thus, ordinary carbon has two fewer neutrons, two fewer protons and two fewer electrons than oxygen. The incredible emptiness of the inside of an atom can be appreciated through the following analogy. If the center of the carbon nucleus was John F. Kennedy Airport in New York State, then the outer orbiting electron would be the size of a small sports plane flying over Bermuda. When a neutrino flies through the Earth, it is this incredible emptiness of the atoms making up the planet that accounts for the fact that most neutrinos travel straight through. It has been calculated that a neutrino passing through one hundred light years of lead only has a 50% chance of colliding with an atomic particle and of being captured.

The terms ordinary carbon and ordinary water are used above because many elements have atoms in which the number of neutrons in the nucleus can vary slightly. Ordinary hydrogen, as pointed out, has one proton, one electron, and no neutrons. Because the variations in the number of neutrons in an atom of an element do not effect its chemical properties, the various forms of atoms of the same element are said to occupy the same position in the table of elements. The Greek word "iso" means "the same" and "topos" means "place"; therefore the atoms of the same element with different nuclear structures are said to be *isotopes of the element*. This name was invented in 1913 by the British chemist Frederic Soddy. If we add an extra neutron to the structure of an oxygen atom, the mass changes by a ratio of 16 to 17. This does not result in a drastic change in physical properties and the two types of atoms are simply referred to as isotopes of oxygen. However, in the case of hydrogen, adding an extra neutron to the nucleus doubles its mass. This changes its physical properties significantly and scientists have thus given this isotope of hydrogen the name *deuterium* from the Latin word for two. There is one other isotope of hydrogen, which has two neutrons added to the nucleus; it is called *tritium* from the Latin word for three. Tritium is actually unstable and is described as radioactive because of the way in which it can spontaneously disintegrate. When both of the hydrogen atoms in a molecule of water are the deuterium isotope, the molecule is described as heavy water. If you were to carry a bucket full of heavy water, you would not notice the difference in weight because the heavier oxygen atoms dominate in determining the molecular weight. A very small fraction of deuterium is present in ordinary water and it can be concentrated in a container by using electrical energy. Heavy water has been used by Canadians in the design of the so-called heavy-water nuclear reactor to generate electrical power. (This nuclear reactor system is referred to as the *CANDU reactor*, which stands for CANadian Deuterium Uranium reactor.) The technical term for heavy water is *deuterium oxide* and its formula is D_2O instead of H_2O, or *hydrogen oxide*. There is a major advantage in using deuterium oxide as the basic detector of neutrino bombardment of the earth. When a neutrino is captured by a deuterium nucleus, a reaction takes place, resulting in the emission of high speed electrons, which initially traveling through water at above the speed of light. (Remember the speed of light in water is less than that in a vacuum.) Once in the water, the electrons must slow down; therefore, when a neutrino capture event occurs, the energetic electrons leaving the reaction area give off Cherenkov radiation.

This Cherenkov radiation is detected by the 2000 *photomultiplier light detectors* placed around the tank of heavy water. As the name implies, photomultiplier tubes are very sensitive detectors of light energy which actually amplify the electrical energy generated by the absorption of photons in their structure to create an electric current which can then be measured by electronic devices. Because the reaction core is surrounded by 2000 of these photomultiplier detectors, scientists can work out the direction in which the neutrino was traveling when it interacted with a deuterium atom. This fact will be used by the scientists of the SNO project to work out where the neutrinos are coming from and hence the type of stellar object which generated them. It will be relatively difficult to replace any of these photomultiplier tubes if they malfunction after installation. For this reason, scientists building the facility took special precautions to ensure that the highest quality tubes were used. It is estimated that the system will continue to function efficiently even if random failure of the tubes leads to up to 10% of them becoming inactive. Above a 10% level of malfunction, steps would have to be taken to replace some of the damaged tubes.

Local blasting in the nickel mine is still proceeding in the general area of the SNO facility. Computer modeling has established that the structure will be able to absorb any local stress generated by distant mining explosions.

As can be seen in Figure 9.4, the outer shield of the SNO detector is made of sulfurcrete. Normal *concrete* is a mixture of gravel, sand, and cement powder. The gravel is referred to as aggregate. In theory, the sand fills in the gaps between the packed aggregates and then the cement reacts with water to seal up the array of sand and gravel to form a solid body. Several types of special concretes are used in various industries; thus, in the building of highrise buildings, the gravel aggregate is replaced with hollow glass spheres to make a lightweight concrete. In the shielding of some nuclear reactors, the chemical barium sulfate is added to the concrete to make it particularly dense to absorb radioactive energy coming from the nuclear reactor. *Sulfurcrete* is a special kind of concrete in which the cement powder binder is replaced by sulfur powder treated with a small amount of plastic as the binder material. The sulfur is substituted for the cement because of its purity and lack of radioactive constituents. The reader should not be alarmed, but the fact is that ordinary concrete does have a very low level of natural radioactivity which can interfere with an experiment as delicate as the SNO project. For the same reason, a very special aggregate composed of crushed dolomite is used in the sulfurcrete. Chemically, *dolomite* is calcium magnesium carbonate ($CaMgCO_3$) which is named after the geologist *de Dolomieu* (1750–1801).

To prevent any remaining local radioactivity in the ground from penetrating the sulfurcrete protective wall, the acrylic vessel and the photomultiplier tube system are surrounded by *ultraclean normal water*.

The construction costs of the SNO are of the order of $47.5 million, the value of the heavy water is placed at $300 million, a contribution to the project in the building of the cavity to house the core by INCO is valued at $150 million. The start-up cost over the first 5 years from 1991 is placed at $61 million. In return, we can anticipate that this expensive search for the tiny neutrino will advance our scientific knowledge about the fundamental properties of the universe surrounding us.

References

[1] For the dynamics of Thor's thunderbolts, otherwise known as lightning, see D.L. Elsom, "Learn to Live with Lightning," *New Scientist*, June 24, 1989, pp. 54-58.
[2] For a discussion of the concept and history of the studies of latent heat in the change of state see any first-year University Physics textbook such as J.D. Cutnell, K.W. Johnson, *Physics*, John Wiley & Sons Inc., New York, 3rd Edition, 1995.
[3] For a good discussion of the role of nucleation in science see A. Walton, "Nucleation," *International Science and Technology*, December 1966, pp. 29-33.
[4] L.J. Battan, "Changing the Weather," in *Modern Science and Technology*, ed. by R. Colborn, Van Nostrand, New York., 1965, pp. 603-610.
[5] C.Knight, N. Knight, "Hailstones," *Scientific American*, Vol. 224, No. 4, April 1971, pp. 97-105.
[6] J. Hallett, "When Hail Breaks Lose," *Natural History*, June 1980, pp. 55-57.
[7] K. Phillips, "Breaking the Storm," *Discover*, Vol. 13 No. 5, May 1992, pp. 63-69.
[8] J. Needham, *Science and Civilization in China, Vol. 5, Chemistry and Chemical Technology, Part 7, Military Technology, The Gunpowder Epic*, Cambridge University Press., Cambridge, 1971.
[9] See the report "Hail Studies Field Program" of the Research Council of Alberta, Edmonton, Alberta, ed. by J.H. Rennick, 1970
[10] L. Wolfenstein, E. W. Beier. "Neutrino Oscillations and Solar Neutrinos," *Physics Today*, Vol. 42 No. 7, July 1989, pp. 28-35.
[11] Hong-Yee Chiu, "Neutrino Astronomy," in *Modern Science and Technology*, ed. by R. Colbourne, Van Nostrand, New York, 1965, p. 502.
[12] K. Zimmer, "Watery Eyes," *Discover*, Vol. 13 No. 3, March 1992, pp. 20-23.
[13] SNO Project Literature is available from Dr. E.D. Hallman, Department of Physics and Astronomy, Laurentian University, 935 Ramsey Lake Road, Sudbury, Ontario, Canada, P3E 2C6.

Chapter 10

Humans as Missiles and Targets

Chapter 10

Humans as Missiles and Targets

The title of this chapter probably conjures up the idea of taking human beings and firing them from a cannon. This of course does happen at the circus; however, such direct use of the human body as a missile will not be our major concern here. Rather, we will be looking at the fact that in several sports and in everyday situations such as traffic accidents, the human body is either a missile or a target. In our discussion of tennis and other racket sports we briefly discussed the use of protection against sports missiles and in a later section of this chapter we will look at how the body can be protected from missiles as diverse as speeding bullets and spinning balls. We will also look at how sports people such as skiers and swimmers are missiles and how scientific investigations are helping to improve the performance of athletes in sports such as skiing and bobsledding. We look at the injuries caused in boxing when the head of the boxer becomes a target for human fist missiles. In road accidents, human beings can be protected with special clothing and/or equipment such as airbags.

Shortly after I came to live in Sudbury, a local teenager was killed in a car accident which she would have survived had there not been an unrestrained portable sewing machine in the trunk of the car. When the car was decelerated in the collision, the sewing machine, acting as a missile, came from the back of the car and killed the driver. Hopefully, the material presented in this chapter will not only be of interest scientifically but may help some readers avoid inadvertent injury if they are unfortunate enough to be involved in a car collision or a sports accident.

Section 10.1 Belly Flops and Bungee Jumping

One of the ways in which individuals can hurt themselves when acting as missiles is when diving into water. Many diving accidents and deaths are caused by drunken diving, but in this section we will confine our discussion to injuries which can occur during responsible diving. Some calculations of the energy involved in sports collisions are given in the physics text by Orear [1]. We will present the results of some of his calculations and the interested reader can check the entire calculations by referring directly to his book.

One of the problems discussed by Orear is what happens to a swimmer who makes a belly flop into water. In a belly flop, a diver enters the water belly first causing a large splash and possible injury. Using the appropriate formula for a typical adult diver, Orear shows that a belly flop from any height greater than 15 meters would involve serious injury. He points out that if a diver attempts to enter water from a height greater than 15 meters, he should try to reduce the area of impact by entering the water head first with hands outstretched above the head to make contact with the water first; alternately, the diver should jump with feet first. Note, however, that when jumping onto dry land, a person should always attempt to allow their legs to bend in order to absorb the impact of landing (something that we do naturally from experience), otherwise severe injuries to the legs can be sustained when jumping from a height as small as one meter [1]. Some humans have survived falls from relatively great heights when they have been drunk. In their drunken condition, their bodies are so relaxed that they act as absorbers of energy. People who are frightened of jumping, and thus become tense and rigid, can suffer severe injury. This is why firemen use a device with a stretchable surface to rescue people from burning buildings. If a person is able to jump into something soft, the material absorbs the energy of the fall and the falling individual does not sustain severe injury. Orear discusses the possibility of someone without a parachute falling from an aircraft and surviving the fall. A free-falling human body stops accelerating when the drag of the air around the body becomes equal to the force of gravity. This occurs at a velocity of approximately 192 kph (120 mph) or 53 meters per second. Sky divers who often take part in acrobatic displays of free fall gymnastics spread-eagle themselves in the horizontal direction to increase the drag around their bodies before they release their parachutes.

Orear has calculated that a person who is in free fall without a parachute can be brought safely to rest by a depth of approximately 3 meters of snow. Orear relates the following story originally reported by R. G. Snyder [2]:

> *During one of the battalion drops from one thousand two hundred feet on a clear relatively warm day, an observer noted what appeared to be an unsupported bundle falling from one of the C-119 aeroplanes. No chute deployed from the object. The impact looked like a puff around an explosion in the snow. When the airmen reached the spot, they found the young paratrooper on his back at the bottom of a 3 1/2 foot crater in the snow which consisted of alternating layers of soft snow and frozen crust. He could talk and did not appear injured.*

Prototype parachutes for breaking the fall of someone jumping from a high tower were first described in Europe in 1500 by Leonardo Da Vinci. The Chinese, however, had used the basic concept of a parachute much earlier. There is a story in Chinese literature dating from 90 BC in which someone who was condemned to death by being thrown from a tower escaped death by using a bunch of cone shape straw hats. A story concerning a Chinese acrobat who parachuted to earth using an umbrella as a float was told by a French writer in his memoirs in 1688. Lois Sebastian Lenormand read about such devices in the memoirs a hundred years later and tried out the technique. It was he who gave the parachute its present name. He suggested to Montgolfier, a hot air balloonist, that he should try the parachute as a safety device [3].

In recent times, a sport called *bungee jumping* has reached the western world from Australia. In this sport, people jump from a great height with stretchable ropes around their legs. The length of the rope is calculated so that the kinetic energy of the falling person is absorbed by the stretching ropes just before the individual would have hit the ground [4]. All of the kinetic energy of the fall becomes potential energy in the stretched rope. The sport appears to have been copied from some Pacific island tribes who test the manhood of their young men by having them jump from high platforms to which they are tethered by stretchable vines.

Section 10.2   Sheepskin Jackets and Metal Helmets

Throughout history, individuals have tried to protect themselves from missiles launched against them. The earliest documents of civilized man show him as an inventor and designer of armor. The subject of ancient armor illustrates that ancient civilizations had some considerable knowledge of "high technology". Thus Blyth has made some interesting studies of the design of ancient weapons [5, 6]. He remarks that Greek artwork depicts heavily armed soldiers wearing body armor and helmets and using shields but carrying relatively light spears as they go into battle. The warrior has always faced a choice between the weight of defensive armor and the ability to use offensive weapons. Blyth states that it would appear from the evidence that, in the Greek world, both arrows and spears were made very light and that the limiting factor was the danger of buckling or fracturing the shaft of a weapon which was being thrust into an adversary. An arrow is obviously designed to be used only once and it is indeed an advantage if it is useless for second shots. The last thing a warrior wanted was for the enemy to pick up an arrow and return it as an offensive weapon. This is part of the explanation for the arrowhead construction shown in Figure 10.1(a). The hardness of the point of a weapon, such as an arrow, is an important factor in determining whether, when thrust with a given amount of energy, it will be able to penetrate a protecting surface. It is the pressure at the arrow tip, and not the overall driving force, that determines whether the penetration will occur at all (see the discussion of the sharpness of a knife in Chapter 2). If the tip of an arrow starts to deform, the pressure exerted on the surface to be penetrated diminishes significantly. Thus Blyth tells us in his account of ancient weapons that the Persians might have had greater success in ancient Greece if their arrows had been tipped with flint stone instead of bronze metal. Bronze-tipped arrows deform more easily than stone-tipped ones [5].

The backward facing points at the rear of the arrow are designed to prevent the arrow from being pulled out of the body after it has inflicted a wound. This fact was demon-

a) The penetration capability of an arrowhead depends on the area of the tip.

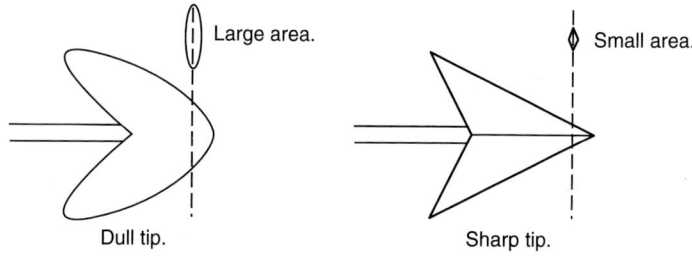

Barbs on the arrowhead prevent the arrow from being easily withdrawn from the target

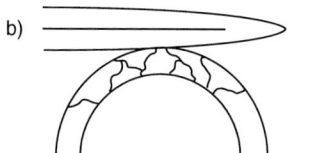
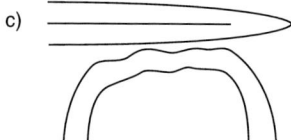

If a helmet is too hard, it will crack open when hit by a sword and only absorb a small amount of the energy of the blow.

If a helmet is made of soft metal, it will deform, rather than crack, when hit by a sword and absorb the energy of the blow.

Figure 10.1 Effective armor relies on absorbing the energy of the offensive weapon. a) The area of the point of an arrow head determines its ability to penetrate the target. b) A helmet made of a brittle material will crack if hit. c) A helmet made of a soft material will deform and absorb the energy of a blow.

strated dramatically in a Clint Eastwood movie "Two Mules for Sister Sarah" in which after he had been wounded by a barbed arrow his companion had to push the arrow through his body to enable it to be removed without creating a terrible wound.

Many people, before undertaking a study of the physics of the situation, would react to the question "What should a shield or a helmet be made of?" by choosing the hardest material available. Their reasoning would be that the hard material would best prevent penetration by the aggressor's weapon. However, in fact, the real task to be carried out by a protective surface is not to resist the weapon blow but to absorb the energy of the blow. Blyth has studied Greek equipment preserved in British museums and has shown that from a technological point of view, the equipment is amazingly sophisticated. To understand one of the surprising properties of the metal helmets that he studied, we need to look at an important property of metals and the process known as annealing. A metal is said to be *ductile* if it can bend relatively easily under an applied force. Pure copper is often quite ductile, but it can be made harder by *work hardening*. You can demonstrate this for yourself by using a simple metal paper clip which is relatively easy to bend open when it is new. If you work a metal paper clip back and forth, it becomes hard and will eventually snap into two pieces. What happens is that as you bend the clip back and forth, you create dislocations in the crystal structure of the original metal and as the metal is continuously worked, the number of these dislocations increases, making the material better able to resist further deformation (by becoming harder and more brittle). It is not as easy to move the dislocations around in the metal as it is to create them. The brittleness of the hard metal is one of the problems which has to be faced by designers of efficient weapons. If you work harden a piece of metal to make an arrow tip hard, you must balance the increasing hardness of the arrowhead against its likelihood of it breaking on impact because of its brittleness. In setting out to fabricate a protective helmet out of piece of metal, the necessary hammering of the piece of metal not only shapes it into the desired pattern but also work hardens the metal. However, hardness in a helmet, if it is brittle, is not a desirable property. If the helmet is hit hard with a sword, it will crack open and only a small amount of the energy of the blow will be absorbed by the helmet as shown in Figure 10.1(b). Metallurgists discovered early on that a piece of work-hardened metal can be made soft again by heating it to a temperature just below its melting point and holding it at this temperature for a certain length of time and then letting it cool slowly. At the higher temperature, the small crystals in the work-hardened piece grow into larger, softer crystals so that the material becomes more ductile. This process is known as *annealing the metal*. The word *anneal* comes from an old English word meaning to burn or heat. (The term ductile comes from the Latin word "ducere" meaning "to lead.") Blyth tells us that his metallographic investigation of Greek helmets shows that the metal had been annealed *dead soft* (that is, made as soft as possible). This had obviously been done deliberately. This fact shows that the Greeks knew that safety depended on absorbing the energy of a blow, not on resisting it as shown in Figure 10.1(c). Blyth has discovered from his measurements that the bottom of a Greek helmet was about one millimeter thick, but that over the forehead and temples, the thickness was 2.5–3 mm or more. In his thesis, Blyth uses this as evidence that Greek armor was carefully designed to maximize efficiency and minimize weight, and that the choice of metal armor for protection is the reason the Greeks used relatively light weapons in their offensive tactics. Indeed, he points out that heavy reliance on stored energy in attack and immobility in defense is characteristic of barbarians such as the Franks, Gauls, and Saxons. That is why the Barbarians attacking the Romans used the club, the axe, or the broad sword (a large heavy sword in which a great deal of kinetic energy could be stored before it came in contact with an enemy). The more sophisticated Romans relied on armor for defense and on mobility when using offensive weapons. These two styles of fighting resulted in contrasting approaches to war and probably led to the defeat of Harold, the leader of the Saxons, at the Battle of Hastings in England in 1066 against William the Conqueror. William the Conqueror used armored horsemen along with bowmen, both of whom used light mobile weapons, against the Saxons who were much more used to fighting offensively in tight formation with high-energy weapons such as the club and the axe. Harold was killed by an arrow through the eye while waiting to use his high-kinetic-energy axe.

In the television show *Death Valley Days*, which recreated legends of the American West, one of the stories told was about a shepherd who was legendary amongst the Mormons of Utah. It was said that this shepherd had the ability to resist bullets. He is said to have been shot at several times; the bullets fell out of his clothing onto the floor without harming him. The legend probably resulted from the fact that the shepherd, used to spending nights on the mountains, wore a double sheepskin jacket. The mass of the wool fibers in such a garment would be a very efficient absorber of energy and provide effective protection against slow bullets. Probably the first time someone fired at him was from a gun with poor ammunition. As the legend grew, people probably stayed farther away when trying to shoot him. As a result the energy of the bullet was even less than if attempting to shoot him at close range.

Such woollen armor would not be particularly effective today — although during the second world war, aircraft personnel were given jackets made of sheepskin with the leather on the outside and the mass of wool from the original sheepskin on the inside. Sheepskin leather was used because its fibrous mass of wool provides both insulation to keep the flying crew warm and missile protection against flying fragments of metal if the aircraft were to be hit by anti-aircraft missiles. The word *flak* was used by the English-speaking airmen to describe the missiles from anti-aircraft guns and also the fragmented material from any hit on the aircraft by such missiles. The word flak comes from a shortening of the German word "Fliegerabwehrkanone" which literally meant "anti-aircraft gun." The phrase entered modern English in phrases such as "he took a lot of flak from the investigating committee" meaning that the person had been attacked verbally by the committee and had to absorb verbal comments in the same way as the sheepskin jackets absorbed the metal fragments.

It is reported that the Chinese used to employ paper armor because, again, several layers of paper would constitute an effective method of absorbing the energy of a missile. The Sikhs of Northern India are a warrior people who from birth do not cut their hair. In adult life they collect the long hair into a cloth turban made out of long lengths of fabric. Although this is considered a religious practice, some scholars believe that the practice of keeping long hair wrapped with a cloth turban was actually a defensive device against sword cuts to the head.

The effectiveness of a fibrous mass against the penetration of a weapon comes from the fact that before progress can be made by the entering missile, every fiber in the way of the missile must be broken. The breaking of the multitude of fibers is a very effective way of absorbing the energy of a missile. The stronger the individual fibers, the more effective the fibrous mass is as an armor system. In 1935, the Dupont corporation invented nylon which proved to be a very strong fiber and suitable for lightweight body armor. Later designs of the previously metal-lined flak jackets were nylon fiber packed. During the later stages of World War II, it is said that this nylon-fiber type of flak jacket reduced casualties amongst bomber crews by 60% [7].

In 1965, Dupont discovered a way to make a plastic known as Aramid and to spin and weave it into a fabric. Their trade name for this superstrong plastic fiber was *Kevlar*

[8]. This fiber has 2.5 times the tensile strength of nylon. That is, it can withstand 2.5 times the stress that would break a nylon fiber. We now know that one of the factors which determines the effectiveness with which a material dissipates energy is the speed of sound in that material. Thus, J.E. Gordon commenting on the ability of objects to absorb energy makes the following statement [9]:

> *The speed of sound in steel, aluminum and glass is approximately four-thousand-eight-hundred meters per second (eleven-thousand miles per hour). This is much faster than the speed of sound in air. Such speeds are far faster than any hammer blow and considerably faster than the flight of bullets. The result is that a hammer or a bullet is pressing against its target for a period, perhaps of about a hundredth of a second, which is very long compared with the time which is required to conduct the energy away from the point of impact in the form of compression waves which are in fact sound waves.*

McKean [7] points out that because sound travels three times faster through Kevlar than through nylon, the stress of a bullet hitting a vest made out of Kevlar is more rapidly distributed over the entire vest. The effectiveness of body armor made out of Kevlar is illustrated in Figure 10.2. Because lead is a ductile metal, lead bullets are deformed by the collision with the body armor (steel, which is not ductile, is used to make armor-piercing bullets).

Dupont originally developed Kevlar as a replacement for steel reinforcement in automotive tires. In 1971, Lester Shubain of the National Institute of Law Enforcement in Criminal Justice suggested that the fiber might make an excellent body armor [7]. Shubain was the manager for standards at the institute and he was looking for a material to replace the existing nylon and metal body armor. The nylon and metal vests were too heavy for routine work and all nylon fiber armor was too bulky and hot. The Kevlar armor that Shubain developed proved to be very light and effective. It should be noted, however, that the absorption of the energy of the bullet, while preventing bullet wounds, cannot protect the wearer against bruising.

One of the very early success stories of the Kevlar armor illustrates this point. In 1977, David Schafer, a policeman in Bettendorf, Iowa, bought his own Kevlar vest (at that time it cost US$127). He was wearing this jacket when he stopped a burglary suspect who fired a 0.45 caliber bullet into Schafer's stomach from six inches away. Schafer tells us that when they took the vest off, they found the bullet in the outer covering. Apart from a bad bruise and a small burn, Schafer was unharmed. As early as 1981, it was estimated that Kevlar vests had saved the lives of 300 law-enforcement officers. Electron microscope studies at the US Army's Natick Laboratories in Massachusetts show that nylon fibers melt when a bullet strikes them. Glass-like fibers which do not melt are subject to transverse fractures which reduce their effectiveness in body armor. Part of the effectiveness of Kevlar armor is that the threads of Kevlar pull into long slender fibers absorbing energy before they break. Many important people, such as presidents and royalty, are said to be wearing Kevlar armor to protect them from assassins'

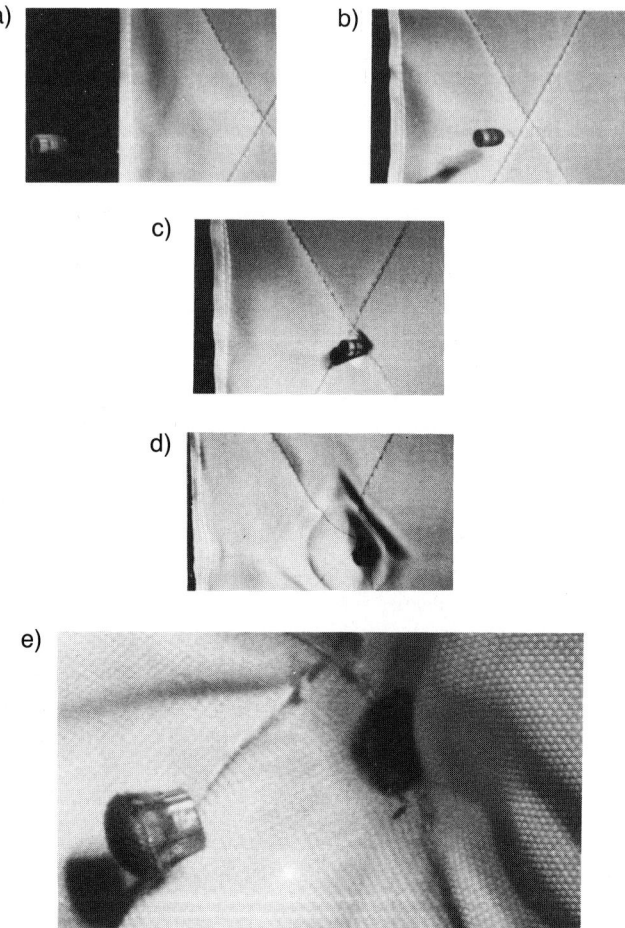

Figure 10.2 The protective power of body armor made from Kevlar is illustrated by the above sequence of pictures. The sequence shows a 0.38 caliber bullet, traveling at 244 meters per second (800 feet per second) being stopped and blunted on impact with seven layers of Kevlar fabric. (Each exposure was less than 1 millionth of a second (10^{-6}s).) Photo courtesy of the Dupont Co. Kevlar® is a Dupont registered trade mark.

bullets. Kevlar is also being used in other safety equipment such as industrial gloves.

Material that can resist the penetration of bullets can also resist the cutting effects of sharp items such as broken glass. Kevlar safety gloves should be worn by anybody picking up broken glass [8]. The edges on broken glass are amongst the sharpest known to man. Slices of animal tissue to be viewed through a microscope are cut with an instrument known as a *microtome*. This word literally means small slice, from the Greek word "temnein", to cut. When we use the word *tome* for a large book, we are using the same word since, originally, a tome was part of a large scroll cut into slices so that each part of the scroll could be handled more easily. The microtome is equipped with freshly broken glass slides to cut the tissue to be studied.

Scientists working for the US Army are studying the possibility of making the armor of the future from silk similar to the threads of a spider's web. It is predicted that spiders' silk would be at least three times as efficient as Kevlar in stopping bullets [10-12].

Section 10.3   Crash Helmets and Safety Visors

In discussing head injuries and their prevention using safety helmets, Gordon [9] makes the following points.

When a projectile is fired into a tank of liquid, such as the fuel tank of an aeroplane, because of the reflection of shock waves, the exit hole is much larger and more difficult to seal because the shock waves which are readily transmitted through the fluid can burst the tank. Unfortunately, the human head is structured rather like a tank of liquid. When it is struck by a bullet, the shock wave creates the same type of wound.

In the J.F. Kennedy assassination studies a great deal of interest centers around the size of the entry and exist holes of the bullet that killed the President.

An essential piece of a modern soldier's equipment is a helmet. Industrial workers also are provided with helmets known as hard hats to protect them from falling objects, bumped heads, and other industrial hazards. Increasingly it is being recognized that many sports people should also wear head protection. For example, in the last ten years, people playing ice hockey and cricket have started to wear protective head gear. However, this equipment is not always as effective as it could be and hopefully some of the information provided in this section will lead people to consider better design of protective head gear both for the work place and for the sportsfield.

In the previous section we discussed how protective armor should be designed to absorb the energy of a blow, not to resist it. To the uninitiated, the inside of a protective helmet used by industrial workers, appears to be designed to ventilate the helmet rather than to make it effective since the hard hat appears to rest sloppily on the head by means of a series of straps with sponge rubber spacings. A typical industrial hard hat is shown in Figure 10.3 [13]. The straps and sponge spacers are there to absorb energy of blows to the helmet [14]. Gordon tells us [9]

The important factor in the design of crash helmets is to cushion the shock waves (from the blow to the head) so as to prevent damage at the back of the skull. This is the reason for the internal headbands in helmets which look as if put there in order to provide ventilation.

On the industrial scene, it is rather surprising to learn that the safety helmets worn by miners in Canada are tested for their ability to resist blows to the top of the head

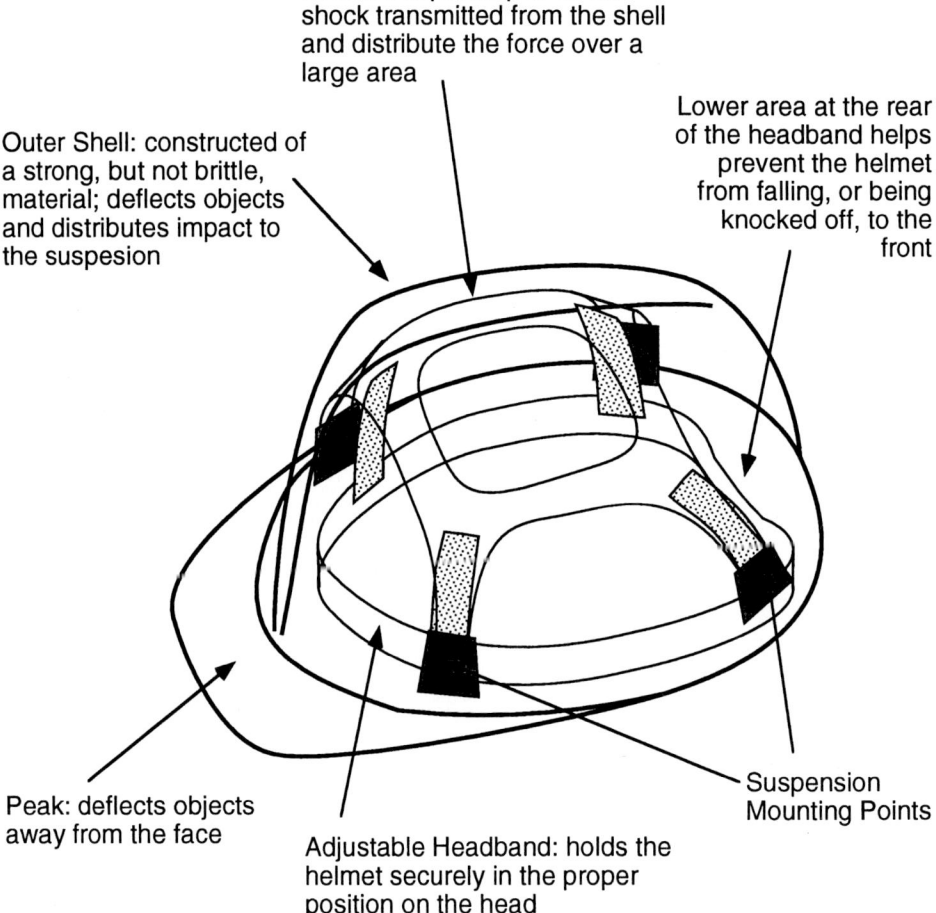

Figure 10.3 Protective headgear is intended to absorb the energy of a blow not to resist it. Shown above is the internal structure of a typical hard hat.

but not to the side of the head and that many accidents in mining situations are actually blows to the side of the head. There is also a lack of research into the design of integrated safety equipment such as a combination of ear muffs against sound, safety glasses for the eyes, and helmets to cushion inadvertent blows [14].

The average motorcyclist is not aware of the fact that crash helmets only protect the head in relatively low speed collisions [15]. Thus Bryan Chinn and his colleagues at the Transport and Road Research Laboratory in Great Britain have been studying the effectiveness of a crash helmet used by motorcyclists. They report that in a head-on impact, a normal helmet will only save the rider from death at speeds of less than 11 kph. However, these helmets are still better than nothing according to Chinn as most accidents involve glancing blows rather than head-on collisions. Seven hundred motorcyclists die every year on the roads of Great Britain and although by law the riders must wear helmets, most of the deaths (about 80%) are still caused by head injuries. It is for this reason that scientists are developing airbags for motorcyclists as discussed in Section 10.5. Chinn [15] tells us that in Britain, surprisingly,

> *The outer shell of current helmets is too strong and the helmets tend to bounce too much. They have a return velocity of 60% of their impact speed.*

He adds that

> *Studies show it is not just a peak acceleration that causes damage to the head during a motorcycle crash but also the length of time that it is subjected to this treatment. Researchers need to find materials which bounce less and absorb energy more efficiently. In British-designed helmets, the energy is absorbed in the polystyrene used inside the helmet which becomes solid when compressed to half its volume. A honeycomb design made of paper and resin would be able to lose 80% of its volume and thus would make more effective use of the space inside of the helmet.*

On a visit to England in 1991, I was pleased to see that cricketers had started to wear protective helmets of the type already common among baseball players, and face protection. When I came to live in Canada in 1968, professional ice hockey players, other than the goalie, did not wear protective helmets. It was considered sissified to wear protective head gear. Thankfully that nonsense situation has changed and by the early 1990s only the occasional foolhardy player could be seen without head protection.

The group of sports people who now need to be convinced of the sense of wearing head gear is cyclists [16]. Already many professional racing cyclists are beginning to wear head protection because they realize the danger of serious head injuries from tumbling off a bicycle. One of my own cousins died from a cracked skull when he fell off a bicycle at the age of thirteen. Two case histories will illustrate the unnecessary suffering and injury created by the failure of young cyclists to wear protective head gear.

In Sudbury, in September 1991, a young boy named Dylan Martin was crossing a local road on his bicycle when a car clipped the front wheel of the bicycle. Dylan was thrown 30 meters in the air before hitting the pavement head first. He was not wearing his cycle helmet. He didn't wake up for two and a half months. When he came out of the coma, his short-term memory had disappeared. He could neither walk nor write. His long-term memory was unaffected. By April of the next year, he had made progress in motor skills such as walking and writing, but he still had to ride around in a wheel chair. Only Dylan's head was injured in the accident and there was not a scratch on any other part on his body. The path to full recovery is still a long hard way ahead for him. A bicycle helmet would probably have saved his brain from serious damage. Dylan, his mother tells us, when he was six or seven, used to wear his helmet, but then peer pressure from the neighborhood kids caused him to leave it at home that fatal day. Fortunately, since October 1, 1995 in Ontario, all cyclists under 18 must wear cycling helmets.

A second story concerns Robby Loftus who was riding his bike down a busy street in Montreal. He swerved to avoid a truck pulling out of a fast-food restaurant and veered into the path of a passing bus. The eight-year-old boy ricocheted off the bus and cracked his head against the curb. At the age of twelve, four years later, Robby is a paraplegic and a recovering head injury patient. His parents are now spearheading a campaign to encourage the use of cycle helmets among children [17].

The reader should contrast these stories of the unfortunate Dylan Martin and Robby Loftus with that of Ryan Norman, reported in the local press in May 1995 [18]. On September 4, 1994 Ryan rode his bike down a hill near his home and into the path of a truck. Ryan, who was 11 at the time, hit the side of the moving truck, was knocked off his bike and lay on the road unconscious. As his mother described it

He was out cold when I got there. I didn't realize it but the helmet he was wearing had broken in three pieces.

Thrown from his bike, Ryan hit his head and other parts of his body on the hard pavement. (Note, in Canada the pavement is the road surface; in England the pavement is the part of the road known as the sidewalk in North America). Witnesses and doctors say that the boy's life was saved by the helmet. He was taken to the emergency department and was observed for a period but then was allowed to go home. His mother reports that since the accident Ryan has suffered some memory problems but doctors are not sure if they are permanent or even if they were caused by the accident.

This successful story of the protection offered by a helmet should encourage all parents to insist that their children should wear cycle helmets. In countries where most medical expenses are covered by insurance held by the individual it is quite possible that insurance companies may refuse to pay all of the expenses incurred in treating injuries sustained by a cyclist who was not wearing a helmet. This would probably have more effect on the wearing of helmets than all the laws and preaching that one can muster for the topic.

A cyclist who survives a crash should replace their helmet if its shock-absorbing material has been crushed in absorbing the energy of the impact. It should also be noted that putting stickers on a helmet or painting it can cause chemical reaction that could make it brittle. Experts warn that placing stickers on a safety helmet could invalidate the warranty provided by the manufacturer.

In some cities in Great Britain, cycle lanes are being provided on existing roads. In Oxford in 1980, when cycle lanes were provided, the injury rate dropped to 0.4 injuries per kilometer as compared to 0.9 injuries per kilometer on ordinary roads. The study also found that only a small number of cyclists, 0.7%, wore helmets. As a consequence, one third of the cyclists injured in accidents had head injuries whereas this figure was only fifteen percent for motorcyclists [15].

As discussed earlier, helmets are now widely used in ice hockey, but players still suffer a large number of eye injuries which could be prevented by the use of visors, i.e., transparent protective shields placed over the eyes. A referee in the Western Hockey League,

Brent Lawson, was not wearing a visor when he was hit by a puck in the eye and his sight was damaged. Another player, Tom Honsberger should have known better: he wore safety glasses at work, but not when playing hockey with his friends. He was hit and blinded in one eye. Doctor Tom Pashby has made a study of eye injuries amongst ice hockey players. He tells us that at least 266 ice hockey players have suffered blinding in Canada since 1972. Some governing bodies are beginning to adopt mandatory visors and it appears that most of the current injuries are to the older players who still resist the use of protective eye equipment. The statistics are summarized in Figure 10.4. A surprising feature of the data summarized in Figure 10.4 is the initial increase in eye

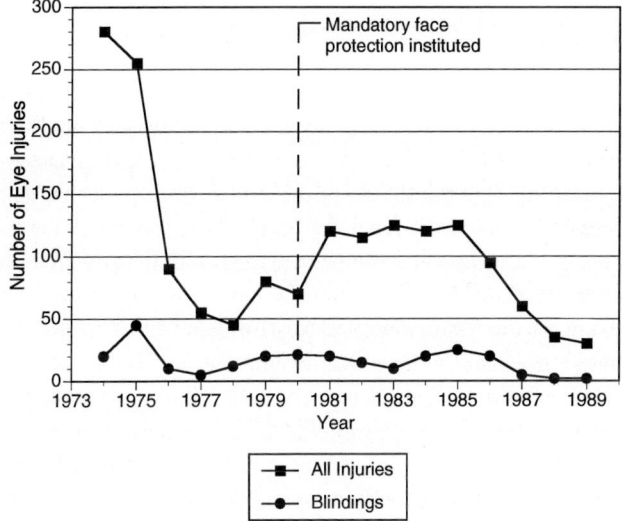

Figure 10.4 Since eye visors have become more widely used in ice hockey, eye injuries are in significant decline.

injuries after face protection was introduced. Scientists believe that this could be a manifestation of the "Superman Syndrome". Because eyes were better protected, players increasingly 'high sticked' and played more aggressively. In fact, after the Quebec major junior hockey league adopted mandatory facial protection, facial injuries went down as predicted but the number of shoulder injuries increased as a result of high-stick play. In general when soldiers and/or sportsmen wear protective gear, the effect can sometimes be counter-productive because of the superman syndrome. The wearers of the protective equipment are sometimes under the impression that they are invulnerable. One army instructor tells us,

> A lot of guys put the (armor) vest on and go running into action. Instead, they should be doing the job the way they were taught to.

The fact that the ordinary citizens can suffer from the superman syndrome is shown by the fact that when drivers of automobiles were made to wear safety belts, there was actually an increase in injuries to pedestrians caused by the more reckless driving of the protected automobile drivers (see Section 10.5).

Section 10.4   Dementia Pugilistica

One sport which some medical authorities would like to see banished to the history books is boxing. The title of this section is the medical term for an illness which many boxers suffer from in later life. The term *dementia* is described in a medical dictionary as a condition of deteriorated mentality that is characterized by marked decline from the individual's former intellectual level and often by emotional apathy (apparent lack of emotional response to events). The term comes from two Latin root words "de", out of, and "mens", meaning "the mind." To be *demented* literally means "to be out of one's mind." The term pugilistica comes from the Latin word "pugnare" meaning "to fight." The same word has given us pugnacious for someone who is aggressive and always fighting. A *pugilist* is the name given to a fighter by those who have a classic education. A more common term for the illness *dementia pugilistica* is "*punch drunk.*" A stock character in some gangster movies is the not too bright, heavily built, bodyguard who used to be a boxer. In recent times, a famous boxer who has been diagnosed as suffering from an illness which causes him to have problems with hand/eye coordination and speech slurring is Mohammed Ali (formerly known as Cassius Clay). The boxing profession is very secretive about what has actually happened to Mohammed Ali. They claim that he is suffering from *Parkinson's disease* which also produces "slurred speech, loss of coordination, reduced muscle strength, and a persistent feeling of fatigue" [19]. A neurologist treating Mohammed Ali denies the ex-boxer is suffering from the effects of blows to the head received over many years as an active fighter; that is, he denies that Ali is punch drunk. He also says that Ali is suffering from Parkinson's disease.

In an article on boxing [20], Gail Vines says that the latest evidence implies that no boxing is safe boxing. Boxing damages the brains of amateurs as well as of professional boxers. From this article we learn that

A blow to the head can exert a force a hundred times the force of gravity (100 g).

Bleeding inside the skull is the most common cause of death in the ring or a few days after a fight. Throughout the world, more than 300 professional boxers have died of injuries in the last 20 years. Defenders of boxing point out that this fatality rate of 0.13 deaths per 1000 participants is lower than that of horse racing (12.8), mountaineering (5.1), or even American football (0.3) according to Vines' article. Such arguments overlook the fact that the other sports do not accumulate lower level damage causing mental and coordination problems in a boxer in what should be the prime of his life. Vines [20] points out:

> *As the soft jelly-like brain swirls within the accelerated bony skull, blood vessels can stretch and tear. Blood then accumulates within the skull and fatally compresses the brain.*

Even with rapid and skillful neurosurgery to reduce pressure and remove blood clots, nearly half of the boxers who develop brain bleeding die. Vines also notes that

> *A variety of studies suggest that a severe blow that causes temporary unconsciousness, a knockout, leads to permanent structural damage to the brain.*

Peter Lamperd at the University of California at San Diego has shown that it is the number of bouts fought by a boxer rather than the number of knockouts that is correlated with chronic brain damage. Professor Nicolas Corssellis of the Clinical Research Center at Northwick, England has studied the brain of boxers who died in the ring or were punch drunk later in the life. He has shown that punch-drunk boxers have lost many nerve cells in specific regions of the brain, particularly parts of what is known as the cerebral cortex, the brain stem, and the cerebellum. Brain cells in the cerebellum that control movement are also lost in the punch-drunk boxer. The ventricular spaces inside the cerebral hemispheres filled with cerebral spinal fluid are enlarged. Most of the studies of the brains of boxers who have died have been criticized by current boxing experts because they claim that many of the earlier boxers suffered from extensive drinking and from neurosyphilis, both of which could generate brain damage. These experts also claim that by the time boxers die they suffer brain damage from diseases such as Alzheimer's disease that are due simply to ageing. But medical experts discount these arguments saying that none of these illnesses show the type of brain damage that all punch-drunk boxers exhibit [20].

Jeffrey Cundy has criticized doctors who will work at the side of a boxing ring saying that the involvement of medics validates the sport and that doctors at the ringside can do no more than provide first aid. Reforms in equipment in amateur boxing are controversial. For instance, Cundy argues that head gear is worse than nothing because it tends to rotate the brain within the skull causing the worst sort of damage [21]. Asymmetrical rotation of the head is both the worst and the most common effect of fist blows to the head. As the brain swirls in the bony skull, the cerebellum is forced into the opening at the base of the skull, while the frontal and the temporal lobes of the cortex are continually bruised and grazed. Boxers often suffer from detached retinas (the network at the back of the eye which enables it to see); detached retinas are an injury which can lead to blindness. In March 1986, the Scottish welterweight champion D. Watt collapsed during a bout and died three days later from a massive brain hemorrhage. The structure of this boxer's brain at autopsy as compared to that of a normal brain was discussed by Vines [20]. At the time that this boxer sustained his fatal injuries, I was in hospital suffering from the effects of a blood clot on the brain which had caused a seizure. The delayed nature of damage from blows to the head is illustrated by my own experience. The doctors treating me believe that I received a blow on the head from some minor activity such as walking into a cupboard door or banging my head as I got into the car. They also believe the original minor injury had been bleeding intermittently over a period of years. The climax only occurred when the blood clot reached sufficient proportions such that the pressure on the brain no longer caused just headaches but actually created a seizure with subsequent unconsciousness. The delay between a blow to the head and the illness from the blood clot can be short or long. In my case there was at least a three year gap. On the other hand, Shuller, a well-known Protestant preacher in the United States, banged his head getting out of a taxi on a visit to Amsterdam. He was later found unconscious on the floor of his hotel room because of the bleeding inside the skull but on the outside of the brain. Yasser Arafat, the leader of the Palestinian Liberation Organization was involved in an air crash and only several months later did he complain of severe headaches which turned out to be caused by bleeding on the inside of the skull.

Modern medical techniques can look inside the skull with X-rays and create a computer-generated section of a living brain. The technique is known as *computerized tomography*. In North America, the technique is generally known as CAT scanning from the fact that the equipment being used to examine a person is rotated about an axis so that the procedure is referred to as *computerized axial tomography*, shortened in everyday speech to *CAT scanning*. CAT scanning has found signs of brain damage even in young amateur boxers before they show any outward signs of brain damage. Helen Grant of Great Britain makes the interesting comment that boxing tends to damage those areas of the cortex (the outer surface of the brain) known as silent areas because of their subtle effects on behavior. Slight damage to the frontal lobe of the cortex for instance might mean that a person is more likely to say outrageous things in public or becomes inept at planning. Mohammed Ali's outrageous statements early in his boxing career may be more than a manifestation of his personality. Could his behavior have been a very early sign of dementia pugilistica? As is also stated in Ref. [20],

> *Particularly devastating is the observation that the boxer's brain and his mental state often continue to deteriorate long after he has stopped boxing.*

A possible explanation for this delayed deterioration of the brain after boxing has stopped is that during the boxing career, the individual uses up what is known as *available redundancy* in the brain's communication systems. The basic meaning of this statement can be understood from the sketches of Figure 10.5(a). Imagine that we are looking at the problem of keeping a military command post A, in communication with a forward gun-firing position at B. If a single straight communication line was laid between A and B, and then if it broke at any point, all communication between A and B would immediately cease. If, however, there were three lines connected between A and B, the presence of the two extra lines would imply redundancy in the communication system which would protect it against breakdown by damage as shown in Figure 10.5(b).

10.4 Dementia Pugilistica

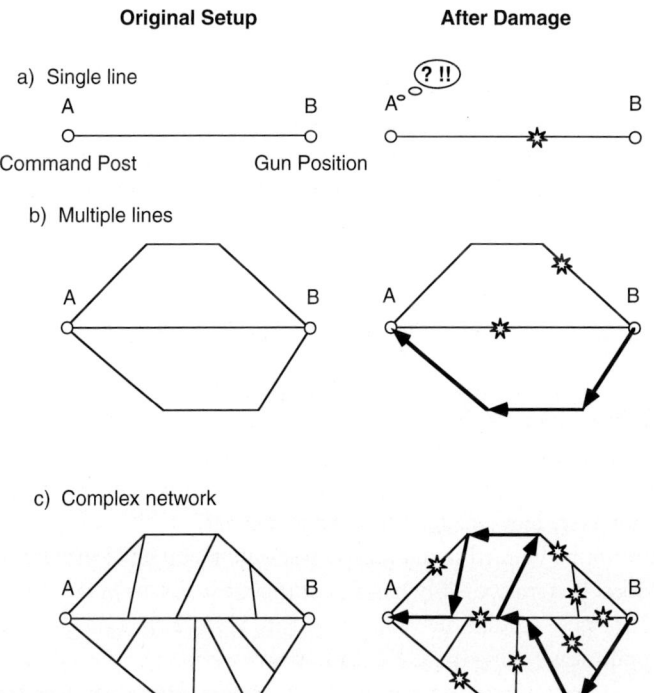

Figure 10.5 Redundancy in a communications network enables the network to sustain a large amount of damage before communication is terminated. a) With a single line of communication, any amount of damage is terminal. b) Several lines of communication allow data to continue to flow with some damage. c) With even greater levels of redundancy, new pathways of communication can be found even with heavy damage.

In fact, the network could be made able to sustain as many as nine impacts if short, connecting communication lines were established between the three main lines as shown in the Figure 10.5(c). Imagine that a commander was at A speaking to someone at B. In a warfare situation as the branches of a communications network were damaged, A would not know that up to nine lines had been severed as long as a pathway still existed, such as the line shown by arrows in the sketch of Figure 10.5(c). It appears to be well established from the study of the brain structure that the networks in the information part of the brain make a great deal of use of redundancy similar to that illustrated between A and B in Figure 10.5(c). Imagine that our boxer had a control network supplying information from the brain to the muscles controlling speech. As the punching he received broke more and more elements of the network, he would still be able to use his motor controls for speaking until the overall network was damaged to a final level.

It appears a reasonable hypothesis that the human brain is supplied with many redundant networks to counteract the fact that, throughout the lifetime of a human, the brain continually sustains small amounts of damage. The available redundancy in the brain is designed to carry a person through the normal life span. The boxer, by allowing his head to be battered, uses up some of this available redundancy at an early stage of life. He may therefore show brain damage much earlier, and certainly before old age. Thus he can become punch-drunk after he has finished his boxing career because he does not have enough available redundancy to withstand the subsequent normal aging process. In 1983, the American Medical Association published an article in a journal stating that

> *The principal purpose of a boxing match is for one opponent to render the other injured, defenseless, incapacitated, or unconscious. Boxing is a throw-back to uncivilized man that should not be sanctioned by any civilized society [20].*

I agree with that statement. I also dislike any form of so-called entertainment in which a performer risks his life to amuse others. However, it is probably unrealistic to think that boxing can be banned since such a move would probably just force the activity underground, into covert meetings in barns and old warehouses where the contestants would not even have medical help after damage.

If boxing cannot be banned then every step should be taken to protect the boxer from his own activities. Britain probably has one of the most advanced set of rules for protecting boxers. The latest regulations were issued in 1995 [22].

In Britain brain scans using the CAT scanner are to become compulsory for all boxers, and will be carried out annually for professional boxers. More advanced techniques such as magnetic resonance imaging (MRI) should be used to check on the brain of the boxer. The Professional Boxers Association has set up a fund to help pay for this more expensive way of inspecting the brain. The association is also to provide funding to be used for research into psychometric testing, the measurement of mental states or processes. It cannot be stated too often that finding brain damage depends on how hard you look for the damage. As Hamlyn points out in a commentary on this subject, there is a need to continue the study of elderly boxers to see if they had deteriorated more than the general population [21]. In Britain boxers who are knocked out may not box for 45 days. Any boxer knocked unconscious while in the view of ringside doctors or who has taken excessive punishment should go to hospital. The main points of the new recommendations to be set up in Great Britain are to be found in the article by Bunce [22].

Perhaps the last comment that we should make on boxing is to describe the experience of the junior lightweight champion of the United States, Deve Ruelas after he knocked out Billy Garcia in a World Boxing Council Championship at Caesars Palace in Las Vagas in 1995. Garcia died on May 18th, 1995, twelve days after lapsing into a coma from the injuries he suffered in the ring. After the fight Ruelas visited Garcia in the Las Vagas hospital where the the injured fighter was hooked up to a life support system. He describes how he met the mother of the boxer he had injured, in the waiting room. The mother did not say a word she just kept looking at the hands of Ruelas. Finally she said [23],

Those hands killed my son.

Not only boxers are suffering from head injuries due to being knocked out or hitting the ground heavily. The current style of play in the American Football League is leading in increased numbers of players suffering from brain damage. For details see the article by Thitpen [24].

Section 10.5   The Automobile: Convenience or Deadly Missile?

In September 1991, a book was published entitled *Autogeddon* [25].
The title is a play upon the name Armageddon of the final great battle predicted in the book of Revelation in the New Testament. The book Autogedden written by Heathcot Williams contains the poem "Now We've Reached the Half Way House," which contains these lines about the automobile:

Half the world's earnings are auto related,
half the world's resources are auto devoted
and half the world will be involved in an auto accident
at some time in their lives

Not very good poetry but starkly realistic in its facts.

The first pedestrian to be killed by an automobile was a real-estate agent, H.H. Bliss. He was killed on September 13, 1899. On the next day, the front page of the New York Times reported the previous evening's accident under their headline:

Fatally Hurt By Automobile

The story, quoted in Ref.[25], reads:

The electrically powered auto was in the charge of an Arthur Smith who was arrested. Bliss was helping a woman off a horse-drawn trolley after getting off himself, and he was knocked to the pavement and run over.

Since that first accident in 1899, it is estimated that the automobile has killed 17 million people world wide! Because we do not usually have dramatic accidents in which hundreds of people are killed, the death rate on the roads, for example 14 people a day in Great Britain, goes largely unreported. However, the total bill for insurance claims and medical treatment of auto victims is a staggering percentage of the national economy. The automobile companies claim that they are making major efforts to increase the safety of the automobile and we will discuss some of their research later in this section. Society, however, remains amazingly tolerant of drunken drivers and resists technology which could reduce the carnage on our roads. For example, in some cities, experiments have been carried out in which video cameras, hung from overhead support systems, can automatically photograph the license plate of a speeding car and subsequently issue a ticket for speeding. The system is known as photoradar. For a period in 1994 to 1995 Ontario had photoradar and it was generating 16 million dollars in revenue a year. It was claimed that it would reduce speed and accidents on the highways. However an election in 1995 led to a change of government which repealed the use of photoradar because of public opposition to the technique. However within six months the government was beginning to hint that it might be bringing back the revenue-generating scheme.

Recently it was suggested that cars be fitted with black boxes similar to those carried on aircraft so that after an accident, experts could see who was accelerating, who was braking, and, from the information stored in the recorder, decide who was responsible for the accident. But again there has been public resistance to such technology [26].

Probably two major contributors to accidental death involving the automobile are the public's ignorance about the forces involved even in minor collisions and the lack of recognition of the effects of alcohol. Alcohol is a strange substance which, while diminishing the brain's agility and the ability to control a system, boosts a person's confidence in his motor skills. Tests have been carried out with bus drivers asked to repeatedly steer their vehicles through a series of narrow gaps defined by road marker cones, pausing after each completion of the obstacle course to consume an amount of alcohol. It was shown that as the amount of alcohol in the blood increased, the ability to negotiate the obstacles decreased; however, the confidence of the drivers increased, leading to what, in real life, would have been a series of nasty collisions.

Two of the problems worked out by Orear in his textbook [1] involve the forces operative in a car crash. Thus in one problem, he considers the case of a 1.5 tonne car traveling at 20 meters per second (72 kph) which collides with a tree. It comes to rest in 3×10^{-2} seconds (30 milliseconds) which corresponds to a deformation of the car's structure of 60 centimeters. This means that the car undergoes a deceleration of about 70g or, in other words, that the average force acting on the car during this time to achieve this deceleration is 70 times the weight of the car. He then discusses what would happen if the car was driven by an 80 kilogram man held by a seat belt 5 centimeters wide and 2 millimeters thick. He assumes that the breaking strength of the seat belt is 5×10^8 newtons per square meter. He shows that the average force on the seat belt during the retardation of the car as it hits the tree is about half the breaking force. Orear goes on to show that if the man were not wearing a seat belt, he would hit the windshield after the car had come to rest and that the head/windshield collision would be perhaps a hundred times shorter than the time it took for the car to stop. He concludes his simple calculation with the stark information that, in such a collision,

The head would break.

As the speeds of automobiles rose, the damage occurring during the collisions increased tremendously. It seemed to take a long time for society to realize that safety design was as important as speed in the evolution of automobiles. The seat belt is a device intended to make the passenger an integral part of the car. As part of the car involved in a collision, the body restrained by a seat belt does not become a free moving missile which has to be brought to rest by collision with the front part of the car; instead, it can be decelerated relatively slowly by the seat belt. There is a great deal of misinformed folklore about seat belts. Many people swear that it is safer not to be strapped to the car in a collision in case you have to get out in a hurry. The facts are clear to those that have studied the problem. In the three years after Michigan introduced a safety belt law in 1984, hospital stays of a week or more from a car crash dropped 43%; there was also a 19% reduction of the number of car crash victims admitted to hospital and a 20% reduction in the rate of crash victims admitted with injured arms and legs. Fatalities dropped 19.7% [27]. In some countries there has been a report of an increase in the number of pedestrians injured after safety belts became mandatory because of the more reckless driving of drivers who felt more secure [28].

The automobile industry was not quick to realize the design features built into Greek helmets that absorb the energy of a collision rather than resist it. Modern cars, however, are now designed to absorb the collision energy both in the front and back of the car. The fact that modern automobiles are actually designed to absorb energy during a collision to help protect the driver and the passengers who are contained in what is now known as the *passenger cage*, means that the damage sustained by a car is much greater now than was the case for the rigid automobile of the past, which was not built to absorb energy. Unaware of the scientific basis of modern car design, an uninformed driver might get out of the passenger cage of the badly damaged car, not realizing his life has been saved by the new design, and grumble, "They don't built cars like they used to". When told that the collision means that the car has to be written off entirely and that a new car must be purchased, the driver grumbles even more. In an older car, the driver probably would have been impaled upon the steering column as the rigid front end led to enormous deceleration rates and would not have been alive to grumble! The public needs to become more aware of the fact that, although there may be higher insurance bills for modern cars, these costs can be more than balanced by the fact that people are alive to see the damage and that society saves in the areas of hospital and medical costs.

The *airbag* is the most significant new safety device to be included in modern automobile design. To study the optimum design of airbags, companies are using dummies wired with all sorts of transducers so that the forces on them, when involved in various types of crashes, can be measured and appropriate safety mechanisms designed.

The basic trigger mechanism used to inflate an airbag is an *accelerometer*. This is an instrument designed to measure the rate of deceleration or acceleration in a force-driven system. The basic principle of an accelerometer can be appreciated from Figure 10.6. A simple device would consist of a cylinder in which a mass is free to move backwards and forwards, except for a restraining spring. If the accelerometer is part of a system which is slowed down in the direction of the tube, the mass continues to try to move forward until the spring can hold it back. If the deceleration is slow, the mass only

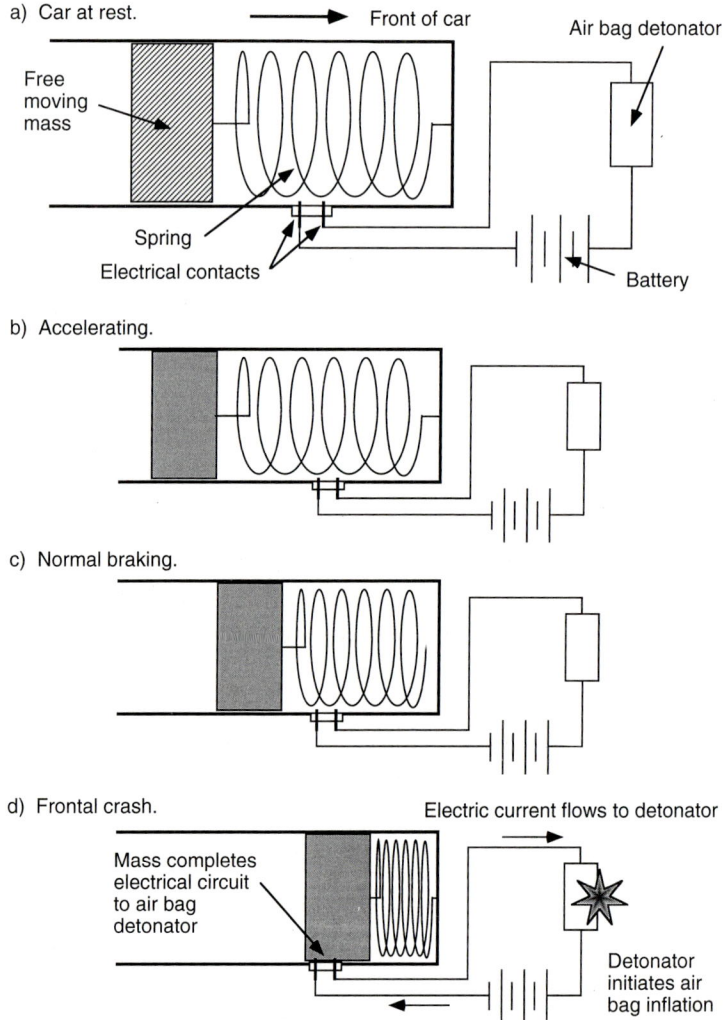

Figure 10.6 An accelerometer activates the inflation of an airbag in a crash. Shown above is a simple accelerometer illustrating the basic principles of such devices.

moves a short distance along the tube; if the deceleration is very rapid, the spring is compressed so far that the mass moves forward until it completes the circuit by connecting the two electrodes in the side wall of the accelerometer (see Figure 10.6(d)). The strength of the spring and the size of the mass are selected so that only the type of fierce deceleration experienced during a crash will trigger the airbag inflation system. In practice there have been some reports of airbags inflating when drivers have had to brake abruptly on the freeway and there have been some reports of drivers being bruised by the inflating airbag. However, as the companies gain more experience in the design of such airbag systems, these problems will diminish. Consumer concern over the way in which an airbag is inflated have been answered in a useful article by Parenti in New Jersey's Courier News [29]. He points out that the actual compound used to inflate the airbag is sodium azide. This has the chemical formula $NaNH_3$. When this decomposes chemically, the main product is nitrogen. Eighty percent of the air we breath is nitrogen and so in the words of Parenti

> *The gases inside the inflated airbag do not pose a health hazard .... Sodium azide is an explosive, but when contained in a metal canister, it is less accessible than the gasoline in the fuel tank of a car or the sulfuric acid in the battery.*

Note that the term *azide* comes from the fact that, in the late 1700s, chemists used two names for nitrogen. The other name was *azote* from the fact that, unlike oxygen, it could not sustain life. One of the Greek words for "life" was "zoe" (a word which has given us zoology for the study of living animals) and the prefix "a" means "without." Although the word is no longer used to describe the gas itself, many nitrogen-rich compounds are described with the root word "aza" from this alternate word for nitrogen.

When the accelerometer is activated, an electric spark ignites a pellet of potassium nitrate which in turn ignites the sodium azide pellets. In a typical system, the airbag inflates in less than one-tenth of a second. When manufacturers install airbags, they lubricate them with cornstarch so that they will unfold properly. In that one-tenth of a second, as the airbag inflates, the corn starch is expelled and sometimes creates what looks like a cloud of smoke around the airbag. The white powder can settle on the interior of a car and the occupants. However, it is no threat either to the people in the car or to rescue workers. Parenti addresses the fact that during the burning of the sodium azide, there is a small amount sodium hydroxide dust produced. In his words [29],

> *This amount is less than one quarter of a gram, equal to about one quarter of a small packet of sugar found in restaurants. Some sodium hydroxide may disperse during the airbag's deflation and, on contact with the atmosphere, converts into sodium carbonate and sodium bicarbonate, a compound better known as baking soda. However, sodium hydroxide is a potential skin and eye irritant and as a precaution, rescue workers should ventilate the interior of the vehicle and wear gloves and goggles during rescue operations where an airbag has inflated.*

Parenti points out that airbags are generally made of nylon. During an accident, the airbag and crash victim come in contact at a rapid speed and with great force. In the event of a severe crash, the driver may suffer friction burns by skin rubbing against the airbag. No evidence indicates that chemical burns are caused by airbags. Emergency workers should follow normal procedures when they come upon accidents where the vehicle has an inflated airbag and should not delay medical attention. Again in the words of Parenti

> *The life saving records far outweigh any possibility of suffering minor skin irritation during an automobile accident.*

Although research into the use of airbags on motorcycles indicates that these would be very effective at preventing injuries, the prospects that motorbikes will be fitted with airbags are not promising. Thus, in an article by M. Hamer, it is stated that discussions at the Council of Europe about the introduction of airbags for motorcycles have reached stalemate and the chances of a manufacturer offering a motorcycle model with an airbag are slim [30]. Chinn who has studied the safety of motorcycles comments on the lack of manufacturers' interest in safety devices with the statement [30] that

> *Manufacturers of motorbikes are remarkably reluctant to fit antilock brakes on motorcycle.*

The article in the *New Scientist* on airbags for motorcycles generated some interesting correspondence in a later issue. One letter by Duncan McKenzie questioned the wisdom of introducing devices which make the driver feel safer [31].

As we have seen, increasing the safety of a driver or an individual in warfare often leads to the superman syndrome; however, the increased hazard from this is far outweighed by the overall benefit of safety equipment. There is an interesting historic perspective on safety devices and over confidence presented by the history of railway technology. In the 1800s it was generally believed that safety precautions encouraged recklessness by creating a feeling of confidence in the workers. These men would therefore be inclined to risk more than they would otherwise do. In addition, if the men believe that danger may be entirely prevented by mechanical inventions, they may be prone to sleep.

However, in spite of their objections, which had a grain of truth in them, the railways went ahead and introduced safety devices. With a century of hindsight, it is clear that British Rail crews and passengers are safer for having block signals and carriage brakes. Their adoption did not produce more accidents due to foolhardy train drivers. Perhaps a sufficient answer to the possible negative effects of safety devices was given a century ago by W. Malcolm, assistant secretary of the Board of Trade in Great Britain:

> *It is not easy to see how the use of defective tire fastenings or inferior couplings could, by stimulating the care and vigilance of railway servants, contribute to the safety of the traveling public.*

Sometimes, even after the manufacturers have done their best to design a safe car, problems are experienced by drivers using new cars. I once had a car which, when I drove it in a snowstorm, suffered from a turbulent air pocket behind the rear window in which snow accumulated and made visibility very difficult. A similar sort of problem was experienced by people driving the station wagon of the type shown in Figure 10.7. Turbulence behind the station wagon created conditions where mud, water, and snow blocked visibility [32].

10.5 *The Automobile: Convenience or Deadly Missile?*   377

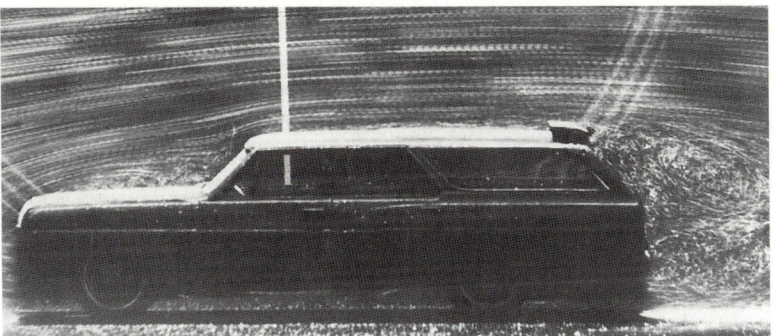

Figure 10.7 Water-tunnel studies with scale models can be used to look at the design of vehicles and to devise corrective measures if problems are found [37]. (Reproduced with permission, National Research Council Canada – Institute of Aerospace Research.)

The photograph of Figure 10.7 was taken when studying a model of the vehicle in a water tunnel. A *water tunnel* is similar in design to a wind tunnel, but much smaller.

Sometimes, people active in the "green movement" or so-called "environmental activists" need to be better informed about the problems they tackle. For example, at one time the use of air conditioners was criticized on the grounds that it caused extra fuel consumption; however, the fuel used by an air conditioner is less than the fuel consumed in driving a car with the windows down, because of the turbulent drag caused by air flowing through the open windows. Even with the advantages of modern design, a car at 110 kph (70 mph) uses up to 70% of its gasoline to overcome wind resistance. As the speed of a car increases, the air drag rises as the square of the speed. That is, if you double the speed, the air resistance increases by a factor of four. This is why in Canada and the United States, during the oil crisis of the 1970s, the speed limits on the major freeways were dropped from 70 mph to 55 mph. The 15 mph difference does not seem to be a significant reduction in speed, but the air resistance experienced by the average car decreased by the ratio of $(55 \times 55)/(70 \times 70) = 0.62$. Thus a 20% decrease in speed resulted in 38% saving in gasoline. The lower speed limits also saved many lives, but finally public pressure caused the US government to relax its speed limits once more because of the public's desire to travel faster rather than safer.

Sometimes car accidents are made worse than they should be because of human failure to use the safety equipment provided. Thus Dr. Garry Dick, a lecturer on biomechanics at the Canadian Memorial Chiropractic College in Toronto, says that it is very common to find that whiplash patients were in cars without properly adjusted headrests [33]. Dr. Dick has made an issue out of headrests because of the large number of patients he treats with avoidable whiplash. He quotes studies which show that between 74% and 90% of male drivers have their car headrests set too low to be effective. The sudden force of an accident whiplashes the neck, often causing nerve tissue and vertebral damage which are difficult to treat. Dr. Dick says that many people never adjust a headrest; they just leave it positioned as it came with the car. The headrest should be no more than two inches behind the center of the back of the head. If it is low behind the neck, the force of an accident could bend the neck back over the headrest, causing serious injury. Reclining the seat increases the chance of whiplash injury because it increases the distance between the head and the restraint. Dr. Dick also criticizes some automobile manufacturers who have equipped their cars with headrests that are difficult to adjust. Dr. Dick is surprised that insurance companies, safety councils, and consumer groups do not place more emphasis on proper use of existing safety devices. He also warns individuals that whiplash injury may not become apparent for up to two years after an accident [33].

Specialists attempting to prevent road accidents believe that a major problem is that people are unable to judge their performance limits. Many nasty accidents are caused by people falling asleep at the wheel having driven too far and for too long [34]. Opinions differ as to how many accidents are caused by tiredness, but the sleep research laboratory at Loughborough University, England points out that up to 20% of the road accidents in Great Britain may involve driver fatigue. Studies in the United States conclude that of the 50 000 deaths a year from motor vehicle accidents 13% are caused by drivers falling asleep. What is particularly disastrous about some accidents involving drivers asleep at the wheel is that they involve serious head-on collisions with innocent drivers traveling in the opposite direction.

As mentioned earlier, in Europe in the early 1990s, groups concerned for drivers' safety started to push for the fitting of cars with a device which would work like an aircraft black box [26]. When an aircraft crashes, major efforts are made to retrieve the flight data recorder, or black box, which can tell investigators the last words of the flight crew and the flight conditions at the time of the accident. A similar device that safety groups would like to see fitted to automobiles would cost approximately $200 per car. Fitting such devices to cars to automatically register speed, acceleration, and use of brakes and lights at the time of the crash would give more information than investigators can gather from inspecting the scene of a crash. The Association of British Insurers says that its members pay out more than $8 billion a year for crash damage, including damage suffered by the 2.7 million commercial vehicles. Scientists tell us that fitting the device would definitely change driver behavior. Accident investigators point to the important human factor of reckless drivers worrying about the devices giving them away. He says that some truck drivers whose vehicles were equipped with speed-recording devices have been found at the scene of accidents trying to eat tapes from the recording instruments in their trucks. Probably the major problem in getting drivers to accept such electronic surveillance would be resistance to what they regard as invasion of privacy. It is amazing how much disaster the public will accept in defense of their freedom. They would rather have their freedom and be killed by a drunken driver than allow an electronic monitor to record the pre-crash movements of a deadly automotive missile. This kind of emotional resistance is similar to that of people who do not want any restrictions on their freedom to carry guns to shoot ducks and as a result prevent reasonable legislation to stop criminals from carrying deadly automatic weapons. Some people fail to realize that in an advanced technical society we must accept some limitations of individual freedom for the greater good of the total community. The motorcyclist who defies the law and drives without a helmet not only destroys himself but creates enormous medical bills for the community to absorb. Does he have the right to such "freedom"?

A very confusing debate about automobile safety started to grab newspaper headlines in the early 1990s. Thus, one story [35] reported in the New Scientist read:

Green cars will cost life says U.S. industry.

Of course the story did not deal with cars painted green but was concerned about possible legislation in the United States to mandate the development of smaller cars to save fuel. By the early 1990s, it was obvious that America was vitally dependent upon imported oil as demonstrated by the US involvement in the Gulf War. Officially, the war was to preserve the political freedom of Kuwait, but many people felt that unofficially the real concern of the United States was the protection of fuel supplies. It was widely pointed out in early 1992 that no rush was made by the Western powers to protect Croatia from the Serbs because no oil was involved. One of the things that can make the United States much less dependent on foreign oil is mandatory increases in the average fuel efficiency in new cars. However, the car industry has formed a group called "coalition for vehicle choice" which fights the proposal to save fuel by building smaller and lighter cars. They have broadcast television advertisements showing two cars in a head-on collision, one weighing 1450 kilograms and the other weighing 870 kilograms. In the advertisement, the entire front of the small car collapses under the impact while the larger car remains relatively intact. Environmental groups, small car advocates, and others say the test and the film are misleading. At present the whole question of the safety of small cars versus large cars is a confused debate. First of all it should be pointed out that if everybody had small cars there would be no big cars to collide with. Second, although the statistics show that the chance of being killed in a small car is twice that of being killed when driving a large car, smaller and lighter cars can be made much safer through the use of seat belts and airbags. Not enough time has elapsed yet to judge the increased safety achieved with these modern devices. It should also be noted that statistics are distorted by the fact that high-risk drivers are more attracted to small sporty cars, thereby biasing the statistics.

Section 10.6 Sports Aerodynamics

Anyone who has watched professional cyclists racing will notice that they adopt "goose style" aerodynamic tactics. The group name for an assembly of geese is a gaggle. Anyone who has watched a gaggle of geese in flight know that they adopt a V-shaped formation. What is not so obvious is that the leading goose changes every now and again because flying as the leader of the V-shaped group is a tiring occupation. As the lead goose forges ahead, it creates a slipstream in the air and the following geese are wise enough to take up their positions in this diverted air, allowing them to fly with less effort. I had a conversation with a professional racing cyclist who told me that the lead cyclist has to pedal about one third harder than his colleagues snuggled into his slipstream. In professional racing, groups of cyclists will move together, taking advantage of this reduction in air drag, with the lead cyclist changing places with a colleague every now and again. This goose-like behavior of cyclists demonstrates the importance of understanding air drag when the human athlete behaves as a missile moving through the air.

It is interesting to note that Lloyd Jenkinson of the Aeronautical Engineering Department of Loughborough University has suggested that airliners should copy geese [36]. He has shown that if aircraft were to fly 400 meters apart there would be significant savings in fuel. Jenkinson points out that airplanes are already flown by computer for most of the time. Thus a plane from London to Los Angeles may be under the pilots direct control for only 5 minutes during take off. He suggests a system in which electronic links between the two planes would be made in triplicate to prevent any technical failure causing an accident. Under his scheme a second plane flying in the shelter of the first plane would still carry a crew for takeoff and in case of emergencies. Airports would have to build parallel runways but in most cases this could be done within the existing boundaries. Jenkinson estimates that the fuel consumption of the second plane would be reduced by 10%. One of the problems facing the adoption of such a system is that air traffic controllers regard any two planes coming within 2 kilometers of each other as a near miss. Jenkinson is trying to persuade the controllers to consider the two planes as one unit. The University of Loughborough has patented the idea and in 1995 was looking for an industrial partner to exploit the technology.

In Figure 10.8 some earlier studies of the wind drag experienced by skiers and snowmobile riders are shown. Note how the windshield's effective height was double its physical size in forcing air over the top of the snowmobile rider and note the turbulence behind both the skier and the snowmobile rider. When reading the literature on modern winter sports, one finds that the conflict of the snow slopes is fought by the windtunnel experts of the respective countries long before the arrival of their competitors at the slopes. In an article on the aerodynamics of skiing, written in 1988, Michael S. Holden of the Calspan Corporation describes the wind tunnel research work carried out on downhill skiing [37]. He gives us much detailed information: for example, a racer in a porous suit will in general have a drag 5% greater than when wearing an externally or internally non-porous sealed suit. He goes on to state that during the ten years prior to 1988, there had been significant changes in the technique of downhill racing, the most important of which was the combining of effective aerodynamics in good skiing. Developments in ski design combined with smoother and more consistent course grooming have resulted in top competitors being able to ski with greater, more subtle control, allowing them to adopt more effective body positions. In another article written at the same time by Chester R. Kyle [38], entitled "The Ski Jump — Flying Without Wings,"

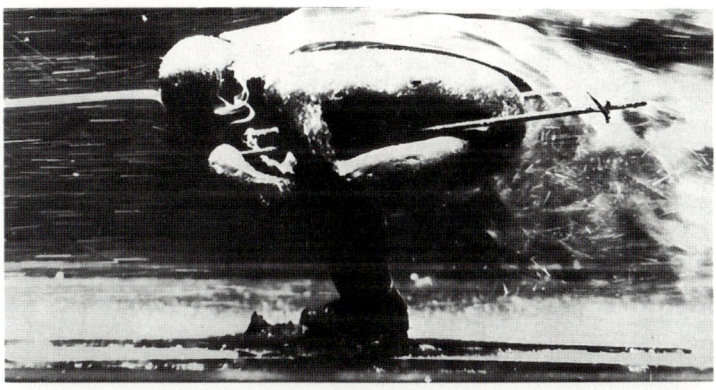

Figure 10.8 Water-tunnel studies have helped athletes to improve performance by reducing air drag at high speed. a) A competitive skier's performance can be improved by studying various body positions. b) Recreational vehicle design can be examined to improve user comfort and vehicle efficiency [37]. (Reproduced with permission, National Research Council Canada — Institute of Aerospace Research.)

one learns that, during an entire career, a ski jumper spends only a few minutes in the air. Special practice wind tunnels can be built to give jumpers several hours to refine their "flying" skills. He tells us that efficient aerodynamics is critical to the ski jump:

> *Earlier jumpers used bad posture on the ski run and windmilled their arms in flight. They also wore sloppy clothing and used narrow skies. All of these produce air drag and less lift, making jumps much shorter than today. Present ski clothing design is very efficient but it is carefully controlled by the rules. At one time, the Austrians used a balloon suit which allowed air to enter the front, trapping it in the back and forming the skier's body into an airfoil shape. These suits were outlawed when the Austrians began out-jumping everyone. The Austrians also put a flap on the tail of the skis to produce a higher lift; again, this was quickly banned. Helmet form is also specified to prevent exotic airfoil shapes being used.*

The importance of computerized wind-tunnel studies becomes even greater for sports such as bobsledding and the luge. (The word *luge* is French for a sled and a luge in competitive sports is a single-person sled.) An account of how these studies have altered the performance of bobsledding has been written by Valkenburgh [39]. He describes the various studies which led to a 40% reduction in the drag on the bobsled models used in the Calgary Olympic Games of 1988 as compared to the models used in the 1984 Olympics. The way in which aerodynamic studies are influencing modern sports is perhaps best summarized by the comments given in a series of articles on the technology of winning winter sports which describe innovations in luge competition [39]:

> *Luge sleds are ridden at terrifying speeds through a serpentine ice run by athletes lying on their backs with feet forward. A luger hangs on for dear life while steering with foot levers attached to the runners. The sled is carefully formed to the human shape and is streamlined in every respect. Lugers dress includes gloves, a safety helmet, a face shield, and slick rubberized clothing that clings perfectly to the human body.*

The top US woman luger, Bonnie Warner, was third overall in the 1986/87 World Cup competition. She was going into the Calgary Games with a time within 0.3 seconds of the West Germans. She thought she would improve 0.1 second in the start which would still leave her 0.2 seconds shy of winning. Bart Hibbs of aeroVironment Inc. was working to improve the aerodynamics of body, sled, and clothing. He calculated that a 9% air drag reduction would gain the missing 0.2 seconds. Wind tunnel tests showed that fairing the hands into the sled along with other revisions lowered air drag by 10%. A fairing is a smooth covering built over irregularly shaped parts (in this case around the hands) to ensure a smooth, aerodynamic outline that will reduce drag. Hibbs also planned to put riblets on the sled and on Bonnie's clothing. (Riblets as we have seen earlier are manufactured by the 3M Corporation.) If Hibbs' calculations were correct, the riblets should have reduced air drag by a further 2%. Riblets or not, snowshoeing is more my kind of sport. I prefer not to use my body as an aerodynamically encapsulated, high-speed missile!

Section 10.7 Roller Coasters as Missile Systems

It is estimated that there are over 200 roller coasters operating in the United States. In an amusing account of the terrors of being a missile when riding a roller coaster Jeffrey Kluger makes the following statement [40]:

> *Modern coasters fall into two categories, wooden or steel and are designed by people who fall into one kind of category, terrorists.*

The sensation one gets on a roller coaster involves both high g forces (remember g stands for the acceleration due to gravity) and the empty stomach feeling caused by low g forces.

Jet pilots risk blacking out if they attempt tight turns in their aircraft which generate forces of 10 g. They suffer what is known as a mental brown out at about 8.5 g. Curtis Somers who designs roller coasters said that beyond 3.5 g ordinary people become frightened and, as Kluger states, become plaintiffs in legal action. As a roller coaster accelerates downhill, the effect of gravity is reduced producing low gs or a feeling of weightlessness. Astronauts orbiting the Earth experience zero g. but a roller coaster is designed to go no lower than 0.2 g. This is enough to give people the thrill of feeling airborne but, in a worse-case scenario, keeps them in the car if the lap bars or seatbelt fails! Those interested in the dynamics of roller coasters should see the articles by Kluger, Pearce, and Pready [40-42].

References

[1] J. Orear, *Physics*, MacMillan, New York, 1979.
[2] R.G. Snyder, "Journal of Military Medicine," Vol. 131, 1966, p. 129,
[3] C. Ronan, J. Needham, *Shorter Science and Civilization in China, Vol. 4*, Cambridge University Press, Cambridge, 1993.
[4] For a discussion of the safety of bungee jumping and other activities at the fair ground see F. Pearce, "Not All Fun at the Fair," *New Scientist*, August 29, 1992, pp. 25-29.
[5] P.H. Blyth, "The Technology of Ancient Warfare," *Science Spectrum*, Number 1-7, 1975, Issue 2. A series of notes on current science activities in Great Britain.
[6] P.H. Blyth, "The Structure of a Hoplite Shield in the Museo Gregoriano Etrusco, Estrato da Bollettino," Abstract of the Bulletin of the Pontif's Museum and Gallery, Volume III, Vatican Press,1982.
[7] K. McKean, "Longer living Through Chemistry," *Discover*, Vol. 2 No. 7, July 1981, pp. 80-82.
[8] Trade literature on Kevlar fibers available from E. I. Dupont de Nemours and Co. Inc., Textile Fibers Department, Kevlar special products, Center Road Building, Wilmington Delaware, 1988.
[9] J.E. Gordon, *The New Science of Strong Materials*, 2nd Edn., Princeton Paperback Printing, 1984.
[10] J. Beard, "Warding Off Bullets By a Spiders Thread," *New Scientist*, November 14, 1992, p. 18.
[11] D. Graham, "Synthetic Spider Silk," *Technology Review*, Vol. 97, No. 7, October 10, 1994, pp. 16,17.
[12] See general discussion of body armor in Chapter 5 of *Science and the Detective* by B.H. Kaye, VCH, Weinheim, 1995.
[13] Diagram of the hard hat is based on the pamphlet available from Scriptographic Communications Ltd., 150 Consumers Road, Suite 404, Willowdale, Ontario, M2J 1P9.
[14] B. Cook, "Head Protection, Are New Hard Hats Really Necessary?" *Occupational Health and Safety*, Vol. 10, No. 1, Jan/Feb. 1994, pp. 28-31.
[15] News story, "If You Want to Get Ahead, Get a Helmet," *New Scientist*, April 28, 1988, p. 27.
[16] News story, "A Bike Helmet Could Save Your Child's Life," *Sudbury Star*, April 21, 1992, p. B3.
[17] News story, Canadian Press, "Wearing a Helmet Encouraged as Part of a Bicycle Safety," *Sudbury Star*, May 17, 1991, pA1.
[18] "Helmets Save Lives," *The Sudbury Star*, May 18, 1995, p. B1.
[19] News story, "Ali Fights a New Round – a Brain Disorder Focuses Attention on the Dangers of Boxing," *Time*, October 1, 1984, p. 79.
[20] G. Vines "Boxing Takes a Battering," *New Scientist*, June 19, 1986, pp. 30,31.
[21] S. Connor, "Boys Box on as Pros Agree to Talk," *New Scientist*, March 15, 1984, p. 6.

[22] S. Bunce, "Board point way forward with better brain scans," *The Daily Telegraph*, October 26, 1995, p. 34
[23] R. Seltzer, "Boxing Ruelas Grapples with Guilt of Killing Opponent," *Sudbury Star*, June 28, 1995, p. A8.
[24] D.E. Thitpen, "Chin Music," *Time*, December 12, 1994, pp. 59,60.
[25] H. Williams, "Autogeddon," Jonathon Cape, London, 1991.
[26] E. Geake, "Black Box Could Put Brakes on Bad Driving," *New Scientist*, November 9, 1991, p. 26.
[27] News Item "Safety Belts," *Technology Review*, Vol. 92, No. 3, April 1989, p. 13.
[28] M. Hamer, "Seat Belts Save Lives, But Pity the Pedestrian," *New Scientist*, October 24, 1985, p. 18.
[29] The article by A.J. Parenti who is chief of police in New Brunswick, New Jersey and president of the New Jersey Traffic Officers Association appeared in the Courier News, New Jersey, October the 14, 1991. The address of the Courier News is P.O. Box 66001201, Route 22, Bridgewater, New Jersey, 08807-0600.
[30] M. Hamer, "Motorcycle Safety is in the Bag," *New Scientist*, April 27, 1991, p. 30.
[31] Correspondence in the *New Scientist*, June 8, 1991.
[32] See article "Pint Size Water Tunnels Serves Broad Areas of Research," Science Dimension, Volume 5, Number 1, February 1973. Science dimension is published by the National Research Council of Canada, Ottawa.
[33] News story, "Properly Adjusted Head Rest Can Prevent Head Rest Injury," *Sudbury Star*, February 7, 1989, p.A3
[34] J. Horne, "Stay Awake, Stay Alive," New Scientist, January 4, 1992, pp. 20-24.
[35] D. Charles, "Green Cars Will Cost Lives Says U.S. Industry," *New Scientist*, November 9, 1991, p. 15.
[36] M. Hamer, "Formation Flying for Future Planes," *New Scientist*, January 21, 1995, p. 8.
[37] M.S. Holden, "The aerodynamics of skiing," The technology of winning. A special advertising section of *Time Magazine*, 1988, pp. T4-T8.
[38] C. Kyle, "The Ski Jump – Flying Without Wings," The technology of winning. A special advertising section of *Time Magazine*, 1988, pp. T17,T18.
[39] P.V. Valkenburgh, "The Aerodynamics of the Bob Sled," The technology of winning. A special advertising section of *Time Magazine*, 1988, pp. T10-T14.
[40] J. Kluger, "Ticket to Ride," *Discover*, Vol. 13 No. 8, August 1992, pp. 78-80.
[41] F. Pearce, "License to Thrill," *New Scientist*, August 29, 1992, pp. 23,24.
[42] R.E. Pready, *Roller Coasters and Their Amazing History*, Published by R.E. Pready, 54 Park Edge Close, Leeds, LS8 2LP, England, 1992.

Chapter 11

Micro and Miscellaneous Missiles

Chapter 11

Micro and Miscellaneous Missiles

Section 11.1   Volcanic Ash and Birds as Engine Stoppers

In this section we will look at some interesting missiles that can cause problems for aircraft. One source of potential problems is active volcanoes which can send ash clouds billowing up to 20 kilometers into the atmosphere. These ash clouds pose a serious danger to aircraft. If the engines of an aircraft suck in volcanic dust, they may malfunction and cut out, causing aircraft to fall out of the sky! [1].

Since 1973, at least 18 civil aircraft have had trouble after flying into volcanic ash clouds. In some of these encounters all of the engines have cut out. There have been cases in which aircraft have dropped more than 6000 meters before attempts to restart the engines have succeeded. It is believed that volcanic ash prevents the flow of air through jet engines when it solidifies as a glassy substance on turbine blades. Specialists speculate that the engines can be restarted because the deposits of silicate (a glassy substance) on the turbines break up as the engine shudders during the plunge towards the ground. The ash fine-particles can also cause abrasion to the cockpit's windshield impairing the pilot's vision. In such encounters, the volcanic ash behaves like a sandblaster because the high-speed impact of ash fine-particles (which are mainly silicate rocks) scours the windshield. In April 1991, the Australian aviation authorities tested a device called AVADS (Airborne Volcanic Ash Detection System) which when mounted on aircraft can detect ash clouds. Radar and Doppler devices can measure the presence of fine-particles in the path of the aircraft but cannot tell the pilot whether they are water droplets or volcanic ash. Fortunately volcanic ash fine-particles absorb sunlight, like the rocks of the Earth, and warm up. They then re-radiate the energy at infrared wavelengths which are unique, and can be detected with appropriate instruments.

It is estimated that a major cloud of volcanic ash injected into the stratosphere not only causes spectacular sunsets, because of the light scattered by the ash fine-particles, but also cools the Earth because the fine-particles intercept the Sun's radiation and radiate it back into space before it can reach the surface of the Earth. After a major volcanic eruption, scientists have been able to measure a 2°–3° cooling in the Earth's atmosphere leading to localized climate modification [2]. A device which measures radiant energy from a cloud is called a *radiometer*. The AVADS radiometer being tested by the Australians measures thermal radiations from clouds at various wavelengths in the infrared spectrum. Research scientists have found that at two particular infrared wavelengths, eleven and twelve micrometers, the difference in the radiation signal from volcanic clouds and from water droplets is at a maximum. Radiation entering the radiometer passes through two optical filters to screen out all of the information on the unwanted wavelengths and then the radiation level is measured at these two sensitive wavelengths. High levels of radiation indicate volcanic clouds and lower levels show that water and ice are present. In March 1991, research workers tested the device by flying near an active volcano in southern Japan. They were able to examine the radiation levels from volcanic ash clouds, water clouds, and clear sky. In these flights, the results matched theoretical calculations. The Australian scientists report that they can detect volcanic ash clouds at a distance of 100 kilometers, giving the pilot several minutes to take evasive action. Remember, commercial aircraft at their cruising altitude can be flying at approximately 1000 kph; therefore, even a 100 kilometers early warning gives the pilot only about 6 minutes to take the necessary steps to avoid the cloud [1].

Another major problem faced by commercial airlines is birds getting into the engines during flight. In the early 1960s, chicken was such a regular meal on airlines that I used to joke that they must be serving the remains of chickens that had hit their engines and came out already roasted. Seriously, however, one of the safety regulations of the US Federal Aviation Authority (FAA) is that engines must be capable of rapidly recovering to at least 75% of their full power after ingestion of small (85 grams) or medium-sized (680 gram) birds. The FAA also insists on engines that can ingest an 1800 gram bird and be shut down without disintegrating [3, 4].

When a bird hits the blades of the compressor fan at the mouth of a jet engine, damage to blades can take two forms: they can either fracture or become distorted, losing their aerodynamic configuration. Distortion of the turbine blade results in a loss of compressive power. This loss of power is the most important and common type of damage. For many years, airlines tested new designs of engines with real birds — typically a humanely killed seagull fired with an air cannon into a scaled-down or full-sized turbine. These tests were time consuming and expensive. In an effort to improve their ability to design turbine blades which will resist bird damage, research workers from the turbine manufacturer Pratt and Whitney have started to use computer simulations thus reducing the development time of stronger, lighter and more efficient engines. The researchers have tested their new computer program by making movies of computer simulations of what ought to happen during a bird strike and comparing these with high-speed photographs of a gelatine ball used to simulate a real bird hitting a real jet engine.

In order to improve the safety of engines, any bird strike on a commercial airline is investigated by experts. For instance, when a jet aircraft traveling from Shannon, Ireland landed in Georgia, and feathery remains were found in its engine, aviation officials called upon ornithologist Roxie Laybourne from the Smithsonian Museum of Natural History. (An *ornithologist* is an expert on birds from the Greek word "ornis" meaning "a bird.") From a study of the feather fragments Leybourne was able to tell the officials that the plane had collided with a kestrel which, from its feather structure, she identified as being the European rather than the American variety. From the feathered remains she was able to deduce that the bird strike had happened during takeoff in Ireland. Leybourne began her investigations of bird strikes in 1960 when a flock of starlings was suspected to have caused an accident during take-off at Boston's Logan International Airport. Again from the feathers, she was able to confirm that the starling was indeed the guilty bird. Bird strike problems in commercial aircraft usually occur at take-off and landing since birds do not fly at the higher cruising altitudes. As a result of her studies of bird impacts, Leybourne has shown that many of the incidents involve birds larger than those usually suspected of causing trouble. She found that herring gulls, red-tailed hawks, and mallard ducks were causing most of the damage and this in turn led manufacturers to redesign their engines to withstand bigger impacts.

Various strategies are used at airports to clear birds away from runways and so reduce the chance of bird strikes during take-off. These strategies include using hawks trained to kill smaller birds, and recordings of the birds' own alarm calls. At some airports it has been possible to frighten birds away by mounting decoy hawks on wire supports. Most recently it has been suggested that gulls can be discouraged at airports by leaving the grass between runways uncut [5].

Section 11.2   Deer Flies and Bumblebees

The title of this section comes from an interesting discussion in an article by John H. McMasters on the flight dynamics of various types of insects [6]. In particular he reports on two myths which became widespread in scientific literature. The first myth claimed that deer flies in New Mexico could travel so fast that their impact could cause dangerous wounds. The other is that, theoretically, a bumblebee should not be able to fly. In discussing people's ability to judge the flight dynamics of observed objects, McMasters [6] makes this comment:

> *People frequently overestimate the speed of moving objects but perhaps the most amazing example of this was the claim by an entomologist that a deer botfly could reach supersonic speed.*

An *entomologist* is a specialist who studies insects. The word entomologist is related to the words tomography and microtome discussed in Chapter 10. The root word from the Greek means "cut in two" and so entomology comes from the fact that if you look at insects, they often appear to be almost cut into two, or more, parts.

As McMasters tells it, Townsend, the author of a scientific article in the Journal of the New York Entomological Society, apparently made the assertion [7] that on 3600 meter (12 000 foot) summits in New Mexico,

> *I have seen pass me, at an incredible velocity, what was quite certainly the male Cephenemyra (a type of deer fly). I could barely distinguish that something had passed. I saw only a brownish blur in the air of about the right size for these flies and without sense of form. As closely as I can estimate, their speed must have approximated 400 yards per second.*

McMasters tells us that the story was widely reprinted, appearing in various books of world records. The story annoyed Irving *Langmuir*, the winner of a Nobel prize for chemistry [8]. Langmuir pointed out that the first problem with the entomologist's estimate of the speed of the deer fly was that 400 yards per second (365 meters per second) at 12 000 feet happens to be about 110% of the speed of sound at that height (Mach 1.1). If it really had been flying at that speed, Townsend should have noticed a sonic boom! Langmuir calculated that to be able to fly at that speed would require a muscular engine equal to half a horsepower. Furthermore, at the speed reported, which was equal to 1300 kph (800 mph), the pressure would crush the head of the fly. In the words of McMasters [6],

> *Deer botflies (another name for this type of fly) tend not to be very graceful fliers even running into things on occasion while zipping around deer and people. Langmuir calculated that the impact of a 0.3 gram botfly traveling at Mach 1.1 would produce a wound equivalent to that of a large calibre pistol bullet. This would make hiking on the summits of New Mexico, a risky business!*

Langmuir checked the circumstances under which the observations were made by attaching a small weight the size of a botfly on the end of a thread and whirling it around to find the upper and lower bounds of speeds at which such an object appeared to be a brownish blur in the air. The mean value turned out to be about 40 kph or 10 yards per second or about 2.5% of the speed estimated by Townsend. Moreover, Langmuir was able to show that the speed of 40 kph was consistent with the energy consumption capacities of the actual botfly. Thus in the words of McMasters [6],

> *The botfly was eliminated from the ranks of fastest flying machine although the entomologist's rather than Langmuir's estimates continued to appear in popular reference books for several years thereafter.*

In the average garden, people face potential attacks from missiles with yellow and black stripes — otherwise known as wasps or yellow jackets. There are also abundant bees waiting in the flowers with stings in their tails. A full discussion of the flight of such missiles is beyond the scope of this text, but we will deal with the question which is likely to be presented to the unwary teacher or parent by the precocious child who has read through books on popular science. A few years ago, out of the blue, I was presented with a problem which I now know is widely spread in the mythology of science but which at the time was new to me. A fresh-faced, would-be scientist, asked me the question during one of our open-day demonstrations. The question was,

> *Didn't aerodynamicists prove that bumblebees can't fly?*

The bumblebee is somewhat larger than the ordinary honey bee and its name comes from the fact that it appears to bumble around the garden. The name bumble came originally from the sound that the bee makes, but the apparently random wanderings of this large bee has now given us "to bumble: to mishandle in a disorganized manner the task facing us".

Commenting on this myth, McMasters, who is himself a aerodynamicist, states [6]:

Whoever this notorious individual was (who predicted the inability of the bumblebee to fly), he has left his legacy for all of us aerodynamicists who follow him to wear about our necks like an albatross.

He also tells us,

The discovery of who this individual was and how the myth originated has provided a sometimes frustrating diversion from my serious inquiries.

McMasters exploration of the problem has led him to describe the origin of the legend as follows [6]:

It is known that the bumblebee story was already circulating in German technical universities in the early 1930s, apparently beginning in the circle of students surrounding Prandtl at the University of Göttingen. The identity of the specific aerodynamicist continued to elude me until recently when I learned from a reliable source that a possible candidate may be a Swiss professor, now deceased, who became famous for his pioneering work in supersonic gas dynamics in the 1930s and 1940s. The story which is told about the origin of the flightless bumblebee is that the aerodynamicist was engaged one evening in a light dinner table conversation with a biologist who asked in passing for enlightenment about the aerodynamic capabilities of the wings of bees and wasps. Intrigued by the question, the aerodynamicist did some preliminary calculations based on the assumption that the wings were more or less smooth flat plates. He also assumed that the flow over the wings would be that associated with laminar boundary layers and thus prone to easy separation or stalling. A situation similar to that which leads sporting goods manufacturers to make golf balls with dimpled surfaces. The resulting calculations based upon this model proved that the bee could not fly..

McMasters points out that the assumptions made in the calculations were widely wrong and that the aerodynamicist himself later discovered part of his error by examining a bee's wing under a microscope but not, alas, before the myth was born and passed into the hands of overeager journalists. Those interested in a full technical discussion of why the real bumblebee can fly will find McMasters' article a fascinating entry into the world of insect flight. As the real bumblebee flies through the air, as always, empirical reality triumphs over theoretical inadequacies [6]. To appreciate the wonder of the way in which insect wings are built and function see the review by Wootton [9].

Not only are insects capable of colliding in missile-like fashion with people and other earthly objects, but some insects have their own protective missile systems. One such insect which uses missiles to protect itself is the bombardier beetle. When disturbed, it sprays its attacker with a jet of boiling chemicals which it turns on and off about five hundred times a second. Thomas Eisner and his colleagues at Cornell University have used high-speed photography at four thousand frames per second to show the jet turning on and off [10]. They forced the beetle to direct its spray at a piezoelectric force transducer to record the impact of each pulse. The *bombardier beetle* produces its hot spray using a system of reservoirs and a reaction chamber. It secretes hydroquinones and hydrogen peroxide (the oxidizer) into a pair of reservoirs each of which is linked to a reaction chamber holding oxidized enzymes. The reservoirs are separated from the chambers by a tight valve. To initiate a spray, the beetle contracts the muscles around a reservoir and forces a small amount of chemicals into the reaction chamber. Back pressure from the explosion prevents more chemicals from entering until the chamber empties. When the pressure drops, the cycle begins anew. Eisner states that by producing a pulse, the beetle is able to control the length of the spray and keep the power of the jet constant. From a mechanical point of view, it appears that a pulsed jet can be produced with a simpler mechanism than can a continuous jet [10].

Even plants can use sprays to protect themselves. The Central American bursera tree can squirt a jet of unpleasant chemicals at anyone who plucks a leaf off the tree [11]. Apparently this squirt-gun response discourages the attention of cattle and goats. This tree has been studied by Judith Becerra and Lawrence Venable of the University of Arizona. They showed that the *bursera tree* contains chemicals known as terpenes distributed in a network of channels in the stems and throughout the leaves. When a leaf is removed, the plant generates a fine spray. The sticky spray may persist for 3 or 4 seconds and travel up to 15 centimeters. The tree can also use what is described as a "rapid-bath" response when a portion of the leaf is damaged. In this response, terpenes are released which flow across the damaged leaf. Within a few seconds they cover at least half of both surfaces. Becerra and Venable believe that this rapid-bath response is aimed at *micro herbivores*, a technical term for small animals such as beetles that eat leaves. The squirt-gun response is more effective against the *macro herbivores* (e.g., goats and cows) [11].

Many plants have chemical weapons that act against those animals that might eat them. For example, it is believed that aspirin and similar compounds are widely distributed in some groups of plants to deter animals from eating them because of the bitter taste. Aspirin was first isolated from the bark of willow trees.

In an interesting article on how some plants use missile technology to spread pollen and seeds, Paul Simons describes in detail certain missiles launched by plants [12]. Thus he tells us that a whole group of *fungi* (plants which are frequently parasites on other plants because they do not contain any chlorophyll to synthesize sugars from carbon dioxide in the air) have squeezy tubes called the "asci" which shoot out their spores like miniature pop guns. Simons tells us that one particular type of fungus can shoot out *spores* (the basic seed of the fungi) up to 40 centimeters which is a distance of several hundred times the spores' own size. Some plants apparently use electrostatic forces to launch their spore missiles relying on the fact that during the day there is usually an electrostatic potential established near plants. Apparently the same type of electrostatic ballistics are used to discharge spores from the sills (the brown fin-like objects) on the underside of mushrooms. One type of parasitic fungi actually appears to use a little harpoon which is fired into its prey with subsequent injection of a spore through a tube. The spore then develops into a fungus that feeds on the unwary victim. George Baron of the University of Guelph discovered this particular form of missile attack. He tells us that he noticed a nematode (a type of worm) writhing in pain after a brief encounter with a fungus. Detailed study revealed that the nematode had been attacked with a harpoon, the launching of which is triggered when the prey touches the fungus. It is ejected at great speed by a mechanism that resembles the finger of an inside-out rubber glove suddenly inflated. In many marginal terrains there are plants known as broom and gorse. These plants launch their seeds by the tension built up in the pod by its differential drying. Simons describes this effect in the following way [12]:

> *On a hot sunny day on heath or moorland you can hear the snap, crackle, and pop of broom and gorse seed pods splitting under the tensions created by uneven drying. Noisy as these flora are, they still only manage to fling their seeds a meter or two. The tropical tree Bauhinnia can send its seed using this type of mechanism a distance of 15 meters.*

Apparently some flowers have developed ballistic techniques to ensure cross pollination using insect accomplices. Thus again Simons gives us the following fascinating account of a missile attack on a pollinating insect [12]:

> *In one plant a favorite mechanism is a spring loaded stamen (the male sex organ which carries the pollen). In flowers such as gorsenbloom the stamens are pinned down with keel shaped petals. As the flower develops, the petals grow larger and the stamens inside become stretched until eventually they are primed like an old fashioned flint locked gun. When an insect lands on the flower it dislodges the stamens from their present holsters. They spring up and punch their pollen onto the underside of the insect where it hitches a lift to the next flower.*

Simons tells us that such encounters can be so painful for the insect that they never visit another male flower in their pollinating meanderings. Thus Simons tells us that, in a study in 1986 by Gustaf Valero and Craig Nelson of Indiana University,

> *It was found that a bee which has experienced this violent explosion is highly unlikely to visit another male flower. Hence forth the intimidated bee will visit female flowers only. Here the pollen rubs off resulting in cross pollination and a selective advantage for the original male flower. A clear case of explosive aggression paying off.*

Paul Simons has also written a book describing many other fascinating details of ballistic warfare in the plant world [13].

Section 11.3 The Nuts and Bolts of Abrasive Cleaning

In our consumption of everyday items we sometimes forget that the producers face literally mountains of waste products that must be disposed of. The people who provide us with packets of nuts are left with mountains of nutshells. In an article discussing the problem of disposing of nutshells, Bradley tells us that nutshells are almost indestructible; they pose a waste-disposal problem in that each year 50 thousand tons of pecan shells alone are discarded in the United States [14]. One use of nutshells is that, when pulverized, they make a relatively gentle abrasive that can be used in oil drilling muds and in "sand" blasting.

Sand blasting is a branch of missile technology that is widely used to clean old buildings. In this technique, abrasive sand particles are fired at the surface to be cleaned where they wear away the deposits of grime accumulated on the surface of the stone and brick. Sand blasting however tends to generate a dirty slurry, a thick suspension of particles, which then has to be disposed of. Recently scientists have developed a new form of sand blasting which does not use sand [15].

Solid forms of carbon dioxide are known as dry ice. In the new cleaning method, crushed dry ice is used as an abrasive in a jet of air. Clive Curtis, a cryogenicist who manufactures a device of this kind, says [15]

The equipment looks like a sand blaster though we can tailor the jet by varying the size and densities of the grains of dry ice. The mixture of residual dry ice and grime blasted by the device falls to the ground and only the dirt is left as the carbon dioxide evaporates.

The stripping of old paint from aircraft is another task that we give little thought to but which turns out to be an immense problem when we come to look at the details. For example the US Airforce used to have to repaint its aircraft every five years. The fleet of aircraft was over 4000 in number and to clean a C130 transport cost $9000 using ordinary chemical paint removers. One of the alternative techniques that the US Airforce has adopted to clean the aircraft is to use high pressure water jets containing ordinary domestic baking soda as the abrasive [16]. The baking soda is incorporated into the water and blasted through a nozzle at the aircraft surface where it acts as a mild abrasive and loosens the paint. This process takes a little longer than the old one (120 instead of 80 hours) but the cost is lower: $4000 per aircraft as opposed to $9000 for the older technique.

Water jets loaded with abrasive powders can also be used to cut objects. For example, a 14-inch concrete slab can be cut at a rate of 1 inch per minute. For a review of water jets and water jet abrasive combinations used in industrial fabrication see Refs. [17, 18].

Section 11.4 Measuring the Size of Dust Fine-Particles

Scientists who are concerned with the possibility that air pollution is altering the Earth's climate are very involved in measuring the size of dust at all levels in the atmosphere. One type of instrument that measures the size distribution of fine-particles in the atmosphere uses missile theory. This involves the firing of dust fine-particles across a gap. The time required for a dust fine-particle to move across the gap is measured using two lasers. [19, 20] The physical principles employed in the design of this type of instrument are illustrated in Figure 11.1.

The system shown in Figure 11.1 is described by its manufacturer in the following terms [20]:

> *The aerosol to be measured expands through a nozzle into a partial vacuum. The air leaves the nozzle at a near sonic velocity and continues to accelerate through a measurement region. The smaller the fine-particle, the faster the acceleration through the measurement zone. The fine-particle velocity is measured using the two laser light beams. As they pass through the laser beams, they scatter light which is detected and converted into electrical signals by two photomultipliers. The time of flight between the two beams is measured with an accuracy of 1% or better. The computer system attached to the equipment uses the known fine-particle density and the known parameters of the acceleration to measure the size of the fine-particle. The equipment measures fine-particles at the rate of 10 000 per second.*

As the dust fine-particle emerges from the injection nozzle, the aerosol is surrounded by a sheath of clean air that channels it into the measurement zone as shown in the enlarged view of the measurement area at the nozzle. The surrounding air keeps it directly on line through the center of the instrument. The use of a sheath of clean air to focus the aerosol being characterized in the instrument is a widely used technique known as *hydrodynamic focusing*. Originally this technique was used with instruments employing liquid streams to examine a series of fine-particles lined up in the device and hence the name hydrodynamic. The term was extended without modification to systems using air, even though the hydro part of hydrodynamic focusing would lead one to suspect there was some water used in the equipment. It is this type of extension of vocabulary which can act as a *semantic trap* for the reader unfamiliar with a given area of scientific literature. (Note the term semantic is the general term used for a discussion of the meaning of words. It comes from the Greek word "sema" meaning "a sign.") The word aerosol is used scientifically to describe any cloud of fine-particles or liquid droplets. Unfortunately, in the everyday language aerosols have come to refer to spray cans using high pressure gas as a propellant. Again the reader should note the difference in scientific and general usage of a word.

The type of calibration graph used by the manufacturer of the instrument in Figure 11.1(a), is shown in Figure 11.1(b). The performance of the instrument can be checked

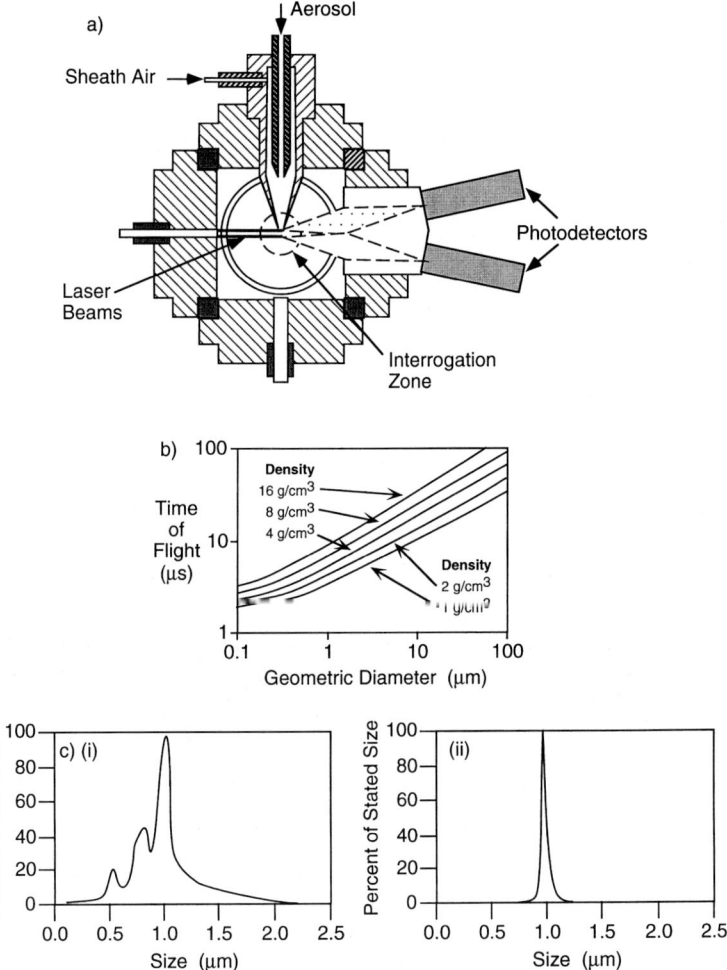

Figure 11.1 One method of measuring the size of dust fine-particles involves firing them over a measured distance between two laser light beams. a) Basic scheme of the Amherst Process Instruments system for measuring the size of aerosol fine-particles. b) Calibration curve for materials of various densities. c) The machine is able to differentiate between fine-particles of various sizes. (i) Size distribution of a mixture of three known sizes of latex spheres. (ii) Size distribution of a suspension of "monosized" latex spheres.

using aerosols of known characteristics as demonstrated by the data of Figure 11.1(c). When measuring dust in the atmosphere, the density of the individual dust fine-particles varies and is often unknown. In such situations the density of the material is assumed to be 1 g/cm³ and the resultant size is called the *aerodynamic diameter* of the dust fine-particles. This aerodynamic diameter is often larger than the physical size of the fine-particle, but scientists take this into account when using the aerodynamic diameter for size distribution data.

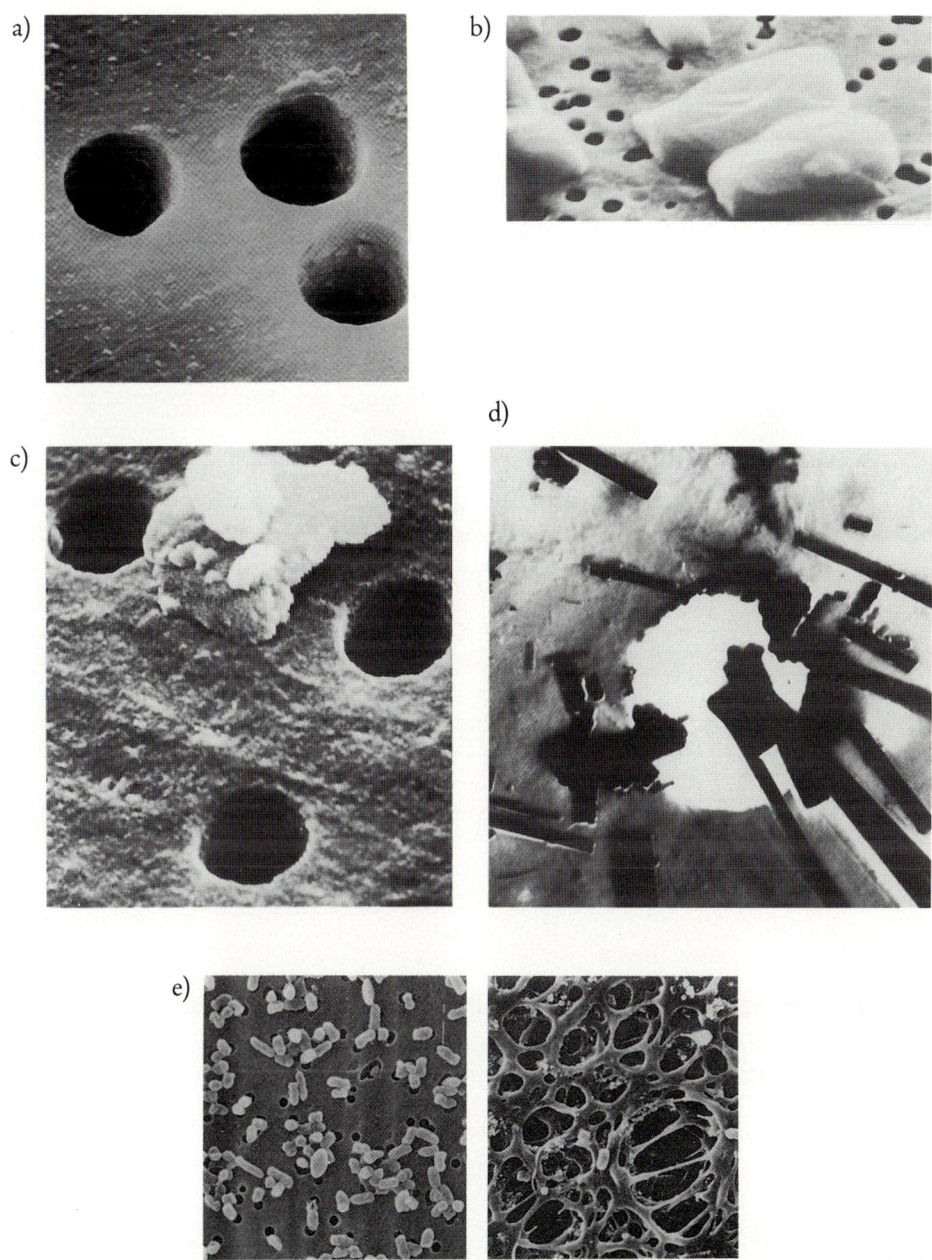

Figure 11.2 Nuclepore filters and other similar systems have revolutionized the art of filtering tiny fine-particles [21–24]. a) Appearance of the surface of a Nuclepore filter. Note the precise, cylindrical holes. b) Ammonium sulfate crystals on a Nuclepore filter. c) Simple dust on the filter surface. d) Asbestos fibers retained by the filter. e) Comparison of the appearance of bacteria captured by a Nuclepore filter and a depth filter.

If scientists are interested in the actual physical size of the dust fine-particle, they can pass the emerging stream from the instruments onto a filter for subsequent examination under a microscope. In recent years our ability to filter dust fine-particles, other tiny pieces of material, and living items has been revolutionized by a new type of filter created by means of some missile technology. A widely used brand name of this type of filter is the Nuclepore filter [21]. In Figure 11.2(a), the structure of a Nuclepore filter is illustrated. Several photographs of material filtered from air and liquid are also shown in Figure 11.2(b),(c), and (d) to illustrate the efficiency of this type of filter. Prior to the availability of the Nuclepore filter, the main type of fine filter in use was the spongelike porous plastic filter. The structure of these two types of filters are compared in Figure 11.3 (b). In contrast to the structure of the sponge filter, the Nuclepore filter has a set of precise, random holes passing straight through the sheet of material through which the liquid or air is passed. A major advantage of the Nuclepore-type filter is that the items being filtered are retained on the surface of the filter and can be viewed easily with image analysis equipment rather than being trapped within the body of a piece of porous material. The second advantage is that the holes can be made very precise:

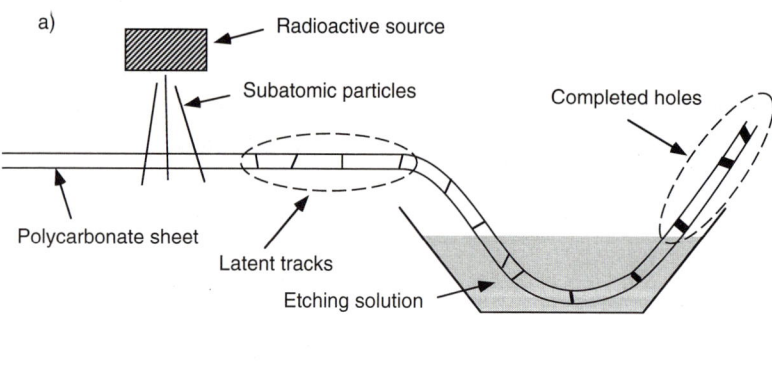

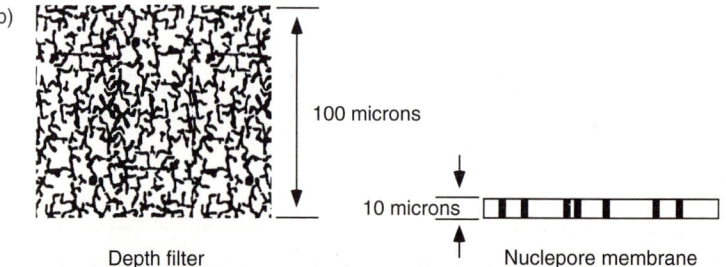

Figure 11.3 Manufacture of the Nuclepore filter and comparison of the structure with a depth filter. a) The precise holes of Nuclepore filters are produced by bombarding a thin piece of plastic with neutrons then etching the tracks of the neutrons to make holes in the sheet. b) The difference in structure of the two types of filters.

the manufacturer quotes a maximum deviation from nominal size of -20%. Thus if the holes are specified as being 2 micrometers in diameter, then the smallest aperture there can be in the surface of the filter will be 1.60 microns and, moreover, there will be no holes bigger than the nominal size.

The holes in the Nuclepore filter are essentially cylindrical and they stop anything larger than the hole. The size of the fine-particles that can be stopped with a sponge filter is more difficult to predict than for a Nuclepore filter. Figure 11.2(e) shows a comparison of the appearance of the surface of a Nuclepore and a sponge-type filter, both rated as having the same stopping power, after filtering a particular bacterium from a fluid [22]. Obviously the bacteria have sunk into the body of the sponge filter, making it difficult to determine what it is retaining and its ultimate maximum size. In contrast, the bacteria are clearly retained on the surface of the Nuclepore filter.

The manufacturing process used to create Nuclepore filters is illustrated in Figure 11.3(a). A thin film of a plastic known as polycarbonate is bombarded by energetic subatomic particles from a nuclear reactor. These missiles from the atomic reactor leave weakened paths in the plastic. Although these paths are not visible to the naked eye, the plastic along the track can be attacked by a weak, warm, sodium hydroxide solution whereas the unbombarded plastic resists attack by this chemical. Consequently, the liquid eats away a cylindrical hole along the pathway of the energetic particles. As can be seen from the enlarged view shown in Figure 11.2(a), the holes are perfectly cylindrical [23, 24]. Examination of the holes after sectioning the plastic shows that they are also essentially perpendicular to the surface of the film. It is these properties of the holes that ensure that the fine-particles or other materials filtered through the Nuclepore filters stay on the surface of the filter. Although the holes are cylindrical, the manufacturers have no control over the geometry of the flux of subatomic particles hitting the surface of the plastic and, as a consequence, the holes produced by the process are spread randomly over the surface of the filter. Manufacturers must keep the density of holes per square centimeter low; otherwise, as neighboring holes are etched, they will grow into each other. If you look carefully at the holes underlying the crystal of ammonium sulfate in Figure 11.2(b), you can see that in the commercial product some of the adjacent holes do grow into each other and of course this means that there exist apertures in the filter surface which are larger than the nominal value of the filter. Studies have shown that it is virtually impossible to eliminate overlapping adjacent holes and the technology has to keep a balance between having enough holes to carry out the filtration process and maintaining a very low level of overlappng holes.

The first *bacteria* seen by scientists under the microscope were rod shaped like those of Figure 11.2(e) and hence the name which comes from the Greek word "bakterion" for "small rod". Not all of the bacteria to be found in this world are rod shaped but the general name has been adopted for this class of organism. In the early days of the study of bacteria and viruses, the filters available were not as precise and were coarser than the present day Nuclepore-type filters. In fact, some older textbooks describe the difference between bacteria and viruses as being that bacteria could be filtered whereas

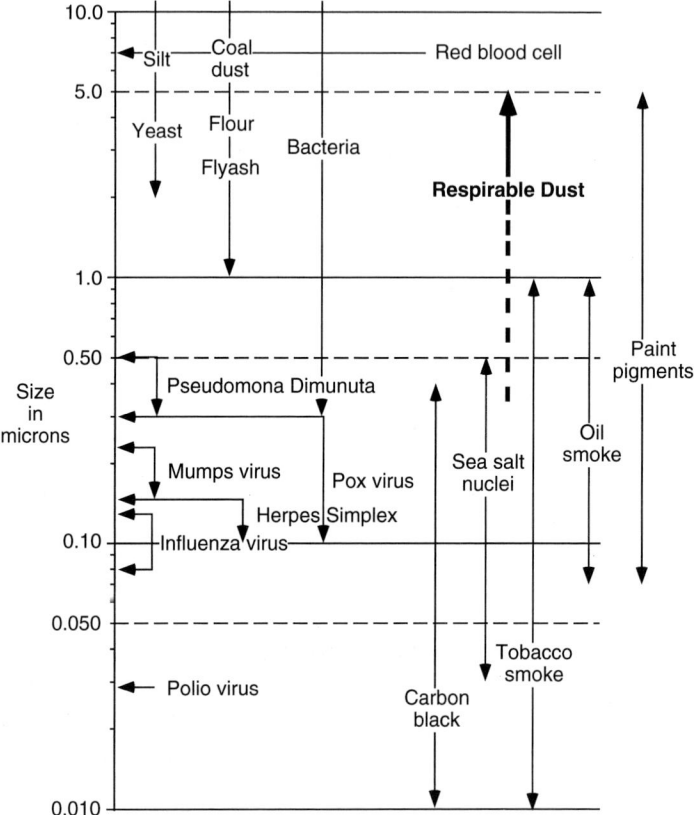

Figure 11.4 The range of filtration problems encountered in modern technology is very wide.

viruses could not. Today biologists have Nuclepore-type filters with holes as small as 0.015 micrometers and in these viruses can be filtered. In Figure 11.4, some data are presented to illustrate the wide range of technological studies in which special filters are used to look at the items being studied. The missile-created Nuclepore and other similar filters are revolutionizing these technologies [22].

"To etch" is defined as to make designs on metal, glass, etc., by eating out the pattern lines with an acid. It has the same root as the German word "essen" meaning "to eat." In modern technical English the word has been extended to describe any chemical method of "eating away" a substance. Thus in the manufacturer of Nuclepore filters, the solution being used to etch the holes is actually not an acid but the chemical sodium hydroxide or caustic soda. This is an example of an alkali, which is actually the opposite of an acid.

Section 11.5   Hammers Without Handles

It is estimated that three percent of the electricity generated in North America is used by the mining industry to pulverize ores, a process which enables the extraction of a valuable mineral by processing in furnaces and other equipment. It is also widely stated that most of the pulverizing machinery being used in the mining industry is only two percent efficient when viewed from the perspective of the energy invested in the powder produced as compared to the energy used in the pulverizing process. Although these estimates are open to discussion, their basic magnitude is correct and therefore it is not surprising that research into better ways of pulverizing ore is being actively pursued in the mineral processing industry [25].

The simplest way to crush a piece of ore is to hit it with a hammer. However, the act of hitting a piece of rock with a hammer is not always fully understood, even by those who make a living of pulverizing rocks. Dr. Gordon who has studied the strength of materials has described the act of hitting a piece of rock as follows [25]:

> *When we strike a solid with a hammer, a whole series of stress waves radiate from the point of impact and move off into the body of the material. They reach the furthest boundary of the solid in a time which is probably between one ten thousandth and one hundred thousandth of a second. They are reflected back as a kind of echo with very little loss of energy. What happens next depends upon a great many things such as the shape of the piece of rock and exactly where the blow was struck. What may happen is that the returning reflected stress waves repeatedly meet the outgoing ones at some critical point and as the stress is piled up at this point, fracture occurs.*

Figure 11.5(a) is a remarkable photograph taken by Dr. Cross of a steel ball 4.5 mm in diameter hitting a glass plate of area 5 cm^2 and thickness 6.35 mm [26]. At the moment of impact, the ball was traveling at 200 meters per second. The photograph was taken 20 microseconds (20 μs) after impact (a microsecond is one millionth of a second). It can be seen that the area near the colliding sphere is black; this is because there is a multitude of cracks in the glass. After 20 microseconds, the shock waves in the glass have already traveled across the plate and have been reflected. The interactions of the direct and the reflected shock waves have started to create cracks on the top and near the surface of the glass plate. Dr. Gordon tells us that a crack in glass travels and grows at about 30% of the speed of sound in the material; therefore, the growing cracks shown in Figure 11.5 are moving at approximately 1700 meters per second. It should be realized that the energy from the tip of a growing crack will itself generate stress waves which in the words of Dr. Gordon [25],

> *... are probably racing about in the material in all directions at the speed of sound being reflected on both old and new surfaces so that we are likely to end with not one crack but many. In other words the material shatters.*

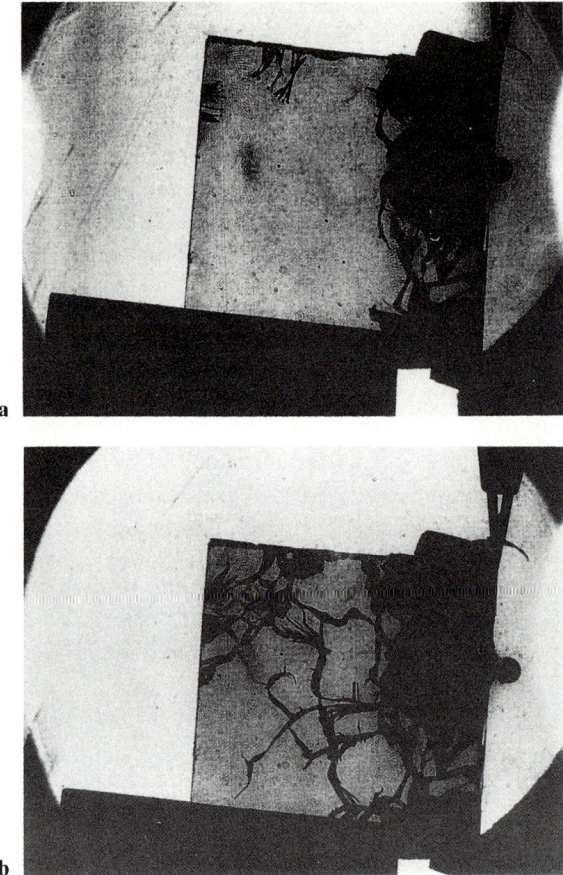

Figure 11.5 High-speed photographs of the failure of a glass plate hit by a rapidly moving steel ball, illustrating the stages involved in ballistic fracture. a) Cracks in the glass plate 20 microseconds after impact. b) Cracks present in the plate 70 microseconds after impact. (Courtesy of Dr. Lee A. Cross, Dayton, Ohio; photographs furnished by Special Illumination Systems, Inc., Dayton, Ohio.)

The second photograph of Figure 11.5(b) was taken 70 microseconds after impact. By this time the cracks have spread throughout the glass plate.

Figure 11.6(a) shows a tracing from another picture taken by Dr. Cross. To create the system of this figure, a steel ball was fired through a small piece of phenolic plastic board. The ball was traveling at 200 meters per second on impact. By the time the picture was taken, the ball had moved a measurable distance. The ball was 4.5 mm in diameter and can be seen clearly to the left of the piece of board. The size distribution of the fragments of the board produced by the steel ball are shown in Figure 11.6(b).

Beginning in 1977, scientists started to study a new form of geometry called *fractal geometry*. Fractal geometry was the brainchild of Benoit Mandelbrot. It concerns itself

11.5 Hammers Without Handles

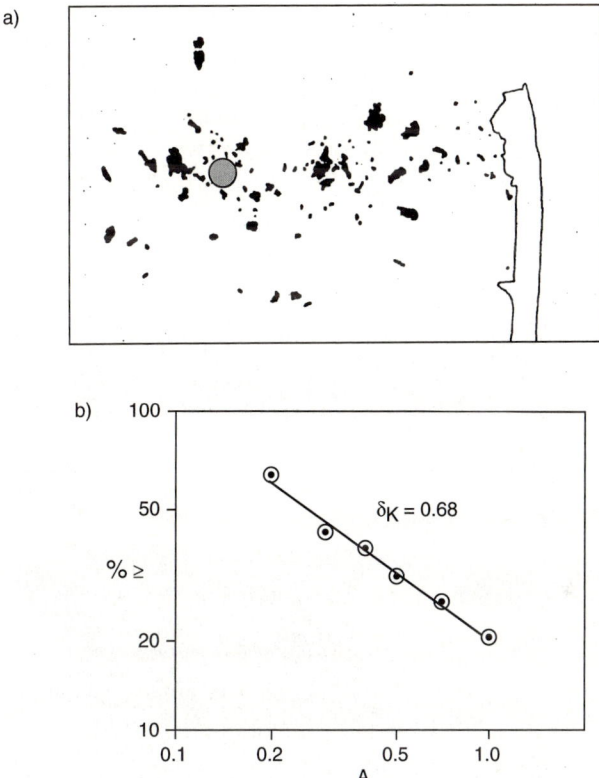

Figure 11.6 The fragments produced by a shattering process look like islands in a sea. a) A missile passing through a piece of phenolic plastic board. b) Size distribution of the fragments produced by the impact of (a).

with the difficult task of describing rugged surfaces such as the fragments of Figure 11.6(a). The traditional geometry most people learn at school is called *Euclidean geometry* after *Euclid*, a Greek mathematician. In ordinary, or Euclidean geometry, we concern ourselves with the dimensions of an object. Thus we describe a point as having no dimensions; a line, one dimension; a square, two dimensions; and a cube, three dimensions. Fractal geometry lets us describe the ruggedness of an object by adding a number to the traditional dimension of a system to describe its ability to fill space. (The name, fractal geometry, comes from the fact that it concerns itself with rugged surfaces such as those produced by fracturing a piece of material.) Thus in Figure 11.7(a), a series of lines of various ruggedness are shown and the way in which the ruggedness is described by a fractional number added to one, the dimension of a straight line, is shown. The number such as 1.45 is called the *fractal dimension* of the line. Describing just how we measure the ruggedness of the lines is beyond the scope of our studies in this course, but the three profiles of Figure 11.7(b) taken from the fragments of Figure 11.6(a) can

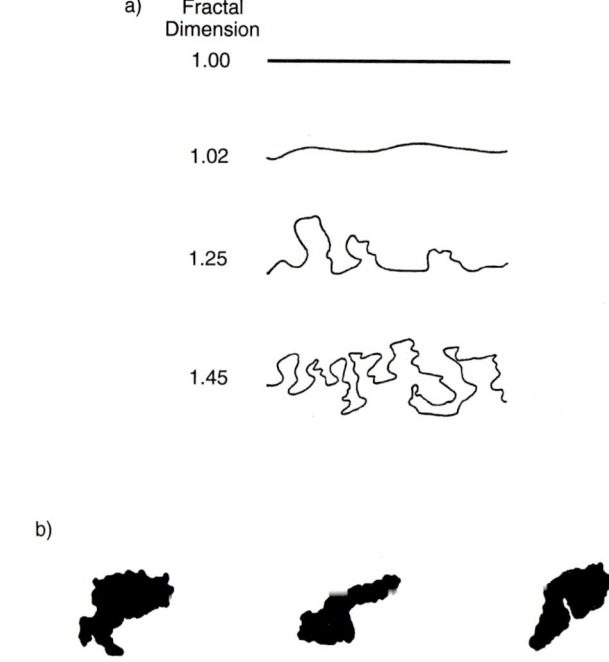

Figure 11.7 Modern science uses the concept of fractal dimensions to describe the structure of rugged boundaries. a) Lines of various fractal dimension. b) Three profiles from the fragmentation process of Figure 11.6(a).

be shown to have fractal dimensions of 1.17, 1.22 and 1.20. This gives a value of 1.20 for the average boundary fractal dimension of the profiles.

One of the exciting theorems in Mandelbrot's book on fractal geometry is that the size distribution of the fragments of a system such as Figure 11.6(a) created by a single ballistic impact should be related to the ruggedness (that is, the fractal dimensions of the profiles) of individual fragments such that if we divide the average boundary fractal dimension of the profiles by 2, it should equal the slope of the line of the cumulative size distribution of the fragments. For the system of 11.6(a), we note that the slope of the size distribution line of Figure 11.6(b) is 0.68 and half of the fractal dimension of the fragments is 0.6 (1.20/2). This is close enough to be an encouraging result, although not a proof of the existence of the relationship. The experiments summarized in Figure 11.6 and 11.7 are stimulating research in laboratories around the world in attempts to predict the type of fragments produced by a missile in such ballistic pulverization. By studying the way in which materials fragment during ballistic impact researchers hope to be able to improve the efficiency of pulverization machinery [27].

The title of this section takes its name from the idea that many of the pulverizing devices in use in the mineral industry consist of balls tumbled in a cylindrical container where, as the cylinder rotates, the balls are lifted up and thrown down upon the material to be crushed at the bottom of the cylinder. Thus these balls are hammer heads without handles. A lot of the inefficiency of traditional pulverizing equipment arises from the lack of efficiency in the collision between the ball and the material being pulverized, because the individual fragments hit by the ball are cushioned by the surrounding fragments and the impact of the blow is dissipated throughout the bed of powder.

Sometimes mineral-processing engineers do not use ceramic or metal balls to achieve the crushing but larger lumps of the material that is itself being pulverized. This process is described as *autogenous pulverization*. The word literally means self-generated grinding, although its meaning is often obscured by the fact that scientists pronounce it au-tó-gen-ous. An autogenous grinding device which brings together many of the physical theories explored in our study of missiles is the Trost mill, originally developed to pulverize spices [28]. The equipment is shown in Figure 11.8. The material to be pulverized is fed as a coarse powder from a storage bin into the machine. Air, used to make pieces of material encounter one another at high speed, flows under the bottom of the supply bin. The supply entrainment system is a Venturi throat! An opposing jet of air coming from the other side of the equipment meets the jet of entrained fragments in the collision zone. At the start of the operation of the mill there are no fragments in the second jet of air and the entrained material is simply swept up into a cylindrical chamber by the flow of air. In this chamber the air is made to spin with most of it leaving through

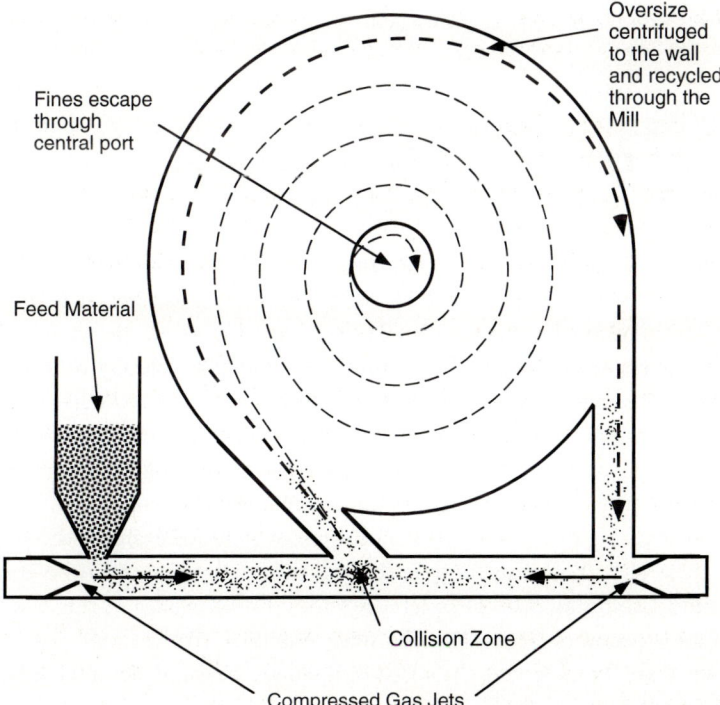

Figure 11.8 The Trost mill used the concepts of missile physics to achieve efficient pulverization [28].

the discharge hole at the center of the cylindrical chamber. Centrifugal force sends the large fragments to the outside of the wall where they tumble around and back down into the opposing jet — opposite the feed air jet. After a few moments, fragments of the material from the two jets meet each other in the impact zone and are pulverized by the encounter. The pulverized material is then again swept up into the cylindrical chamber. Now the smaller fragments of "fines" are captured by the spinning air stream and swept out the middle of the chamber whereas any unpulverized larger fragments are tumbled around and back into the collision zone again. The speed at which air is taken out of the top chamber controls the fineness of the product being produced by the mill. Thus if the air flow out of the middle is slow, only very fine material can exit with the stream of air. The cylindrical chamber acts as a fractionation device and a feedback loop sending larger material back to be pulverized some more. The fact that the whole process is operated by air means that any heat produced in the pulverization process is continually taken away by the air stream This is useful when pulverizing spices in that the aromatic flavor is not driven off by the heat of the pulverizing process. Indeed, the material being fed into the system can be chilled in liquid nitrogen both to pre-crack it and also to preserve the flavor and taste chemicals. This type of mill can also be used to pulverize material which is inherently explosive since the jets can also consist of nitrogen or carbon dioxide gas instead of air. As engineers realize that they need to be more efficient in creating pulverizing encounters in their machines, we can expect to see devices such as that in Figure 11.8 becoming more widely used in the pulverization industry.

Section 11.6   Fat Bodies and Magic Bullets

In popular discussion of the drug industry, a new drug which appears to be able to attack a disease is sometimes described as a magic bullet. However, it is important to point out that missile technology is actually being used to create anti-cancer drugs that can truly be called magic bullets. If a chemical intended to destroy cancerous cells is administrated to the whole body, the chemicals kill healthy cells as well as cancer cells. The drug industry is working very hard to develop what is known as *targeted delivery* systems to avoid the problems associated with body-wide application of a drug. One targeted delivery system which is being investigated makes use of micromissiles which are themselves made using missile-based technology. Consider for example the system illustrated in Figure 11.9. In this technique, the solution of a chemical to be encapsulated is sprayed onto a spinning disk. Droplets are thrown off the edge of the disk when the centrifugal force becomes larger than the surface tension of the liquid which causes it to cling to the disk. By increasing the speed of the disk, we can make smaller and smaller droplets. The droplets produced in this technique are very uniform in size. The disk is spun inside another cylinder which has holes around its periphery. These holes are

11.6 Fat Bodies and Magic Bullets

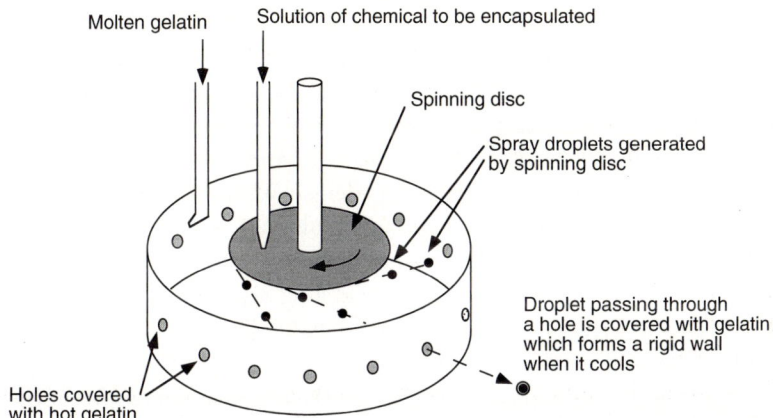

Figure 11.9 In one method of making microcapsule droplets, the chemical to be encapsulated is fired through holes covered by molten gelatin. The cooled microcapsules have a solid gelatin coating with a core made up of the chemical. (This technique was developed by the South West Research Institute in San Antonio, Texas.) Reproduced by permission of the Council of Powder Technology and the Hosokawa Powder Technology Foundation, Osaka, Japan. C1995, KONA. All rights reserved.

covered by a molten gelatine or other suitable coating liquid. The droplets leaving the spinning disk are fired through the holes in the outer cylinder. As they pass they are covered with hot gelatine or another coating liquid. They immediately start to cool and form little beads filled with liquid [29]. In one technique the *microcapsules* are covered with a waxy coating that melts at 41.7 °C (107 °F). The chemical used to destroy cancer cells is encapsulated with this material and the tiny capsules are placed in the blood stream. The body's defense mechanisms do not attack these tiny beads because they appear to be harmless waxy spheres. However, when they arrive at a cancerous tumor, this region can be heated with infrared rays from outside the body which melt the microcapsule coating, thus delivering the chemotherapy drug directly to the tumor.

In other experiments using the microcapsules, the outer coating is made of fat. The technical term for a fat compound is *lipid* and the little globules of fat carrying chemotherapeutic drugs to a tumor are called *liposomes* from the Greek word for fat and from "soma" for "body". These liposomes can be very efficient magic bullets. Growing cancer tumors are often very rich in blood vessels and so blood carried microcapsules can be efficiently delivered to the tumor without attacking other areas of the body. The function of the liver is to break down fats in the blood and so it removes the fatty coating from the microcapsules, freeing the drug to do its work [29].

Scientists have mixed magnetic materials into the suspension of a chemical to be microencapsulated. A magnet can then be used to concentrate the tiny bullets at the site of a tumor to be destroyed. J. Andrews and colleagues at the University of Michigan in Ann Arbor have incorporated a radioactive material into tiny glass spheres [30]. When these spheres are injected into the artery feeding the liver, the microspheres lodge in the tiny peripheral branches of the blood vessels where they deliver radiation to any cancerous tissue in that area. Some of the microspheres lodge in healthy parts of the liver, but, as Andrews points out, there are roughly three times as many small arteries in a tumor than in the surrounding tissue so you can deposit more glass spheres and hence more radioactive material in the tumors. Andrews and his colleagues used these magic bullets of radioactive glass to treat 24 patients with inoperable liver cancer. Without this treatment, the life span of the patient would have been 5 to 6 months. In nine patients the cancer was actually reduced and it stopped growing in another seven patients. After 18 months, seven patients were still alive and two survived for more than 32 months after the therapy.

In the new field of biotechnology, scientists are trying to create modified plants and animals which have the advantages conferrred by the introduction of foreign genes. One of the ways that biologists are modifying living cells is to fire small bullets directly into living cells. These bullets carry the genetic material to modify the living cell. Thus John C. Sanford and colleagues have built a device which can fire small tungsten fine-particles (1-4 micrometers in diameter) at velocities of between 300 to 600 meters per second [31]. They have shown that these tungsten bullets can penetrate cell walls and membranes in a non-lethal manner. Thousands of cells can be penetrated simultaneously as they grow in living tissues. The tungsten bullets have been able to deliver genetic material into a variety of plant species including onion, tobacco, and rice.

The genetic material that is implanted is known as *DNA*. This is an abbreviation of the word *deoxyribonucleic acid*, the chemical name for the substance in chromosomes that determines the genetic information used to build living creatures. It consists of long molecules composed of two interwound polynucleotide chains. Each element or nucleotide is chemically a sugar combined with one of four compounds known by the names adenine, cytosine, guanine, and thymine. The pattern of these compounds along the chain carries the information invested in the chromosomes. *Chromosome* is a word meaning "colored body". When scientists first started to look into living cells, they discovered thread-like bodies which absorb dye to appear colored. The name chromosome was invented in 1879 by the German scientist W. Flemming. In human beings, there are normally 46 chromosomes with half of them coming from the mother and half from the father. Damage to the chromosomes is often the source of birth defects in children.

P. Christou and colleagues have developed another ballistic procedure which shoots foreign genes, that is DNA, into plant cells. Their genetic gun has been used to alter the genetic constitution of a type of rice known as Indica [32]. Varieties of this type of rice account for four-fifths of global rice production and are the basic food for two billion people. Christou and colleagues are trying to make varieties of Indica which will be resistant to widespread pests including the tungro and the beetle called the rice water weaver which damages rice plants by burrowing into and colonizing the roots. They have already used their technique to transform commercial Japanese rice varieties grown in the United States and Japan. After having stripped the rice cells of their protective walls they suspend them in solution. The cells are then bombarded with tiny gold beads coated with loops of DNA containing the gene to be transplanted into the rice plant. In the first step, when using the bombardment technique, the beads to be used to invade the plants are laid on a piece of aluminum foil of about one square centimeter. The "gun" is pointed upward at the underside of the foil. The nozzle of the gun is 13 millimeters across and contains a water droplet suspended between two tiny electrodes. When a voltage is applied to this droplet, it is vaporized with such force that the shock wave propels the foil upwards. A mesh above the aluminum foil blocks the foil but not the accelerating beads of gold which have gained high kinetic energy. The gold beads continue upwards to reach the rice cells which are embedded in an upturned plate of culture medium. Christou went on to prove that foreign genetic material was indeed implanted into rice cells in this way. The implanted genes conferred resistance to certain antibiotics and herbicides. Later, the modified plants survived exposure to these toxic compounds, whereas normal plants died.

We have come a long way from our original inspection of the dimpled golf ball and the effect of its dimples on its flight characteristics. In beginning to write this book, I had no idea where I would be at the end of the journey as I explored the dynamics of all kinds of missiles and here we have just discussed missile techniques used today to achieve genetic engineering in plants — and perhaps tomorrow in humans! I hope the reader has enjoyed our meandering journey from missile to missile as much as I have.

When looking for a final topic to round off our exploration of missile technology I came across a story regarding the problem of deer at the National Institute of Standards and Technology in Gaitherburg, Maryland. An abundance of deer were causing problems by eating the flowers on the grounds of the research institute and at one point a doe crashed through the window of the ladies' room. After some consideration, the solution adopted to control the deer population was to fire darts, primed with contraceptive, at the deer using blow pipes. The darts, it was hoped, would prevent the deer from breeding and reduce the size of the herd living on the estate. As the story is told in the New Scientist [33]

> *Other more dramatic methods were first considered, including the idea of unleashing a pack of wolves on the estate, but the concensus at the institute was that, 'While we would lose a few deer we might lose a few slow moving chemists and engineers as well.'*

Apparently contraceptive missile technology was a lesser price to pay than losing a few scientists!

References

[1] I. Anderson, "How Pilots Avoid Volcanic Clouds," *New Scientist*, April 13, 1991, p. 22.
[2] D.E. Axelrod, "Role of Volcanism in Climate and Evolution," Special Paper 185, 1981, Geological Society of America, P.O. Box 9140, 3330 Penrose Place, Boulder, Colorado, U.S.A., 80301.

[3] See News story, "When Birds Meet Aircraft, a Computer Simulation Can Help," *New Scientist*, August 4, 1990, p. 51.
[4] News Story, "Engine Trouble," *Discover*, Vol. 12 No. 3, March 1991, p. 16.
[5] C. Putnam, "The Long And Short of Discouraging Gulls," *New Scientist*, September 24, 1994.
[6] J.H. McMasters, "The Flight of the Bumble Bee and Related Myths of Entomological Engineering," *American Scientist*, Vol. 77 No. 2, March/April 1989, pp. 164-169.
[7] C.H.T. Townsend, "On the Cephenemyra Mechanism and the Deer Fly Circling of the Earth by Flight," *Journal of the New York Entomological Society*, Vol. 35, pp. 245-252.
[8] I. Langmuir, "The Speed of the Deer Fly," *Science*, Vol. 87, 1917.
[9] R.J. Wootton, "The Mechanical Design of Insect Wings," *Scientific American*, Vol. 263, No. 5, November 1990, pp. 114-120.
[10] See G. Barnes, "The Bombardier Beetle," *The Physics Teacher*, Vol. 30, January 1992, pp. 26,27.
[11] News story, "Little Squirt Saves Plant from Being Eaten," *New Scientist*, January 5, 1991, p. 48.
[12] P. Simons, "An Explosive Start for Plants," *New Scientist*, January 2, 1993, pp. 35-37.
[13] P. Simons, "The Action Plants," Blackwells, Oxford, 1992.
[14] D. Bradley, "Why Nuts are Hard to Crack," *New Scientist*, December 26, 1992, p. 14.
[15] A. Cogham, "Dry Ice Leaves Dirty Cleaners Out in the Cold," *New Scientist*, November 19, 1994, p. 28.
[16] R. Gould, "Bicarbonate Adds Fizz to Aircraft Stripping," *New Scientist*, October 8, 1994, p. 24.
[17] E. Raia, "Cold Cuts," *High Technology*, Vol. 5, No. 12, December 1985, pp. 58,59.
[18] R. Hamilton, "Water Jets," *Technology Review*, Vol. 92, No. 7, October 1989, pp. 9,10.
[19] One of the instruments of this type is manufactured by TSI Incorporated, 500 Cardigan Road, P.O. Box 43394, St. Paul, Minnesota 55164. It is known as the APS Instrument.
[20] An instrument used to measure dust sizes by looking at the time of transit of the dust fine-particles between two laser beams is manufactured by Amherst Process Instruments Inc., Mountain Farms Technology Park, Hadley MA 01035.
[21] M.C. Porter, "A Novel Membrane Filter for the Laboratory," *American Laboratory*, November 1974, pp. 63-76.
[22] See discussion of Aerosol filtration in *Direct Characterization of Fine-particles* by B.H. Kaye, John Wiley, New York, 1981
[23] M.C. Porter, H.J. Schneider, "Nuclear Membranes for Air and Liquid Filtration," *Filtration Engineering*, January/February 1973.
[24] Information on the Polycarbonate Filters is available from the Poretics Corporation, 151.I Lindbergh Avenue, Livermore, CA.
[25] See Chapter 9 entitled "Fracture, Fragments and Fractals," in *A Random Walk Through Fractal Dimensions*, by B.H. Kaye, VCH, Weinheim, 1989. See also B.H. Kaye *Chaos and Complexity, Discovering the Surprising Patterns of Science and Technology*, VCH, Weinheim, 1992.
[26] The photographs used in this section were provided by Dr. Lee A. Cross, Vice President, Special Illumination Systems Inc., P.O. Box 501, Dayton, OH., 45409-501, and are used with his permission.
[27] B.H. Kaye, "Fractal Dimensions in Data Space; New Descriptors for Fine-particle Systems," *Part. Part. Syst. Charact.*, Vol. 10, pp 191-200, 1993.
[28] The Trost mill is manufactured and sold by Garlock Inc., Plastomer Products, 23 Friends Lane, Newtown, PA 18940-9990.
[29] For a discussion of the various methods of making microcapsules and their various uses in a range of industries see B.H. Kaye, "Microencapsulation: The Creation of Synthetic Fine Particles with Specified Properties". *KONA*, No. 10, p. 65-82, 1992.
[30] News story, "Radioactive Beads Combat Liver Cancer," *New Scientist*, January 5, 1991, p. 94.
[31] J.C. Sanford, T.M. Klein, E.D. Wolfe, N. Allen, "Delivery of Substances into Cells and Tissues Using a Particle Bombardment Process," *Particulate Science and Technology*, Vol. 5, 1987, pp. 27-37.
[32] See News story, "Genetic Gun Makes Rice Growers Day," *New Scientist*, November 2, 1991, p. 23.
[33] See "Feedback" section of *New Scientist*, October 14, 1995, p. 84.

Index

A

ablation 237
Aborigines, battle tactics 119
absolute zero, determination 244
acceleration
– due to gravity (g) 36
– maximum sustainable by a human 91
accelerometer
– airbag trigger 373
– operation of 374
acetylene 171
aerobatic loop, weightlessness 91
Aerobie ring 155
aerodynamic
– diameter 396
– drag reduction 380
aerogel, insulation 248
Agincourt, Battle of 52
agriculture in mines 291
airbag
– inflation 375
– trigger, accelerometer 373
– automobile 373
– motorcycle 376
– operation 375
– possible injury from 374, 375
Airborne Volcanic Ash Detection System (AVADS) 387
aircraft
– ceramic armor 207
– engine, bird strike regulations 388
– formation flying 380
– pilotless, V1 (Buzzbomb, Doodlebug) 213
– reduction of turbulence over a wing 97
– speed of sound 186
– use of riblets 97
airfoil 151
aliens 61
aluminum oxide 235
Alvarez, Louis and Walter (K-T boundary) 317
Amazons 51
ammonia 179
ammonium perchlorate 231
ammunition
– cartridge 183
– caseless 191
– G11 assault rifle 191
– percussion cap 185
– round of 183
angina 175

angle of attack 152
Ångstrom 176
Ångstrom, Anders Jonas 176
angular momentum
– conservation of 89
– toy top 93
angular velocity 89
antimissile-missile 193
apogee 273
Apollo objects 294
– speed 294
Archemedes 182
– density of a crown 209
– principle 211
– steam powered cannon 182
archery 50
armor
– ballistic 207
– ceramic 207
– disadvantage in battle 53
– distribution of impact energy 359
– Kevlar 358
– nylon 359
– paper 358
– Philistine 110
– piercing bullets 192
– sheepskin jacket 358
– speed of sound 359
aromatic compound 171
arrow
– aluminum 54
– barbed, purpose 355
– construction of 35
– feathers 43
– oscillation 44
– sharpness 355
– spin stabilization 43
– spin versus oscillation 45
– tuned 45
– vanes 43
arrowhead 49
artificial gravity 92
asteroids 292
– Apollo objects 294
– Apollo, speed 294
– Geographos 320
– kinetic energy 294
– (K-T disaster) 318
– number and size 293, 294

- valuable content 321
- watching 320
astronomical unit 292
atmospheric pressure 145
atom 346
attitude angle 152
autogenous pulverization 405
automobile
- aerodynamic drag 377
- airbag 373
- air conditioning 377
- black box 372, 378
- crash 379
- driver fatigue 378
- energy absorption in a crash 373
- first fatality 371
- fuel consumption 377
- head restraints 377
- passenger cage 373
- seatbelts 373
available redundancy 368

B
backspin 21
bacteria 399
- genetically engineered 337
badminton bird *see* shuttle cock
badminton 81
ball mill pulverization 404
ballast 211
Barnard's Star, 280
barometer 146
Barringer crater 287
basalt 296
baseball
- aerodynamics 129
- ball characteristics 123
- ball construction 124
- bat 26
- batter 127
- coefficient of restitution 124
- coefficient of restitution, consequences 124
- game play 127
- pitcher 127
- protective equipment 132
- related deaths 131
- safer ball 131
- scuffing 129
- spitball 129
battering ram (ballistic pendulum) 190
bazooka 228
belly flop, danger of 354
benzene ring 171
Bernoulli, Daniel 143
Bernoulli effect 22

Bernoulli's principle 143, 148, 151
bicycle
- helmet 363, 34
- paths 364
- racing 780
Big Splash Theory 324
black hole 258
black powder 166
black-body 250, 251
- peak wavelength 251, 254
Blakey, Rusty 338
blowpipes 165
bobsledding 382
bola/bolas 87
bolide 299
bomb 169
bombardier beetle 392
boomerang
- flight characteristics 158
- gyroscopic stabilization 156
- history 156
- precession 158
- purpose 156
- rules of competition 159
- use 157
bouncing bomb 120 ff.
bow 33 f.
- composite 45 ff., 56,
- compound 54, 56
- crossbow 48
- early use 50
- effect of pulley 57
- energy transfer 42
- force-distance graph 42
- machine 54
- modern construction 53
- recurve 54
- reflex 46
- relation to stringed instruments 61
- stabilizer 91
- stringing 47
- Yew wood 35
boxer 368
boxing 366 f.
Brahe, Tyco 306
breccia lens 296
brecciated rock 290
Bren gun 191
Brent Crater (Algonquin Park) 294, 297
Brownlee 313
bullet
- alternate materials 192
- biodegradable 196
- calibre 183
- devastator 196

– dumdum 195
– gyroscopic stabilization 185
– hollow-point 197
– incendiary 193
– pass 100
– plastic 197, 198
– proof vest 358, 390
– rubber 197
– speed 186, 188, 221
– tracer 193
Bumblebees 390
bumper 309
bungee jumping 355
buoyant weight 211
bursera tree 392

C
calorie 39, 40, 334
calorimeter 39, 334
Canadian Shield 291
cancer 407
CANDU nuclear reactor 348
carbon dioxide 171, 256
catapult *see also* slingshot 111
cellulose 178
Celsius, Andros 39
Centaurs 50
center of gravity 13 f.
centrifugal force 88 ff.
cgs system 36
chemical warfare 167
chemotherapy *see* cancer
Cherenkov radiation (cosmic ray) 267, 348
chromosome 408
circular motion 88
cleaning technology 394
cloud 339
– seeding 338 ff.
coefficient of restitution 8
– baseball 124
– golf 10 f., 28
– hardwood 46
– peas 13
– soccer 12
– tennis 68, 73
– track surface 13
coherent light 199
collision 10
comet 304 ff.
– coma of 305
– dead 301, 307, 309
– Halley's 306
– jets 309
– nucleus 304
– structure 304

– origin 308
– tail 304
– visibility 304
Computerized Axial Tomography (CAT scan) 368
conduction of heat 245
Congreve, Sir William 227
convection 248
– convective growth 339
conversion of energy
– kinetic to heat 309, 324
– kinetic to potential 189, 355
– potential to kinetic 18, 33, 38, 42, 111
cork 123
cosine wave 106
cosmic drifter 314
cosmic dust 308 ff.
– collection 313
– recoverage 312
– effect on spacecraft 308
– observing 309
cosmic missile 289
– chondrite 314
– footprint of 296, 310
cosmic rays 266 f.
crash helmet 362 f.
crater 290 ff.
– (K-T extinction) 319
– complex 297 f.
– determining age of 287
– ejecta 296
– impact 287, 297
– formation 294
– Moon dust 323
Crecy, Battle of 52
Creighton Fault 291
Cretaceous-Tertiary (K-T) disaster 317
cricket 125 ff.
crossbow 48, 52
cryoblation 236
cryogenics 236 f.
cryonics 237
Cupid 58
cybernetics 207

D
darts 140
David and Goliath 108
deerfly 390
degree 104
dementia pugilistica 366
density 27, 211
deoxyribonucleic acid (DNA) 408
deuterium 348
Dewar flask 258
Dewar, Sir James 258

diamond 171, 214
diatomaceous earth (kieselguhr) 174f.
dimensions 403
dinosaurs 315, 319
discus 152
dolphin 96
Doppler effect 78, 278
Doppler, Christian Johan 78
dry ice 338
dynamic pressure *see* Bernoulli's principle
dynamite 18, 174
dyne 36

E
earth
- age 315
- asthenosphere 260
- atmosphere of 261
- crossing asteroids 294
- development of life 315
- escape velocity 269
- explanation of nutation 326
- geologic history 315
- hydrosphere 261
- internal structure 260
- mantle 261
- mesosphere 261
earthquake 303
Einstein, Albert 3
electromagnetic spectrum 252
elements, periodic table of 346
energy 18
- absorption 373
- dissipation 23
- storage 41, 47, 91, 111, 327
- transfer 42
- kinetic 18
- potential 18
erg 38
ergonomics 17
erosion 290f.
ether 264f.
ehtymology 65
Euclidean geometry 403
explosives 171ff.
- food powders 177
- metal powder 233
- plastic 180
- sawdust 179
- Semtex 180
extraterrestrial visitors 312
eye, evolution of 252, 267

F
Fahrenheit, Gabriel Daniel 39
Fawkes, Guy 225
fireworks 282f.
First Law of Motion 35
Fizeau, Armand 264
fluid drag 96f.
fluid flow 23
flywheel 91, 98
focal point 169
football 100
force 37, 49
- gravity 36, 91
- impulsive 20
- reactive 223
formation flying 379
fossil 315
Foucault, Jean 94
fractal dimensions 402ff.
free fall 354
Frisbee 153f.
fuse 169

G
galaxies 279
Gatling, Richard Jordan 191
Gay-Lussac 244
genetic engineering 337, 408
geyser 337
gimbal, 95, 213
Giotto mission 308, 311
glucose 177
Goddard, Robert 228
golf 4ff.
golf ball 5, 20
- dimples 4f., 23ff.
- acceleration 4
- Aquaflyte 26
- atomic 28
- biodegradable 26
- coefficient of restitution 10
- electronic 28
- floating 27
- guttie 5
- light weight 27
- modified 28
- Polara 25
- professional drive 4
- regulations 10, 25, 27
- spin rate 23
- structure 5
- true flight 14
- turbulence 4
golf club 20
golf course 11, 27

Index

golf green 11
Goliath 110, 114
graphite 173
gravity 13, 35f., 92
greenhouse effect 256
greenhouse 256, 291
grenade, origin 168, 170
gun 181ff., 223
guncotton 173, 177, 179
gunpowder 166, 170
guttapercha 5
gyroscope 95f.
gyroscopic compass 96

H
hail 340ff.
hail abatement 342ff.
Halley's comet 306ff.
Hastings, Battle of 49, 357
Headrick 154
heat 245ff.
– conduction 245
– exchange 250
– infrared radiation 248
– insulator 245
– loss 245, 248, 250
– shield 262
– transfer 245
heavy water (D_2O) 345, 348
helmet 357, 363
Hess, Victor Francis 266
Hiroshima 303
hockey 364
holography 101
hydrazine rocket fuel 236
hydrodynamics 395
hydrofoil 151
hydrosphere see earth 261

I
iceberg 336
impact, power of 294
impulsive force 20
Indian Mutiny 184
insulation 246
isotopes 348

J
javelin as a sport 114, 139, 142
Joule, James Prescott 38
joule 38, 40
Joule-Kelvin effect 234

K
Kekulé 172
Kelvin, Lord 244
killing stick 156f.
kinetic energy 18, 19, 27

L
laminar flow 23
Langmuir, Irving 390
laser 102, 199ff.
"lead" pencil 173
lens 168
Ley, Willy 229
light year 279
light 199
– coherent 199
– colors 199, 252
– incoherent 199
– monochromatic 199
– speed of 264, 279
lipid 407
liposome 407
liquid propellant, cryogenic fuels 231, 236
longbow 33, 41, 52
luge 382

M
Mach number 188
Mach, Ernst 188
Magnetic Resonance Imaging (MRI) 370
magnetic spheres, cosmic dust 312
Magnus effect 21, 120, 126
Mandelbrot, Benoit 402
mandrel 231, 233
Manicouagan Québec crater,
manometer 145
mass 19
mass extinctions 317, 320
Mercury 325
– barometer 146
– thermometers 39
metal
– annealing 357
– annealing dead soft 357
– burning 282
– dislocations in crystal structure 357
– ductility 356
– powder 233
meteor 289, 327
Michelson, Albert 266
Michelson-Morley experiment 266
Microabrasion Foil Experiment (MFE) 309
microcapsules 407
microencapsulation 406
micrometer 176

Milikan, Robert Andrew 267
missile
– ballistic 230, 238
– heat-seeking 212
monomer 178
Moon 321, 324
– dust 323
Morley, Edward 266
musket 181, 183, 185

N
nanometer 176
Napier, John 241
Nemesis 320, 344, 348 f.
neutron 346
Newton, Sir Isaac 35
Newton's First Law of Motion 35
Newton's Second Law of Motion 36
Newton's Third Law 223
newton 38
nitrocellulose *see* guncotton 177
nitroglycerin 173 ff.
Nobel, Alfred Bernhard 174
nuclear structure 346
nuclear winter 319
Nuclepore filter 396, 399
Nullarbor Plain 326

O
Odysseus 47
Oort, Jan 308
orbit 269, 271 ff.
oscillation 44, 74
oscilloscope 72
ozone layer 261

P
π (pi) 106
Paleozoic 316
parachute 354
Parkinson's disease 366
Parthian shot 51
Pascal Blaise 146
Pascal, 146
passive satellite 273 ff.
peashooter 165
periodic table of the elements 346
phalanx 114
Philistines 108 ff.
photo "radar" 372
photomultiplier tube (SNO) 349
photosynthesis 177
Pioneer 10 278, 282
Pitot, Henri 150
Planck's constant 255

Planck, Max 254
planetoid 293
planets, characteristics due to collision 279 f., 324
plants with deep roots 291, 393
polymer 178
– polysaccharide 178
– polyurethane 231
polymerization 178
potential energy 18, 34
pressure wave 72
pressure 49, 355
proton 345
pulverization technology 401
pumice 296
pyrometer 256

Q
quantum theory 250, 255
quoit 139

R
radiation 106
RAdio Detection And Ranging (RADAR) 79
radius 105
railway safety 376
rain drops 338
rainmaking 338
Rayleigh, Lord 254
recycling plastics 237
relativity, special theory of 266
revolver 191
riblets 97, 382
Richter, Charles F. 301
rifle 185
Robin Hood 33, 41
rocket engine 236
rocket fuel
– hydrazine 236
– liquid 231
– solid 231
rocket 170, 225, 227, 342
Rutherford, Ernest 143

S
saccharine 178
Sagittarius 35, 51
sandblasting 394
satellite 271 ff.
– active 273
– air resistance 276
– communications 271
– footprint of 271
– gravitational effects 276
– passive 273 ff.
seatbelts 373

seismograph *see* earthquake 301
sharpness 49
ship 211
shock wave 187, 238
Shrapnel, Henry 195
SI system of units 38
silver iodide) 338
sine wave 106
ski jumping 381
sling 111 ff.
slingshot 111, 118
SNO photomultiplier tubes 349
Sobrero, Ascanio 173
soccer 12
Soddy, Frederick 348
sodium azide 375
solar system 276, 309, 324
solar wind 304
solid fuel 231 ff.
soot 319
sound wave 72, 78
sound, speed of 186, 359, 401
space junk, dangers of 326
space travel 92
spacecraft 262, 308 f.
– ablative shield 239
spaghetti racket 75
spear 114
spin to stabilize a missile 100
sport injuries 130
sports ergonomics 18
Stefan's Law 252
Stefan, Josef 252
Stefan-Boltzman constant 252
stone skipping 119, 120
stratosphere 261
streamline 23
stroboscope 19
Sudbury Basin *see* Sudbury
Sudbury Neutrino Observatory (SNO) 345
Sudbury 287, 290
sugar 177
sun 252
"superman syndrome" 365, 376
sweet spot *see* tennis 71

T
tangent 88
tank 213
target 61
targeted delivery 407
Teflon 100
Tell, William 48
temperature scale
– absolute 242

– Celsius 39
– Centigrade 39
– Fahrenheit 39
– Kelvin 242
tennis 65 ff.
– ball speed 80
– ball, coefficient of restitution 66 ff., 77, 80
– court 80
– elbow 70
– officiating 68
– racket 66, 69 ff., 76
– robotic umpire 68
Thompson, William *see* Lord Kelvin
thuderstorms 333
time-of-flight 395
toluene 173
top (toy) 93
topspin 21
torpedo 96
torque 89
Torr 146
Torricelli, Evangelista 145
Torricellian vacuum 145
tortional pendulum 90
toxophilite 51
transducer 71
transponder 273
trigonometric functions 107
trinitrotoluene (TNT) 173
tritium 348
Trojan horse 190
troposphere 261
Trost mill 405
Tunguska event, description 299 f., 303
turbulence 23, 97
turf science 11
twisting force 89
Tyrannosaurus Rex 315

U
U-tube manometer 145
ultraviolet catastrophe 254
ultraviolet radiation 248
universe, age of 281
Uranus 326

V
V1 *see* pilotless aircraft
V2 *see* rocket
vacuum 144
Venturi throat 143, 148
Venus 326
volcanic ash 387 f.
volcanic rock
– basalt 296

– magma   296
– pumice   296
von Braun, Werner   229
vortex spoiler   154

W
Wallis, Sir Barnes   120
water hazard   27
water jet   394
water tunnel   377
water   27, 261, 334
waterfall   38
Watt, James   199
Watt   199
wavelength   73, 199, 251
weight   19, 36, 211
weightlessness   91
Wein's Law   254
Wein, Wilhelm   254

Whipple Meteor Bumper   309
Whipple, Fred   308
Whitehead, Robert   96
wind tunnel   22, 240, 380 f.
wing   151
work   18, 36

X
Xenophon   115
xylophone   74

Y
yo-yo   98 ff.
– professional use   100
– real   98
– rotational speed   100
– sleeping   99
– Teflon axle   100